Uranium Matters

Uranium Matters

Central European Uranium in International Politics, 1900–1960

by
Zbynek Zeman and *Rainer Karlsch*

Central European University Press
Budapest   New York

© 2008 by Zbynek Zeman and Rainer Karlsch

Published in 2008 by
Central European University Press

An imprint of the
Central European University Share Company
Nádor utca 11, H-1051 Budapest, Hungary
Tel: +36-1-327-3138 or 327-3000
Fax: +36-1-327-3183
E-mail: ceupress@ceu.hu
Website: www.ceupress.com

400 West 59th Street, New York NY 10019, USA
Tel: +1-212-547-6932
Fax: +1-646-557-2416
E-mail: mgreenwald@sorosny.org

All rights reserved. No part of this publication may be reproduced,
stored in a retrieval system, or transmitted,
in any form or by any means, without the permission
of the Publisher.

ISBN 978-963-9776-00-5 cloth

Library of Congress Cataloging-in-Publication Data

Zeman, Z. A. B. (Zbynek A. B.), 1928–
 Uranium matters : Central European uranium in international politics, 1900–1960 / by Zbynek Zeman and Rainer Karlsch.
  p. cm.
 Includes bibliographical references and index.
 ISBN 978-9639776005 (hardcover)
 1. Uranium industry—Germany (East)—History. 2. Uranium industry—Czechoslovakia—History. I. Karlsch, Rainer, 1957– II. Title.

HD9539.U72G38 2008
338.2'749320943—dc22

2007052047

Printed in Hungary by
Akaprint Nyomda

For Kačka and Ms. Crawford

Table of Contents

List of Tables vii
Preface xi
Part 1: Unparalleled Power 1
Part 2: The Erzgebirge Region 35
Part 3: The Politics of Czechoslovak Uranium 61
Part 4: Wismut AG: a State Within a State 159
Concluding Remarks 271
Appendix 1 275
Appendix 2 279
Archives 281
Bibliography 283
Name Index 299

List of Tables

Table 1. Production and Export of Uranium from
    Czechoslovakia to the Soviet Union, 1945–1989   72

Table 2. Uranium Production in the Soviet Block,
    1946–1950   76

Table 3. Employees in the Czechoslovak Uranium Industry,
    1945–1990   114

Table 4. Export of Uranium to the Soviet Union by
    Wismut AG, 1954–1990   170

Table 5. Fluctuation in the Numbers of Directed Labor
    at Wismut in 1947   187

Table 6. International Comparisons of the Incidence
    of Lung Cancer   254

Preface

The present study concentrates on the development of uranium industries in the Soviet zone of occupation in Germany and Czechoslovakia during the crucial years after the war, and touches on how and why Soviet influence was extended far into the center of Europe. Apart from the book before you, the uranium project resulted in an edition of contributions to a conference,* a double issue of *Der Anschnitt, Zeitschrift für Kunst und Kultur im Bergbau* (2–3, 1998), articles, research papers, several conferences in Germany, a seminar in Washington DC and exhibitions in both Germany and the Czech Republic. Indirectly, Dr. Karlsch's interest in uranium matters led him to the subject of his last book, *Hitlers Bombe*, which was published, to much acclaim and some controversy, by Deutsche Verlags-Anstalt in Munich in 2005.

It summarizes the findings the authors considered important. Both the English and German versions were based on the same material, although they were drafted by Dr. Karlsch and Professor Zeman independently.

The Erzgebirge, or Ore Mountains, mark the border between Bohemia and Saxony. In Czech, the mountain range is called Krušné hory, or the Cruel Mountains.

Zbynek Zeman

---

* Rainer Karlsch and Harm Schröter, eds. *Strahlende Vergangenheit, Studien zur Geschichte der Wismut.* St. Katharinen, 1996.

Part 1

Unparalleled Power

Terminal

On 24 June 1945, Joseph Stalin reviewed a great victory parade in Moscow. Regiments of Red Army infantry, cavalry and tanks threw innumerable banners and standards, taken from Adolf Hitler's armies, at Stalin's feet. The day of torrential rain was charged with symbolism: once again, through endurance and suffering, Russia had won a great war. Mikhail Kutuzov's soldiers had once thrown the standards of Napoleon's armies at the feet of Tsar Alexander in the same way. In that summer of 1945, Marshal Zhukov, the victor of Stalingrad and Berlin, stood next to Stalin. A month later, at Potsdam on 24 July, Harry S. Truman told Stalin that America possessed an atomic weapon.

The third and the last conference of the Second World War's Big Three was given the code name Terminal. It was to take place in Berlin, where few buildings escaped damage during the Red Army's final sweep into the city. Cecilienhof, the residence of the former Crown Prince of Hohenzollern in nearby Potsdam, was proposed as the most suitable venue for the conference. It had a large ballroom, and the Berlin suburb of Babelsberg offered accommodation for the delegates and their staff. On their way to Schloss Cecilienhof, the delegates were able to avoid the center of Potsdam, which had been heavily damaged, and cross the Jungfernsee on a pontoon bridge. The journey took about ten minutes by car.

The eighth plenary session in Potsdam started on 24 July, at 3:15 in the afternoon. Winston Churchill immediately complained about the situation of British military and diplomatic missions in Romania and Bulgaria. He told Stalin that he "would be astonished to read the catalogue of incidents to our mission in Bucharest and Sofia. They were not free to go abroad. An iron curtain had been rung down."[1] The session in Potsdam was, however, memorable for another reason.

When it ended late in the afternoon, the delegates stood around in twos and threes, chatting. Churchill saw Truman go to Stalin, and the two men were left alone with their interpreters. Churchill, standing about five yards away, watched them with close attention. He agreed with Truman that Stalin would have to be told about the successful test of the atom bomb in New Mexico, which had taken place ten days before the meeting in Potsdam. In his memoirs, Churchill noted:

I knew what the President was going to do. What was vital to measure was its effect on Stalin. I can see it as if it were yesterday! He seemed delighted. A new bomb! Of extraordinary power! Probably decisive on the whole Japanese war! What a bit of luck! This was my impression at the moment, and I was sure that he had no idea of the significance of what he was being told. Evidently in his intense toils and stresses the atomic bomb had played no part. If he had had the slightest idea of the revolution in world affairs which were in progress his reactions would have been obvious. Nothing would have been easier than for him to say "Thank you so much for telling me about your new bomb. I of course have no technical knowledge. May I send my expert in these nuclear sciences to see your expert tomorrow morning?" But his face remained gay and genial and the talk between these two potentates soon came to an end.

As we were waiting for our cars I found myself near Truman. "How did it go?" I asked. "He never asked a question" he replied. I was certain therefore that at that date Stalin had no special knowledge of the vast process of research upon which the United States and Britain had been engaged for so long, and of the production for which the United States had spent over four hundred million pounds in an heroic gamble.[2]

Franklin D. Roosevelt and Churchill were convinced of the need to shield the nuclear project from German intelligence. The requirement for absolute secrecy rebounded on Stalin as well as on most of the members of Churchill's own cabinet. Neither the British war cabinet, nor the Defence Committee, had discussed the bomb before it was dropped. In a telegram to Eden on 18 April 1945, Churchill proposed that the Americans should advance into the region south of Stuttgart before the French troops reached it. Churchill believed that German nuclear research installations were to be found there, and he asked Eden to treat his suggestion as "background in deep shadow."[3]

Truman had lunched with Churchill on 18 July, and the first part of their conversation concerned the atomic bomb. It was left out of the record for the cabinet, and only Anthony Eden and Sir John Anderson received a note of it. Truman asked Churchill what should be done about telling the Russians. The President was determined to tell them, but was uncertain about the timing. Should he do so at the end of the conference in Potsdam? Churchill advised him to link the news with the experiment in New Mexico, "a new fact of which he and we had only just had knowledge. Therefore we must have a good answer to any question 'why did you not tell us this before?'"[4]

The Race for the Ultimate Weapon

The race for the atomic bomb was, in the end, won by America. The explosions over the New Mexico desert, and then over Hiroshima and Nagasaki in the summer 1945, marked the beginning of a new age. The scientists who had fled from fascist Europe, where Hitler made a point of distrusting "Jewish physics," helped to draw the attention of the governments in Washington and London to the potential threat of Nazi Germany developing the "ultimate weapon." Leo Szilard helped persuade Albert Einstein to write, on 2 August 1939, a letter to president Roosevelt, warning him of the destructive potential of nuclear fission. Einstein's letter also contained a reference to the importance of the uranium mines at Jáchymov, which had come under Nazi control after the Munich agreement in September 1938. In response to the letter, Roosevelt established a uranium commission in October 1939, in order to determine the feasibility of building the bomb. Other, similar impulses came from Britain.

Between the two wars, nuclear scientists had formed a small international group: they knew each other and worked closely together. When Otto Hahn and Fritz Strassman concluded that they brought about nuclear fission in their experiments, they reported on their findings in a well-known scientific journal, *Naturwissenschaften*, at the beginning of 1939. Research in the instability of matter was marked, however, by bad timing. Important advances were made about the time of the outbreak of the Second World War. The international group of scientists fell apart, and articles on their work disappeared from academic journals. As politicians grasped the possibility of developing the ultimate weapon, secrecy and military necessity came to dominate the scientists' work.

After the defeat of France in May 1940, Joliot Curie, the leading French nuclear physicist, stayed in Paris and sent two of his assistants to England. They brought heavy water in tin cans, and continued their work on slow neutrons in Cambridge. The team at the Cavendish Laboratory discovered that, in the course of neutron reactions, plutonium emerged. Virtually unknown in nature, it behaved like uranium 235.

In April 1940, Rudolf Peierls and Otto Frisch, refugee scientists in England, wrote a memorandum suggesting that a lump of pure uranium

235 would create chain reaction necessary for the bomb. They also proposed an industrial method for separating uranium 235 from uranium 238. They also forecast the horrors of the bomb, and explored its strategic and moral implications. A five kilogram bomb would liberate the energy of several thousand tons of dynamite, causing radiation fatal to living beings long after the explosion. The British government established the "MAUD committee," which concerned itself with the feasibility of producing the nuclear weapon. In July 1941, the committee reported that it would be possible to make a bomb, which was "likely to lead to decisive results in the war." Churchill made the MAUD report available to Roosevelt.

The Japanese attack on Pearl Harbour took place on 7 December 1941, and the German declaration of war on the United States followed four days later. Roosevelt was anxious to receive British cooperation and wrote to Churchill about a joint atomic project. The British response was measured, because they wanted their own project—code-named Tube Alloys—with American cooperation.

In the summer of 1942, the Americans began to construct a large technological venture around research hitherto confined largely to university laboratories. The atomic project was given the code name "Manhattan," and Lt. General Leslie R. Groves, a West Point graduate who had supervised the building of the Washington Pentagon, became its manager. There was some hesitation over the appointment of Robert Oppenheimer as the head of research. A university teacher of theoretical physics with an international reputation, he was more interested in pure research than in its application. He was also regarded as something of a political risk. When his appointment was confirmed, Oppenheimer began bringing together the best available scientists in America. He assumed that about 150 of them could successfully complete the project; eventually, their number grew to some 2,500.

By the end of 1942, the advantage of the American project became apparent. The British knew that the US had cornered the Canadian market in uranium and heavy water for the next two years, and that their own project would have to move to North America. The British tried to come in on the US project, but the Americans were reluctant to accept them. The exchange of information ceased, and the western allies remained united by the fear that Germany would beat them in the atomic race.

On 27 February 1943, Churchill telegraphed Roosevelt that "My whole understanding, was that everything was on the basis of fully sharing the results as equal partners. I have no record, but I shall be very much surprised if the President's recollection does not square with this."[5] Churchill was convinced that the possession of the bomb would provide the key to national power after the war, and he tried hard, in August 1943, to convince Roosevelt to sign the Quebec agreement. It helped the British to participate in the Manhattan Project, and led to joint purchase and utilization of uranium supplies. The partners in the agreement promised never to use the bomb against third parties without the other's consent, nor to pass on atomic information to third parties. All available British scientists joined the US project, while the Anglo-French team from Cambridge moved to Canada. They could not go on working in Britain, and the Americans would not have them. The US government had little liking for the Free French, while the leader of the Anglo-French team was unacceptable to American scientists.

After the Quebec agreement was signed, the Americans underwrote the British project in Canada. John Cockroft, the English Nobel Prize-winning physicist, became its head. Canada joined Britain and America in the nuclear project as an additional partner and the main provider of uranium. Canadian officials sometimes played the role of an intermediary between Britain and America.

On 2 December 1942, Enrico Fermi successfully set up a self-sustaining chain reaction in a reactor built into a squash court at Chicago University. In addition to the reactors built by DuPont, the Manhattan Project had six other reactors, which produced uranium 235 by three different methods. All of this—as well as the construction of the first experimental bomb—was achieved at a time when the US armament industry was nearly stretched to its limit producing conventional weapons, as well as advanced new equipment such as radar. The atomic project cost the American government some $2 billion.

The Manhattan Project scientists fulfilled their objective and delivered the bomb to the US government. The first plutonium bomb was tested in the New Mexico desert on 16 July 1945. The Germans had signed unconditional capitulation and the Potsdam conference was to start the following day. In the morning of 6 August, a uranium bomb was exploded over Hiroshima, and on 9 August a plutonium bomb destroyed Nagasaki. Stalin declared war on Japan on 8 August, and on 14 August, the Japanese emperor declared his willingness to capitulate.

Throughout the war, the Allied intelligence services' fear that Germany would develop its own atomic weapon was never quite allayed. The Germans had the necessary technology and scientific strength. They had Jáchymov under their control and, after the occupation of France and Belgium, came into possession of large stores of uranium. Yet early in the war, and at least until the first months of 1942, it seemed that Germany would win the war with conventional weapons alone. The political will of the Nazi leadership to create the weapon did not remain constant, and the German scientific effort was not centralized on the scale of the Manhattan Project.

On 29 April 1939, the president of the Physikalisch–Technische Reichsanstalt, Professor Abraham Esau, organized the foundation meeting of the Uranium Society (Uranverein). Several research groups joined as members: Nobel Prize winner Werner Heisenberg and his colleagues in Leipzig and Berlin; the scientists from the Ordnance Office (Heereswaffenamt), who were working under Kurt Diebner in Gottow-Kummersdorf to the south of Berlin; a group around Paul Harteck in Hamburg; and the physicists at the Kaiser-Wilhelm Institute for Medicine in Heidelberg, with Walter Bothe at their head. Nikolaus Riehl, a pupil of Hahn, also joined the Uranium Club. As head of the research division of the Auergesellschaft, Riehl offered the Heereswaffenamt the services of his department.

Early in December 1939, Heisenberg delivered a report on the "possibility of using nuclear fission as a technical source of energy" to the Heereswaffenamt. It outlined plans for the building of a nuclear reactor. An effective moderator, which would slow down neutrons without absorbing them, was required. Graphite and heavy water were considered: a complicated electrolytical process was required for the production of heavy water. The technology became available to the Germans after the occupation of Norway and the merger of Norsk Hydro in Rjukan with IG Farbenindustrie. Norsk Hydro was obliged to increase production, making it available exclusively to the German Reich.

In the summer of 1940, the Wehrmacht overran France, the Netherlands and Belgium, acquiring large supplies of uranium. Though the director of the Union Minière, Edgar Sengier, had ordered the shipment of uranium and radium reserves to New York in September 1939, some remained in Belgium and fell into German hands.

Beginning in July 1940, the German companies Auergesellschaft and Degussa began purchasing uranium compounds from the Union Minire. The largest lot (1,244 tons) was secured by Roges GmbH, a German war materials company, in May 1942. It also made purchases in occupied France and bought further 200 tons of uranium compound from Union Minire. Until the summer of 1944, the Belgian company assisted the Auergesellschaft with the purification of uranium oxide.[6]

Germany's own production of uranium remained relatively small during the war, and uranium mining enjoyed no special priority. The

Germans took little interest in the known reserves of uranium in Bulgaria and in Portugal. In Jáchymov, the mining of uranium lingered on and no attempts were made to increase production. Only the mines at Schmiedeberg in Silesia were enlarged during the war. In autumn 1939, the Heereswaffenamt requested the Auergesellschaft to help with the processing of uranium and, within a few weeks, the company built a plant at Oranienburg, with the capacity of about one ton of uranium oxide a month.[7] The Degussa plant in Frankfurt am Main followed, and from the end of 1944, another plant was built at Berlin-Grünau.[8] Initially, the military was more concerned with maintaining an adequate supply of luminous paint than with the nuclear project. Early in 1942, the Supreme Command of the Wehrmacht assumed the requirement of 7 grams of radium, in addition to 1.5 grams for the Italian and the Japanese armies. Three grams of radium were produced by Joachimsthal; the rest of the requirement was met with deliveries from France and Belgium. This covered the uranium needs of the Wehrmacht for three years.[9]

The Germans had enough uranium to make the bomb.[10] After the occupation of France, they also acquired the cyclotron at the Joliot Curie institute. The cyclotron was to stay in Paris, and German scientists were to have access to it.[11] Cyclotrons were subsequently constructed in Germany at the institutes of Walter Bothe at Heidelberg and Gerhard Hoffmann in Leipzig, and at the Reich Post Office laboratories at Miersdorf and Berlin-Lichterfelde.

In the meanwhile, experimental work was started in Hamburg and Berlin. Heisenberg and Karl Friedrich von Weizsäcker had a laboratory built on the site of the Institute of Biology and Virus Research at the Kaiser-Wilhelm-Gesellschaft at Berlin-Dahlem. The laboratory was known as the "Virus House." Their experiments, based on Fritz Houtermans' work on the release of nuclear chain reactions, confirmed that the only two suitable moderators were graphite and heavy water.* The Ura-

---

* Even by the standards of an abnormal era, Houtermans led an extraordinary life. A half Jewish communist, he fled from Hitler's Germany to the Soviet Union in 1935. He was arrested during the great purges and delivered to the Nazis after the Ribbentrop-Molotov pact in the autumn 1939. He was then put into the care of the Gestapo and released after Max von Laue's intervention. Houtermans found employment in Manfred von Ardenne's team at Berlin-Lichterfelde and worked for the Nazi regime in order to save his life. In the spring 1941, he warned an American agent of the German nuclear programme and in the autumn he took part in the plunder of the Soviet laboratories in Kiev and Kharkov. After the war, his role in nuclear research fell into oblivion. (cf. Paul Lawrence Rose, Heisenberg and the Nazi Atomic Bomb Project, 135 et seq.)

nium Society put its faith in heavy water. Heisenberg established that the chain reaction of natural uranium created element 94, or plutonium. In September 1941, he knew that the way to the construction of a nuclear bomb was clear. The reason why he did not complete the work are still disputed by scientists and historians. Did Heisenberg hesitate to deliver the bomb to Hitler? Did his project simply lack resources after having been put on the back burner by the Nazis? Or did it suffer from conceptual difficulties in connection with his estimates of "critical mass?" At the end of October, the legendary meeting between Heisenberg and Niels Bohr took place in Copenhagen. The two men were unable to establish common ground. While Bohr feared that Heisenberg wanted to pick his brain and convince him to cooperate with the Germans, his former protégé believed that Bohr was unaware of his reservations with regard to the Nazis.

In the autumn of 1941, it seemed as though a German victory was assured, and they had no need for any miraculous products of science, later known as "Wunderwaffen." The military lost interest in nuclear research and the Heereswaffenamt passed on responsibility for the project to the Reich Research Office (Reichsforschungsrat). On 26 February 1942, the office held a conference of nuclear physicists in Berlin. The captains of the armaments industry kept their distance, and the Reich leaders also seemed rather lackadaisical in regard to the military uses of nuclear fission. On 21 March 1942, Joseph Goebbels wrote in his diary that: "Researches in the field of destruction of the atom have succeeded so far that they could be possibly considered for the conduct of the current war. The tiniest input has such an immense power of destruction that one looks forward with some horror to the course of the war, should it last still longer, or to a later war."[12]

The German nuclear project was divided among several institutes, among which two were outstanding: Heisenberg's group at the Kaiser-Wilhelm Institute of Physics in Berlin, and the Heereswaffenamt group under the direction Kurt Diebner at Gottow, near Berlin. The key theoretical question concerned how to achieve critical mass, i.e. the correct amount of fissionable material (U 235 or plutonium) needed to maintain an explosive chain reaction while avoiding a spontaneous explosion. Estimating critical mass would be a deciding factor in determining whether the construction of the bomb should be attempted.

A meeting took place in Berlin on 4 June 1942, in which Albert Speer, the minister for armaments, General Leeb, the head of the Heereswaffenamt, General Field Marshal Milch and other top military officials took part. The generals showed some interest in the possibility of building a nuclear weapon. Questioned by Milch, Heisenberg stated that an atomic bomb, with enormous destructive power, should not be larger than a pineapple. After Heisenberg referred to the enormous investments required, Speer wanted to know what kind of a sum the scientists had in mind. Weizsäcker mentioned a sum that seemed ridiculously low and it became clear to Speer and to the military present that the project could not be significant for the conduct of the war.[13] There were no more meetings between Speer and Heisenberg. Although Speer was not convinced of the practical value of nuclear research, he went on supporting it in a modest way. In 1943 he set aside 3 million Reichsmarks (RM) for it, and 3.6 million RM in 1944.[14] Hitler himself remained skeptical about the nuclear project. In addition, he suffered from deep distrust of "Jewish physics" and dismissed anything having to do with Albert Einstein out of hand.

Still, the leaders of the Third Reich may have known more about the US Manhattan project than had been assumed. The Sicherheitsdienst had tapped the transatlantic telephone connection and broke the radio traffic code between Moscow and the Soviet embassy in Washington.[15] The information thus acquired was passed on to scientists close to the SS. In addition, the Germans and the Japanese received reports on the Manhattan project from Spanish agent Alcazar de Velasco early in 1944.[16] We have as yet no evidence on the uses the Nazis made of the fragmentary information they acquired on the British-American program.

Prominent scientists and heads of the armaments industry remained convinced that nuclear explosives could not be manufactured in time to influence the outcome of the war. After the spring of 1943, Heisenberg made no more references to the "explosive." The Allies, on the other hand, interpreted the situation differently, believing they were in a technological race with Nazi Germany. Assaults by British commandos and the Norwegian resistance in February, and the attacks by the RAF in November 1943, paralyzed Norsk Hydro's production of heavy water near Vemork. A ferry was sunk which was to bring the remaining heavy water to Germany in February 1944. In any case, the heavy water plants in Norway were dismantled in the middle of 1944 and shipped to Germany. New methods of producing heavy water were developed and new plants were built at the Leuna Werke near Merseburg.

In the meanwhile, experimental work in Germany had advanced. Kurt Diebner's group made important discoveries, leaving the achievements of Heisenberg's group behind. At the end of 1943, Professor Walther

Gerlach of Munich took over the running of the Uranium Society, and attempted to make peace between the various research groups there. At the same time, the growing intensity of Allied air attacks prompted the relocation of nuclear research projects under the Reich Research Office to Freiburg, Hechingen, Heidelberg in South Germany and Stadtilm in Thuringia.

Heisenberg and his colleagues believed that Germany's defeat was within sight. During 1944 and 1945, they concentrated on one task: to set off chain reaction in a uranium machine (nuclear reactor). In a cave at Haigerloch, near Tübingen, the final experiments were begun. Early in March 1945, the Heisenberg group was on the threshold of achieving self-contained chain reaction.[17] Other research groups passed beyond that stage, and historians, who had focused their attention on the activities of the most eminent member of the Uranium Society, were long loath to acknowledge this fact. Apart from Heisenberg and Kurt Diebner in Gottow and later at Stadtilm, there were at least two more groups working independently of each other in the atomic field.[18] They were the research groups of the Reichspost at Berlin-Lichterfelde, under Manfred von Ardenne; and at Miersdorf, with Dr. Georg Otterbein at its head; as well as a little-known SS research group divided between Thuringia, Austria and Bohemia.

In addition to its routine work, the research department of the Reich Post Office took over some important military research, including enciphering and radar development. Dr. Wilhelm Ohnesorge, the minister of post and a close party comrade of Hitler since 1920, wanted to give his department extra political weight by supporting research. Ohnesorge had at his disposal a scientific think-tank based in Kleinmachnow, near Berlin.[19] In December 1939, Manfred von Ardenne drew his attention to the "unusual importance of the discoveries by Hahn and Strassmann."[20] It was a shrewd political move, as the post office had a large fund for basic research and an expert minister who carried weight in the Nazi hierarchy.

In January 1940, Ohnesorge decided to support the project "for the technical development of process and production in the field of atom disintegration."[21] Ardenne started building a powerful apparatus for the production of radioactive isotopes and a cyclotron, which became operative at the beginning of 1944. New Post Office research institutes were established at Kleinmachnow and at Miersdorf. Equipment belonging to the main institute at Berlin-Lichterfelde, such as its high tension apparatus and heavy cyclotron, was duplicated at Miersdorf.

Ohnesorge was determined to deliver the first "uranium bomb" to his Führer. He reported to Hitler on several occasions on the work in progress; in June 1942, Hitler was amused by the thought that his minister of posts should be engaged in the development of new weapons. The research conducted by the Reichspost, Heereswaffenamt and SS in the autumn of 1944 was further advanced than it has been hitherto assumed. After the failed assassination attempt on Hitler in July 1944, Himmler's SS became considerably more powerful. Himmler pressed for the transfer of all important armaments and research projects under SS control. Nuclear weapons fell into the same category as the V-Waffen (i.e. rockets and jet fighters). SS General Dr. Hans Kammler, the head of Gruppe C of the SS economic and administrative directorate, was in charge of the considerable enterprise, including extensive underground projects.[22]

Himmler and the SS began to assert their influence when the life expectancy of the Third Reich was low, right before defeat created total chaos. It is not clear how far Kammler was able to pursue the high technology projects under his control. It is also hard to establish whether the scientists who cooperated with the SS, and who worked in the underground complexes in Austria, Thuringia, Silesia and Bohemia achieved significant results. In his 2005 study entitled *Hitlers Bombe*, Rainer Karlsch threw new light on the dilemma of the German bomb. He argued that when Hitler started making references to the Wunderwaffe in the summer of 1944, he had a nuclear weapon in mind. There exists evidence that, in 1944 and 1945, small test explosions took place on Rügen and in Thuringia, taking the lives of prisoners of war and of other prisoners.[23] Stalin apparently kept the film of a test explosion in his desk in the Kremlin.

The Allies' fear that Hitler might possess a nuclear bomb by no means decreased as the war went on. The influential publisher of a scientific magazine, Paul Rosbaud, was the most important source for the British on the German atomic endeavor. He was in touch with prominent scientists and kept sending alarming reports to London.[24] In 1944, General Groves decided to establish a specialized unit that was to gather intelligence on the German atomic program. Lt. Col. Boris T. Pash became the commander of the unit, which was called the "Alsos." Pash's scientific adviser was Dutch physicist Samuel A. Goudsmit.[25] Pash set up office in London in 1944, with Captain Horace C. Calvert as his liaison officer. Calvert discovered that the Germans had acquired large stores of uranium near Brussels when they occupied Belgium, and that Auer-Gesellschaft processed the ore.

The Alsos mission arrived in liberated Paris on 25 August 1944. They contacted Joliot Curie straightaway, who knew much less about the German effort than the Americans hoped. In the middle of November in Strassburg, Pash discovered a German physical laboratory and some documentary material. Goudsmit thought that it proved that the Germans neither had the bomb, nor were they able to construct it.[26] In retrospect, Pash regarded the finding that the German atomic weapon presented no threat with pride. He expressed the view that this was the most important intelligence discovery of the war, and alone justified the existence of the Alsos.[27] Groves regarded the Alsos mission as completed when the remainder of the Belgian uranium was located. However, the hunt for German atomic scientists continued; many were found at Heidelberg, Hechingen, Stadtilm and in Bavaria.

After the end of the war, ten prominent German physicists, including Hahn and Heisenberg, were interned for six months at an English country house known as Farm Hall, near Cambridge. All their conversations were secretly recorded. The British wanted to know whether the Germans had hidden away nuclear material and research reports, and wanted to prevent German scientists from going to work for the Soviet Union. When the Germans heard that the US bomb was used on Hiroshima, they did not believe the news. After becoming aware of the extent of the US-British project, and of its success, Otto Hahn said to his colleagues: "If the Americans have the atomic bomb then you're all second-raters. Poor old Heisenberg."[28]

In reply to the news of the explosion of the atomic bomb, and in an attempt to give an outline of the history of their own project, the German scientists drafted a press release. They stressed that German nuclear research did not center on the development of a nuclear explosive. This later led to the legend that German scientists slowed down their work to keep from delivering the bomb to Hitler. After losing the scientific and technological race, they moved to occupy the moral high ground.*

The observation that history is often no more than propaganda of the victors is reflected in the historiography of the atomic bomb. There exists a whole library of books on the making of the bomb, and only a few vol-

---

* Robert Jungk's book, *Brighter Than Thousands Suns*, published in 1956, did much to give that view wide circulation. See also the Bibliographical note at the end of Part 1.

umes on the work connected with the production of a small thermonuclear weapon in Germany. The Japanese nuclear program suffered a similar fate. In the 1970s, the American press reported that the Japanese also had a nuclear program. Japan was among the first three countries in the world with a cyclotron.[29] Professor Yoshio Nishina was the key person in the project code-named NI and financed by the military. NI had been started in July 1941, five months before the attack on Pearl Harbour.

Nishina had spent eight years before the war in Europe, studying with Rutherford in Cambridge and Bohr in Copenhagen. In Japan he became the "father of new physics," and in 1931 he was given his own laboratory at the University of Tokyo. An independent research project was started in Japan by the navy, under the code name F-Go Program, around the summer of 1942. The scientific head of the project was Dr. Bunsaku Arakatsu.[30] Arakatsu had also spent a few years abroad, including time at the Cavendish Laboratory in Cambridge under Rutherford. In 1927, Arakatsu studied at Berlin University (today the Humboldt-Universität) under Einstein, becoming a member of Einstein's circle of friends. Next to Nishina, Arakatsu, who taught physics at the University of Kyoto, was the most prominent nuclear physicist in Japan.

Japanese research projects, however, suffered from shortages of uranium and a dearth of large energy plants. The most important resources of uranium were located on the Korean peninsula, and an industrial complex, run by the Japanese navy, was therefore developed at Hungman (Konan in Japanese). A vast fertilizer factory was built at Hungman, as well as a hydroelectric plant, a propellants plant and a heavy water installation. The approximately 300 scientists working there on the nuclear project probably had a cyclotron at their disposal, as a part of the plant was located underground. Raw materials, uranium in particular, came from the occupied territories in Korea, China and Inner Mongolia.

Shortly before the end of the war, US intelligence came across some unexpected information. Japanese scientists were apparently planning to conduct an atomic bomb test near Konan on 12 August, six days after the Hiroshima explosion. The bomb developed for the Japanese navy was possibly intended to be used by kamikaze pilots. However, this rumor could not be verified, as Konan was occupied a few days later by the Red Army. Most of the Japanese installations were subsequently destroyed.

Though most experts are skeptical about a Japanese nuclear bomb test, it seems that the Japanese project was further advanced than had been assumed. After the two nuclear attacks on Japan, and the country's subsequent unconditional surrender, the history of its atomic project was suppressed. Japan came to regard itself as a victim of the American raids. The scientists who had taken part in the nuclear program remained silent, or insisted they had taken part in developing atomic energy for peaceful purposes. Shortly after the war, the Japanese cyclotron was destroyed by the US Army, and the few extant documents confiscated. The Second World War had hardly ended and new conflicts were starting to emerge. The United States had little interest in publishing information regarding the nuclear achievements of the defeated powers.

The post-war history of Europe, and of the world, was dominated by the conflict between two new superpowers: the United States and the Soviet Union. The conflict was driven by the arms race, especially in the field of the nuclear weapons. Stalin's shrewdness and paranoia helped him to conceal his extensive knowledge of his Western allies' nuclear project in Potsdam on 24 July 1945. He had received the first information on the possibilities of an atomic bomb from London as early as the autumn of 1941. The information came from John Cairncross, private secretary to Lord Hankey, a member of the cabinet responsible for overseeing the work of the intelligence services.[31] It was passed on to Moscow by Anatolii Veniaminovich Gorsky (Vadim), who worked under diplomatic cover as the resident in London. Gorsky's message was sent on 25 October 1941, and reached Moscow shortly before the celebration of the anniversary of the revolution on 7 November. Lavrentii Beria, the head of the NKVD, delivered the report to Stalin.

In March 1942, another reminder of the importance attributed by the British to nuclear research reached Moscow. Gorsky reported the details of the nuclear project and the role which Canada, a rich source of uranium, was to play in it. Another report from London concerned nuclear research carried out by Hahn and Heisenberg in Germany. It also referred to the Norsk Hydro plant in Rjukan, Norway, capable of supplying Germany with 4,500 kg of heavy water a year. Beria suggested that consultations with the scientists should take place.

Many leading Soviet scientists were pessimistic about the practical uses of atomic energy. They noted that articles on the subject had stopped appearing in western scientific publications. A young physicist Georgi Flerov, who served as a lieutenant in Voronezh early in 1942, wrote to Stalin in April. It is not certain whether the letter reached Stalin, though it was known in the scientific community. Flerov's letter harshly criticized, in familiar terms of Stalinist invective, the physicists who were pessimistic about the possibility of developing nuclear energy for military purposes.

Beria used intelligence reports from the West to acquaint Stalin and the State Committee for Defense (GOKO, the highest government office concerned with military matters) with the subject of nuclear weapons. The reports showed that research in nuclear physics was considered to have significant military applications, and that the scientific community in Germany, as well as the British and Americans, also intended to use it for military purposes. Beria intended to create a body of scientific consultants and attach it to the State Defense Committee. He named three scientists (Kapitsa, Skobeltsyn and Slutski) as having worked in the field of nuclear fission. As none of them had in fact done so, it seems that Beria was better informed on the British than the Soviet side of atomic research.[32]

Igor Vasilevich Kurchatov, who was then researching the problem of protecting shipping against the threat of magnetic mines, was invited to come to Moscow on 15 September 1942. There he met Mikhail Georgevich Pervukhin, a party technocrat with the rank of deputy prime minister in charge of the chemical industries. Kurchatov possibly met Stalin as well. He moved to Moscow permanently in February 1943, when the State Defense Committee formally established an atomic energy research program. Kurchatov's colleague, Iulii Borisovich Khariton, thought that Kurchatov was "an exceptional leader who organized a strategically correct program from the very beginning."[33]

While Beria was considering how best to use the intelligence material from London, Molotov made it available to Pervukhin. Molotov asked him to find out what Soviet scientists knew about the research carried out abroad, and their opinions about the kind of research that should be undertaken in the Soviet Union. In January 1943, Pervukhin had a meeting with Kurchatov and two of his colleagues, asking them for a memorandum concerning nuclear research. When it was finished, Pervukhin handed it over to Molotov with a warm recommendation. The State Committee for Defense passed a special resolution on the organization of nuclear research, and the laboratory of the Academy of Sciences (which later became known as Laboratory No. 2) was established. On 10 March 1943, Kurchatov was confirmed in the post of the scientific director of the project, thus becoming Robert Oppenheimer's opposite number in the Soviet Union.

After Kurchatov expressed doubts to Molotov about the feasibility of constructing the bomb in the foreseeable future, Molotov decided to pass on the intelligence materials to him. In March 1943, Kurchatov began studying these materials in his room in the Kremlin. Among the scientists, he was the only recipient of information from secret intelligence sources. Though unable to disclose its origin to his colleagues, he abandoned his skepticism about the practicality of the bomb.[34]

In two letters to Pervukhin on 7 and 22 March 1943, Kurchatov compared the work of the Soviet physicists with the intelligence he had from Molotov, wondering whether they were a true reflection of the state of research in the West. He was surprised that Western scientists preferred the diffusion to the centrifuge method of isotope separation. Kurchatov noted that the intelligence reports contained "some fragmentary comments about the possibility of using not only uranium 235 but also uranium 238." In a letter to Pervukhin on 22 March, Kurchatov wrote that "I looked carefully through the latest research in the Physical Review on transuranic elements...and I was able to determine a new direction in the solution of the whole uranium problem...The prospects in this direction are unusually attractive."[35]

Lavrentii Pavlovich Beria knew that Churchill and Roosevelt had discussed cooperating on the atomic project in Washington in June of 1942, and that the main part of the project would move to America. So far, the most valuable information had come from British sources. John Cairncross, who passed the contents of the MAUD report on the feasibility of the nuclear weapon on to the Russians, moved to Bletchley Park, where German radio traffic was deciphered, before getting a job in the Secret Intelligence Service (SIS). John Philby began reporting to the Soviets early in 1944, after the establishment of a section that was to deal with "past records of Soviet and Communist activity." Donald Maclean, a young diplomat and a Soviet spy, dealt with Anglo-American cooperation and the building of the atom bomb at the British Embassy in Washington.[36]

Moscow had several well-placed British sources of political intelligence at its disposal. For scientific and technical information, Klaus Fuchs was probably most valuable. After his move to North America he had his first meeting with the Soviet controller, Harry Gold, on 5 February 1944 in New York. In September 1944, Kurchatov assessed the Manhattan project as "a concentration of scientific and engineering-technical power on a scale never seen before in the history of world science, which has already achieved the most priceless results."[37] The secret service reported to Beria using sources from Los Alamos, apparently for the first time, as late as 28 February 1945. The report contained details of the construction of the bomb, provided by Theodore Alvin Hall, a precocious, nineteen year-old Harvard physicist. Hall was convinced that nuclear arms monopoly would continue to threaten world peace and, together with technical sergeant David Greenglass, became a useful source of intelligence on the Manhattan project. It therefore seems likely that the Russians had two sets of instructions on how to make the bomb before them: Hall may have revealed the implosion method before more detailed instructions came from Klaus Fuchs.[38] When Stalin returned from Potsdam in August, he asked Kurchatov to come and see him. Stalin was impressed with the American achievement, and dissatisfied with Soviet physicists. Kurchatov tried to explain why the Soviet program had made so little progress: many people had died, equipment was destroyed, the country faced famine, and nothing was available. Kurchatov did not say that, during the war, Stalin had hedged his bets as far as the nuclear project was concerned. By the end of 1944, it employed about a hundred scientists.

After the explosions of the US atomic weapons in the summer of 1945, Stalin knew that the power of the Soviet Union and its victory in the war would be called in question. He told Kurchatov that he wanted to know what the scientists needed, and that they would lack for nothing in the future. The Russian people were again compelled to make huge sacrifice. In the midst of post-war devastation and shortages, Stalin began laying down the foundations of the Soviet military–industrial complex.

Uranium Monopoly and the Division of Europe

In the race for the nuclear weapon, availability of uranium ore was of the essence. General Groves tried to secure a monopoly over the purchase and processing of uranium by all available means. On 3 December 1945, Groves confidently reported to the Secretary of War that the Combined Development Trust, a joint British-American government agency, controlled by Groves himself, had cornered 97% of the world's production of uranium, and 65% of thorium.[39]

Sometime in February 1945, Groves recommended to General Marshall that the Auer uranium processing plant near Oranienburg be destroyed by air attack.[40] The role of the Auer Gesellschaft in the supply of uranium was confirmed independently by information from Brussels and from Paris. Auer had taken over the management of a formerly Jewish company in Paris, which traded in rare metals, and thorium in particular. Dr. Egon Ihwe, who looked after this side of the Auer business, had supplied Alsos with the information. By the end of 1944, Groves knew of the central importance of the Oranienburg processing plant. For a long time, his memoirs remained the only source on the true purpose of the heavy air raid on 15 March 1945.[41]

The Auer Gesellschaft plant was not among the targets originally selected for the Eighth US Air Force, which carried out the raid. Instead, the SS Main Ordnance and the Heinkel aircraft plant and airfield were to be targeted. The Auer processing plant was not mentioned in any existing documents or literature, though the neighboring railway junction was. The only mention of the air raid on the Auer plant comes from a report dated 16 March: "The industrial area just West of the target [i.e. the SS depot], including the gas mask factory of [there is an empty space in the report] AUERGESELLSCHAFT AG recieved [sic] many hits, causing at least two explosions and starting numerous fires."[42]

Among the target dossiers of the US Strategic Bombing Survey, there exists a report on the raid where this gap in the text is filled in, by hand, with "GS-5451," the code for the Auer plant. The sentence quoted above is also underlined by hand. It may be assumed that these unusual additions, made in only one of the 243 copies of the report distributed in the offices of the US Air Force, were made by somebody who had special interest in the work of the Auer Company. It was described in the report as a maker of "gas-masks, anti-gas equipment for submarines and air filters." The Americans and the British selected targets on the basis of a detailed list of German industrial plants, including the Auer Company. Its description as a producer of gas masks is hardly plausible.

The operational order named the Mean Points of Impact—that is, places where the bombs were to be concentrated. One was in the middle of the railway station, the second on the rails of the south-eastern exit. This was where the Auer plant was situated, on both sides of the railway lines. The plant therefore lay at the centre of the massive air raid, which was conducted by the Eighth Air Force. 612 aircrafts took part in the raid, and the composition of the strike force was significant: every bomber flight was to include one plane with delayed fuse bombs. The delays were 1, 2, 6 and 12 hours, in order to hinder salvage efforts.

The order for the operation came in late. The Eighth Air Force had planned for routine raids on synthetic fuel plants in the area of Leipzig and Ruhland on 15 March. A high priority order then arrived from the United States Strategic Air Force, giving the composition and tactics of the strike force and moving the target to the area of Berlin.[43] The command of the Eighth Air Force was thus transferred from General James H. Doolittle to the supreme commander of the US Air Force, General

Carl Spaatz. All this tallies with the report on the raid as told in General Groves' autobiography.

About three weeks after the raid, the Auer Company asked Reich authorities for compensation for the damage caused by the raid. On the basis of accounts from June 1944, the total cost of the damage was put at 61 million RM.[44] A few days later, the Red Army reached the town. A Soviet commission of experts started inspecting the plant on 21 April 1945. Ihwe, the manager, was questioned. All documentation and patents had to be handed over to the commission. All usable equipment was stripped and sent to the Soviet Union, as well as the available rare metals. When Nikolaus Riehl, the German physicist, visited the Oranienburg processing plant, he realized that "The Russians who accompanied me appreciated that the bombings were directed not against the Germans but against them."[45] When some of the uranium the Germans had initially impounded in Belgium was located at Stassfurt, deep within the Soviet zone, the Alsos advanced there regardless and impounded some 1,100 tons of ore.

Until the spring of 1945, the Allies had been united in their desire to bring the war to an early end, though often divided on how to best achieve this. The European Advisory Commission (EAC) was a backroom team proposed by Eden and established by the Foreign Ministers Conference in Moscow in October 1943. Based in London, it played the key role in drawing up zones of occupation in Germany, and kept the lines of communication open on matters of strategy between American, British and Soviet military commands.

Preliminary work concerning the division of Germany into zones of occupation was done in connection with the preparations for the Second Front. General J. H. Morgan and his staff worked under the assumption that the Americans, whose bases were located in the western part of Britain, would go through France and Belgium to occupy south Germany, including Bavaria, while British forces would liberate Holland and move to the Ruhr and northwestern Germany. The Russians were to take the East, and Berlin was to come under joint control of the Allies. A committee of the British cabinet, chaired by Clement Attlee, confirmed the division and drew up the western border of the Russian zone.[46] This was late in the summer of 1943, and the boundary proposed in London eventually became the eastern border of the Federal Republic of Germany. It put 40% of German territory and 36% of the population on the other side of the boundary, under Soviet control.

The EAC final protocol on the zones of occupation was signed 12 September 1944, and confirmed by the Big Three at the conference at

Yalta in February 1945. It was a contract on the limits of advance of the Allied armies in Germany. It assumed momentous political significance later, by dividing Germany—and Europe— into two spheres of influence.

Starting in March of 1945, as American and British troops advanced fast into the center of Germany and the Red Army was approaching Berlin, Churchill became increasingly concerned by Stalin's policy in Eastern Europe. The future of Poland worried him, as did the reports of British officers on Allied Control Commissions in the countries under Soviet occupation. He wanted the western commanders to move as far east as they could, in disregard of the zonal protocol. Due to the speed of its eastward advance, the US Army moved into the "tactical zones" to the north of the Erzgebirge, and occupied large parts of Saxony and Thuringia, a region well beyond the zonal limit.

Two days after Hitler's last birthday, on 22 April l945, the US Seventh and the French First Army crossed the Danube. Augsburg and Munich fell to the Seventh Army on 30 April, and some 30,000 prisoners in the Dachau concentration camp were set free. In the north, British and American armies established bridgeheads across the river Elbe, while the Russians fought their way through the streets of Berlin. Hitler was by then dead or dying, and Field Marshal Keitel admitted that there was no way of relieving Berlin.

The armies were everywhere on the move, and the fluid nature of warfare prompted Eisenhower, also on 30 April, to telegraph his strategic intentions to Moscow. He suggested that, in the north, his approximate aim was to reach the line Wismar-Schwerin-Domitz, while in the center he would hold his position along the Elbe and Mulde rivers. Concerning Czechoslovak territory, Eisenhower proposed to pause at the l937 border before advancing, should the situation require this, to a line between Karlovy Vary–Plzeň and České Budějovice. Eisenhower assumed that the Red Army would clear the country up to the Elbe and Moldau (Vltava) rivers. He added that "with knowledge of our mutual plans adjustments of contacts in this area" would be made by local commanders.[47]

On 4 May, one of Patton's armored units was about to run into the Russians in the Linz region. In the evening, General Bradley telephoned Patton to give him the green light for an advance to Czechoslovakia. There were eighteen divisions in Patton's army, comprised of some 540,000 men. He wanted to be through the mountain passes as soon as possible, "before anything hit us." The advance began in the morning on 5 May: "In view of the radio reports that the Czechoslovakian citizens had taken Prague, I was very anxious to go and assist them, and asked Bradley for authority to do so, but this was denied. As a matter of fact, however, reconnaissance elements of the Third Army were in the vicinity of Prague and by that act marked the furthest progress to the East of any western army." On 6 May, Patton received confirmation of the ruling that his troops were not to advance beyond the stop line running through Pilsen for distances greater than required for security reconnaissance. "I was very much chagrined because I felt, and still feel, that we should have gone on to Moldau River and, if the Russians did not like it, let them go to hell."[48]

The Karlovy Vary–Jáchymov region lay within the area of operations of the Vth Corps of Patton's Third Army. It was commanded by General Huebner, who ordered his infantry and armored units to attack Karlovy Vary. The main road between Karlovy Vary, Sokolov and Kynšperk was defended by the German Kampfgruppe Benicke. After some skirmishes, the Americans reached the line Kraslice-Jindřichovice—Horní Slavkov on 7 May. The Red Army got there four days later, on 11 May: the contact line with the Americans stabilized during the summer. It ran roughly from Sachsenberg in Germany across Rotava, Jindřichovice, Dolní Niva, Vintířov to Doubí, and from there to Teplička and Bečov.[49] Czechoslovakia was an allied country, and both the Red Army and the US Army were expected to withdraw soon. No zones of occupation had been agreed upon for Czechoslovakia.

In the meanwhile, developments in Germany were more blurred. In the spring of 1945, Churchill left Washington, a dark vision of the future of Soviet-dominated Europe fixed in his mind. The controversy about the withdrawal of the US Army to the agreed contact lines culminated early in May. Churchill warned Eden that the retreat of the American army would be one of the most distressing events of world history. The Americans were unimpressed by Churchill's fears, as they did not want to risk a breakdown of the Allied occupation policy. Without withdrawal from the territory which was to be a part of the Russian zone of occupation, the future of the Allied Control Commission was threatened, as well as the control of Berlin by the four Powers.

On 11 June 1945 President Truman cabled London that the agreement on the zones should be adhered to, and the British and the American troops withdrew from parts of Mecklenburg, Sachsen-Anhalt, Thuringia and Saxony. The Americans also agreed that the Red Army should establish its headquarters at Karlovy Vary. The territories vacated by the Americans soon proved to be the richest source of uranium for Stalin's atomic bomb.

The brief occupation by the Americans of parts of the future Soviet zone gave rise to speculation. Why did the Americans withdraw and enable the Russians to access the richest sources of uranium in Europe? Did the promise of uranium play any role at this time, when Germany was divided into the zones of occupation? Did the Americans miss a historical chance when they withdrew from the line of their furthest advance to the East?

In Stefan Heym's 1984 novel *Schwarzenberg*, the answer was put simply: "When I imagine, how differently the history of the republic would have unfolded had the Americans known what was hidden under the soil of Schwarzenberg! They could have known it as did the miners, who mined silver and tin in the region in the old times and who struck pitchblende again and again. And were not the radioactive springs in the vicinity a further proof?"[50] The idea of a missed historical chance was recycled in contemporary publicity.[51]

There exists no evidence that the uranium problem played any role in the negotiations on the division of Germany into zones of occupation. The negotiations took place at a time when Stalin was pessimistic about the atomic project, and the whole venture, including the search for uranium, did not enjoy a priority status. The uranium reserves in Thuringia, Saxony and Bohemia played no role in Churchill's correspondence during the spring of 1945, with its insistence on retaining the territories under British and American occupation. At that time, neither he nor Truman were aware of the uranium reserves there.[52]

There nevertheless exists a sharp contrast between the decision of the Americans to withdraw from Saxony, Thuringia and Bohemia on the one hand and, on the other, the decision to bomb the Auer company plant in Oranienburg. Both decisions were made within a period of three months. While General Groves knew enough about the German atomic project to try and reduce Soviet war booty, he was less well informed about the Erzgebirge. He knew only of the oldest, low yield European uranium mines at Jáchymov.

Early in the summer of 1945, the Russians moved into the region vacated by the Americans. They were not aware that, by the time mining was concluded there after 1989, it would produce over 231,000 tons of uranium.

The Erzgebirge and the Soviet Uranium Gap

Some months before the end of the war, the NKVD began preparing itself for the search for German nuclear scientists and their technical equipment. Avramii P. Zaveniagin, Colonel General in NKVD and a mining engineer, became the head of the project. He had managed Magnitogorsk, a metallurgical concern, and the Norilsk mining company. In 1937, he became deputy minister of heavy industry. Zaveniagin became Beria's deputy in 1941, and in the spring of 1945 he began to look after security of the Soviet atomic project. He was described as the "Soviet General Groves."[53]

The position of the Soviet nuclear project and Zaveniagin's role in it were further strengthened on 20 August 1945, when Stalin ordered the establishment of a special section of the State Committee on Defense,[54] inviting Beria to be its head. According to a member of the team of nuclear scientists, the departments under Beria's command worked with the precision of a Swiss watch.[55] Of all the Soviet government agencies, the NKVD alone was capable of securing priority allocation of the resources for the project to build the bomb.

The Soviets were initially more interested in tracing the German atomic program and recruiting German scientists than in the search for uranium.[56] Despite strong opposition from the Communist Party bureaucrats, Zaveniagin sent a group of forty Soviet physicists to Germany.[57] They succeeded in convincing eminent German scientists, including Manfred von Ardenne, Gustav Hertz, Heinz Pose, Nikolaus Riehl, Peter Adolf Thiessen and Max Volmer, to work for the Soviet atomic program.

As early as 15 May 1945, NKVD presented in Moscow the results of their investigations into the German plants and research institutes which concerned themselves with nuclear matters.[58] Among the institutions visited were the Kaiser-Wilhelm-Institut für Physik in Berlin-Dahlem, Manfred von Ardenne's institute in Berlin-Lichterfelde, Institut der Reichspostforschungsanstalt in Zeuthen (Miersdorf), the Siemens cyclotron laboratory run by Gustav Hertz, as well as the plants and warehouses of the Auer company in Berlin-Charlottenburg, Berlin-Grunau, Oranienburg and Zechlin.

The objects came under NKVD control and were soon dismantled. Special units found about 300 tons of uranium oxide and 7 tons of uranium metal in Berlin, Gottow, Zechlin, Kagar and Rheinsberg.[59] In Stadtilm, a small town in Thuringia, the special unit found a uranium processing plant that used to belong to the Degussa Company.[60] The Auer Company's plant in Oranienburg, destroyed in the American air raid in March 1945, was also thoroughly searched; a few tons of pure uranium oxide and several hundred tons of thorium derivates were found there.

According to western estimates, about half of the scientific work on the "uranium problem" in the Soviet Union was carried out in prison laboratories.[61] In addition, prisoners in the gulag system carried out most of the construction and mining work for the atomic project. Alexander Solzhenitsyn, who mapped Stalin's Gulag Archipelago in great detail, made only a brief reference to the camps that contributed to the nuclear project. Prisoners in the top-secret camps had to make special pledges of non-disclosure. According to Solzhenitsyn, these pledges were renewed every three months. When the initial phase of the work on the *atomgrady* (atomic cities) was completed, a large group of the workers were sent to Kolyma in September 1950, as a "particularly dangerous special contingent."[62]

Next to the gulag archipelago, there grew up the "white" archipelago of the nuclear enterprise. Professor Lev Altschuler spent twenty-two years as its employee: "The white archipelago became the centre of thousands of highly qualified scientists, engineers, construction and production workers and many others, who had survived the war and persecution." They were completely isolated from the outside world: "My future workplace was miles away from the nearest railway station. We were transported in buses, wrapped into long fur coats. We passed through settlements which looked like Russia before Peter the Great. When we arrived, we saw monastery churches and houses, cottages scattered in the forest, and the common symbols of this era, the "zones"...[they were] camps inhabited by prisoners from every region of the Soviet Union and of every nationality. According to local people...a large number of prisoners had recently rioted and escaped into the forest. The rebellion was led by an army pilot. Every morning, columns of convicts marched through our village. Such was the stark reality."[63]

Parallel with the development of the nuclear program itself, Beria's agencies tried to secure sufficient supplies of uranium. Russian geologists estimated Soviet uranium reserves at about 2,000 tons, concentrated mainly in central Asia.[64] Its location meant that it could hardly become the basis of an efficient uranium industry. In the last months of the war, the NKVD therefore began exploring the possibilities of acquiring uranium abroad. Behind the advancing Red Army, and among the special units engaged in the search for scientific and technological booty of every kind, groups of geologists were to be found.

The first group came to Bulgaria at the end of November 1944. It followed a German trace: Soviet troops had discovered some German documents concerning uranium reserves in the vicinity of the town of Buchovo. The first report was sent to Beria on 12 January 1945, who informed the former leader of the communist International, Georgi Dimitrov, of the Soviet's interest.[65] In October, a Soviet-Bulgarian mining company was established, and Soviet geologists came to Buchovo. Political prisoners were employed in the uranium mines and, by the middle of 1946, the company had produced 272 tons of pitchblende, which was then sent to the Soviet Union.[66] Dimitrov took a keen personal interest in the project. The Soviet director of the enterprise nevertheless did not inform Dimitrov about the quality of the Buchovo pitchblende. The volume of production was also kept a Soviet state secret.[67]

After Bulgaria, the NKVD turned to Czechoslovakia. At the end of August 1945, an expedition was sent to Jáchymov, including Semion P. Alexandrov and Alexander Orlov, two experts who had taken part in the search at Buchovo. Beria's special committee for the atomic bomb received the first report on Jáchymov on 14 September 1945. The estimated uranium reserves in Jáchymov amounted to 300 tons.[68] It was, however, in the German side of the Erzgebirge in the Soviet zone of occupation, which would eventually become the most important source of uranium for Beria's project.

The Jáchymov mines were not considered to be a rich source of high-grade ore, but the dearth of uranium in the Soviet Union nourished the hopes of new finds; and the Jáchymov trace led the Russians to neighboring Saxony. In May 1945, after Soviet specialists interrogated the geologists in Freiberg, they decided to carry out the necessary exploration themselves.[69] Early in August 1945, an expedition of Soviet geologists, led by Professor Kreiter, came to Saxony.[70] The geologists visited the headquarters of the Sachsenerz-Bergwerks AG in Freiberg and the mines near Schneeberg and Johanngeorgenstadt. In Freiberg, they consulted geological maps of the former silver fields, as well as the documents of the mining administration. Johannes Schmidt, who as chief supervisor (Obersteiger) knew his way around the mines better than anyone else, pointed out the old works where explorations could be started.

Sometime in August 1945, the Russians decided to undertake thorough explorations in Saxony. The geologists, led by Semion P. Alexandrov began with a review of the old mines in September. They pretended to the Germans that they were looking for bismuth and cobalt.[71] Professor Friedrich Schumacher, director of the geological institute at the mining academy in Freiberg, was probably the first German who found out, during a conversation with Professor Kreiter on 10 September 1945, the true purpose of the search in Saxony.

Schumacher agreed with Kreiter to establish an "office for colored metals," which was to carry out commissions for the occupying authority. Together with Professor Gustav Aeckerlein, who had studied the sources of radioactive waters in Saxony, Schumacher was asked to help with the examination of the reserves of uranium.[72] Their first report was ready on 8 October,[73] and it estimated the prospects for uranium mining as being generally poor. They thought that Johanngeorgenstadt, in close proximity to Jáchymov, was more promising than Schneeberg. After Taking into consideration the amount of ore necessary to produce the volume of oxide, Schumacher forecast a "possible reserve" of 80 to 90 tons of uranium oxide there. It was ten times the amount of uranium produced in Johanngeorgenstadt to date.[74]

In view of Soviet interest in uranium and of the uncertain post-war situation, it may be assumed that the German professors were trying to ensure a continued existence for their academy, and possibly even a new future for it in the uranium industry. Although their skepticism regarding the economic feasibility of uranium mining remained unchanged, Schumacher went on working, during 1946, on estimates of the reserves in Johanngeorgenstadt and Marienberg.

The first Russian estimates were even more pessimistic. At the end of 1945, Alexandrov put the Schneeberg reserves at 10 tons and Johanngeorgenstadt at 22 tons.[75] In the spring of 1946, Alexandrov sent a report to Moscow recommending that, despite the low estimates, mining works be started at Johanngeorgenstadt. In the meanwhile, Professor Aeckerlein was commissioned by the Russians to build or acquire the instruments necessary for locating and measuring uranium deposits. He made an emanations meter and then turned to producers in the British zone, where he acquired Geiger counters and other instruments.

The two Freiberg professors did not work for the Russians long. In the course of denazification, Schumacher was suspended as the head of the office for colored metals in November 1946. He moved to Jugoslavia in the spring 1947, where he worked, until 1950, as chief geologist for a lead and zinc mining enterprise. In 1952, Schumacher started to work for the mineralogical institute at the University of Bonn. The value of his work for the Soviets lay primarily in his estimates, which were more optimistic than those initially presented by the Russian geologists. Due to the Allied ban on Germans working in the field of atomic physics, Aeckerlein's radium institute was closed down in the spring of 1948.

The NKVD officers hoped that enquiries among the members of the Kaiser-Wilhelm-Institut in Oberschlema would give them some leads. Since 1937, Professor Boris Rajewsky and his colleagues at the Kaiser-Wilhelm-Institut für Biophysik at Frankfurt am Main had been making examinations of radon content in the air and the waters of the mines at Schneeberg and, later, in Jáchymov.[76] Dr. Adolf Krebs, the head of the Oberschlema branch of the Rajewsky's institute, was arrested by NKVD officers on 1 October 1945. Before being taken to Moscow, he was interrogated in Dresden and Berlin.[77] He was allowed to return to Germany in the spring 1946, from where he immigrated to the United States. In a report drafted early in 1948, Krebs reinforced the low estimates, current in America at the time, of the potential of the Erzgebirge region.

In June 1946, the council of ministers of the USSR gave the go-ahead to uranium mining in Saxony,[78] although prospects were still uncertain. Another search party under N. M. Khaustov established, about the same time, the existence of deposits near Oberschlema.[79] The existence of strong radioactive waters, as well as the evidence gathered by the Kaiser-Wilhelm-Institut, helped the Soviets make the decision to start mining operations at Oberschlema. The town had a well-known spa, which was closed down after the war. In July 1946, the occupation authorities allowed resumption of its service, which was once again suspended only a few weeks later.[80] Geological probes were made in the vicinity of the spa buildings, and large parts of the spa quarter and the famous springs fell victim to the search for uranium. The local people regarded the destruction of the spa as the beginning of a new "mining fever."

From the end of 1946, Soviet estimates of the uranium reserves in the German part of the Erzgebirge kept on going up. Whereas reserves of about 100 tons were assumed at the end of 1945, by January 1947 the estimate of proven reserves increased to 252 tons. While Dr. Krebs was drafting, early in 1948, his report on the poor reserves available to the Soviets in their zone of occupation, the estimates reached the figure of 1,600 tons.[81] The Soviet authorities by then knew that Saxony could provide more uranium than any other part of their empire.

Notes

Bibliographical note: There exists extensive historiography concerning the American and the British nuclear programs, including official histories. Richard G. Hewlett and Francis Duncan wrote the first volume of the US history, entitled *The New World: A History of the US Atomic Energy Commission, vol. 1, 1939–1946*, published first in 1962 and reissued in 1990 by the University of California Press (UCP); Richard G. Hewlett and Francis Duncan, *Atomic Shield: A History of the US Atomic Energy Commission, vol. 2, 1947–1952*, 1969 and UCP 1990; Richard G. Hewlett and Jack M. Holl, *Atoms for Peace and War: 1953–1961*, UCP 1989. Margaret Gowing wrote *Britain and Atomic Energy*, published in London in 1964, and, assisted by Lorna Arnold, *Independence and Deterrence: Britain and Atomic Energy 1945–1952*, sv 1; Policy Making and Policy Execution, sv 2, London 1974. The historiography of the Soviet bomb is somewhat poorer, and no official account of the Soviet program exists. The first study appeared in Munster, Germany in 1992: *Die Sowjetische Atombombe* by Andreas Heinemann-Gruder, and was followed by David Holloway's *Stalin and the Bomb* in 1994, published by Yale University Press. There exists a valuable collection of documents on the Soviet atomic program edited by L. D. Riabev, *Atomnii Projekt SSSR*, volumes 1–4, Moscow 1998–2003; and Arkadii Kruglov's *The History of the Soviet Atomic Industry*, London 2002.

[1] Gilbert, *Never Despair: Winston S. Churchill 1945–1965*, vol. 8, 97.
[2] Churchill, *The Second World War: Triumph and Tragedy*, vol. 6, 579–80.
[3] Gilbert, *The Road to Victory: Winston S. Churchill 1941–1945*, vol. 7, 1302.
[4] Gilbert, *Never Despair*, vol. 8, 66.
[5] Gilbert, *The Road to Victory*, vol. 7, 415.
[6] Bundesarchiv (BArch) Berlin-Hoppegarten, R 121/583, Bc.3. Schreiben von Dillmann und Ihwe, 8 July 1944 (Letter from Mr. Dillmann to Ihwe).
[7] Auer Society of Berlin, Institut für Zeitgeschichte (IfZ) Munich, OMGUS, Adjutant General, Top Secret (AGTS), Nr. 38/1a.
[8] German Reports, G-324, Herstellung von Uran bei der Degussa; Bericht von Dr. Völkel, 9 November 1945.
[9] State Archive of the Russian Federation GARF Moscow, 1458/40/160; Reichswirtschaftsministerium. Abt Chemie, Wochenbericht von März 1942.
[10] Jonathan E. Helmreich, *Gathering Rare Ores*, 254.
[11] Richard Rhodes, *Die Atombombe oder die Geschichte des 8. Schöpfungstages*, 344.
[12] Joseph Goebbels, *Tagebücher aus den Jahren 1942–1943*, 136.
[13] Albert Speer, *Erinnerungen*, 240.
[14] Helmut J. Fischer, *Hitler und die Atombombe*, 56.
[15] Hubert Faensen, *Hightech für Hitler*, 105ff.

16. Robert K. Wilcox, *Japan's Secret War* and David Alan Johnson, *Germany's Spies and Saboteurs*.
17. Mark Walker, *Die Uranmachine*, 185.
18. Rainer Karlsch and Mark Walker, "New Light on Hitler's Bomb," *Physics World* (June 2005): 15–18.
19. Hubert Faensen, *Hightech für Hitler: Die Hakeburg—Vom Forschungzentrum zur Kaderschmiede*.
20. Manfred von Ardenne, *Sechzig Jahre für Forschung und Fortschritt*, 89 et seq.; Manfred von Ardenne, *Ich bin ihnen begegnet*, 117.
21. Thomas Stange, *Die Genese des Instituts für Hochenergiephysik der Deutschen Akademie der Wissenschaften zu Berlin (1940–1970)*, 10.
22. Jan Erik Schulte, *Zwangsarbeit und Vernichtung: Das Wirtschaftsimperium der SS*, 407.
23. L. D. Riabev, ed., *Atomnii projekt SSR 1939–1945*, vol. 2; and a letter from Kurchatov to Stalin of 30 March 1945, on the German atomic bomb.
24. Arnold Kramish, *The Griffin*.
25. Samuel Goudsmit, *Alsos* and Boris T. Pash, *The Alsos Mission*.
26. Goudsmit, *Alsos*, 79.
27. Pash, *The Alsos Mission*, 157.
28. *Operation Epsilon: The Farm Hall Transcripts*, with and Introduction by Sir Charles Frank, Bristol and Philadelphia, 1993, 71.
29. Pacific War Research Society, *The Day Man Lost*, 153.
30. Wilcox, *Japan's Secret War*.
31. Christopher Andrew and Vasili Mitrokhin, *The Mitrokhin Archive*, 150.
32. Holloway, *Stalin and the Bomb*, 84.
33. Interview with Iulii Khariton in *The Bulletin of Atomic Scientists*, May 1993.
34. Holloway, *Stalin and the Bomb*, 91.
35. Interview with Khariton in *The Bulletin of the Atomic Scientists*, May 1993.
36. Andrew and Mitrokhin, *The Mitrokhin Archive*, 165–166.
37. *Idem.*, 173.
38. *Idem.*, 174.
39. Leslie Richard Groves, *Now it Can be Told: the Story of the Manhattan Project*, 69.
40. Groves, *Now It Can Be Told*, 230.
41. Werner Schüttmann and Helmut Schnatz, "Ein erster Schritt zum Kalten Krieg? Der Amerikanische Luftangriff auf Oranienburg am 15 März 1945," *Der Anschnitt*, Zeitschrift für Kunst und Kulture im Bergbau, 50. Jg., Heft–3/1998.
42. *Idem*.
43. Air Force Historical Research Agency, Maxwell, item 520 334, 8 US Air Force Mission, B5023, Planning of the Mission on 15 March 1945, quoted in Schüttmann and Schnatz, op. cit.
44. Landesarchiv Berlin, Rep 05-07, Nr 28.
45. Nikolaus Riehl and Frederick Seitz, *Stalin's Captive: Nikolaus Riehl and the Soviet Race for the Bomb*, 79.

46 John Wheeler-Bennett and Anthony Nicholls, *The Semblance of Peace*, 271.
47 L. F. Ellis, *Victory in the West vol. 2, History of the Second World War*, 331.
48 George S. Patton, *War as I Knew it*, 326–7.
49 A personal communication from Stanislav Kokoška, okresni archiv Karlovy Vary-Rybare, T45155.
50 Stefan Heym, *Schwarzenberg*, 12.
51 For instance Paul Reimar, *Das Wismut-Erbe. Geschichte und Folgen des Uranbergbaus in Thüringen und Sachsen*.
52 Gregg Herken, *The Winning Weapon: The Atomic Bomb in the Cold War 1945–1950*, 107.
53 Norman M. Naimark, *The Russians in Germany: A History of the Soviet Zone of Occupation 1945–1949*, 210.
54 Mark Kramer, "Documenting the Early Soviet Nuclear Weapons Program," *Bulletin of the Cold War International History Project*, Nr 6–7, Washington 1995.
55 Vladislav Zubok and Constantine Pleshakov, *Inside the Kremlin's Cold War: From Stalin to Khrushchev*, 142.
56 Nikolaus Riehl, *Zehn Jahre im goldenen Käfig*, Stuttgart 1988; Nikolaus Riehl and Frederick Seitz, *Stalin's Captive: Nikolaus Riehl and the Soviet Race for the Bomb*; Andreas Heinemann-Gruder and Arnd Wellmann, *Die Spezialisten: Deutsche Naturwissenschaftler und Techniker in der Sowjetunion nach 1945*, Berlin 1992; and Norman M. Naimark, *The Russians in Germany*, Cambridge Mass, 1995.
57 Nikolaus Riehl, *Zehn Jahre in goldenen Käfig*, 5 et seq.
58 Archive for Russian History, Moscow, special file J. V. Stalin, Beria's letter to Stalin of 15 May 1945.
59 Riabev, *Atomnii projekt SSR 1939–1945*, vol. 2, 323–325.
60 Albrecht Ulrich, et al., *Die Spezialisten: Deutsche Naturwissenschaftler und Techniker in der Sowjetunion nach 1945*, 50 et seq.
61 David Holloway, *The Soviet Union and the Arms Race*, 21.
62 Holloway, *Stalin and the Bomb*, 193.
63 Lev Altschuler as quoted in Karel Pacner, *Atomoví špioni*, Prague 1994, 456.
64 W. I .Wetrow, "Die Bildung von Betrieben für die Gewinnung und Aufbreitung von Uranerz," *Ministerium der Russischen Federation für Atomenergie, Die Entwicklung der ersten sowjetischen Atombombe*, 1.
65 Pavel Knyscheweskij, Moskaus Beute, *Wie Vermögen: Kulturgüter und Intelligenz nach 1945 aus Deutschland geraubt wurden*, 78.
66 Vladimir Picugin, "Aus der Geschichte des sowjetischen Atomprojektes," 57.
67 Pavel A. Sudoplatov, *Special Tasks*, 199.
68 *Idem*.
69 Nikolai Grishin, "The Saxony Uranium Mining Operation ('Vismut')," 127 et seq.
70 Rob Roeling, *Der grote Trek naar het Duitse Ertsgebergte. Arbeiders in de*

uraniummijnbouw: dwang, vorlokkingen en sociale omstandigheden, 1946–1954, 16 et seq.

[71] Stadtarchiv Schneeberg, Schreiben des Sachsenerz-Bergwerk und die Landesverwaltung Sachsen, Abteilung Gesundheitswesen, 21.9.45. The authors are grateful to Gotthard Bretschneider for indicating the source.

[72] Norman Fuchsloch, Forschungen zur Uranprospektion an der Bergakademie Freiberg im Auftrag der Sowjetunion, in *Der Anschnitt*, Heft 2–3, 1998; *Idem*, Rainer Karlsch, Der Uranwettlauf 1939 bis 1949.

[73] Wismut GmbH, Geologisches Archiv, Nr 55 187, Bl 1–35.

[74] *Idem.*, Bl 25.

[75] Chronik der Wismut GmbH, Teil 1, CD-Rom Chemnitz 1998, 20.

[76] Werner Schüttmann, "Die Geschichte der 'Schneeberger Lungenkrebses,'" *Der Anschnitt*, Heft 2–3, 1998.

[77] RADIZ Information 11/1996, biography of Adolf Krebs and RADIZ 9/1996; Inge Meutzner, Zeitzeugenbericht.

[78] Nikolai Grishin, "The Saxony Uranium Mining Operation ('Vismut')," 127

[79] Wismut Unternehmensarchiv, Mikrofilm Nr 423/Teil 4, Bl 71, Jahresbericht 1953 (in Russian).

[80] Hoover Institution Archives, William Sander, Box 1, Folder ID, Reports 1946: Demontage des Radiumkurbades Oberschlema.

[81] Chronik der Wismut, 21.

Part 2

The Erzgebirge Region

The Silver Mines and Healing Springs

From the early middle ages onward, the mining of silver, tin and of other ores has been carried out in the border region between Saxony and Bohemia. An impenetrable mountain forest, the Erzgebirge became one of the most densely populated regions in Europe. According to legend, carters from the Harz Mountains found traces of lead on the wheels of their wagons when they passed through the site of present-day Freiberg. The news of the discovery of metal ore reached Siegerland, a rich source of iron in North-Rhine Westphalia, and its miners laid the foundations for the industry early in the 12th century. So, at any rate, goes the legend.

The first three pits, sunk where the town hall now stands, yielded silver ore of high quality. Freiberg became a free mining town in 1218, flourishing during the reign of Margrave Henry the Enlightened (Heinrich der Erleuchtete). The Freiberg mining laws were enacted in 1296, becoming the basis for mining regulation in all Saxon towns and elsewhere in the mining regions of Europe. The local mining school (Bergakademie) became the first institution of its kind in the world.

The initial period of silver mining, lasting from the 12th to the13th century, was followed by intensive tin mining during the 14th and 15th centuries. Tin mining began in Graupen (Krupka), Bohemia circa 1230, and a century later in the region of Neustadt. This extensive mining challenged the English tin monopoly.[1] In addition to tin, bismuth was also mined, and another silver rush began in 1470.[2] "Much commotion in the mountains" (das große Berggeschrei) was set off, as the news of the discoveries spread and drew many miners and their families to the Erzgebirge. New towns were founded in quick succession: Schneeberg in 1471, Annaberg in 1497, Sankt Joachimsthal 1520 and Marienberg in 1521. Their growth reflected faith in the future of the mining industry. There was little agriculture in the mountains and food supplies had to be brought from afar. The situation lasted until the 20th century, accounting for cyclical crises during the mining recessions.

Schneeberg became a free mining town in 1481. New discoveries of silver helped the town, and the ruler of Saxony (Kurfürst), achieve uncommon riches. Frederick the Wise referred to the town as "little Venice." There were iron and copper mines and smelting plants in a nearby valley at Schlema, and silver was mined in Niederschlema. People in the neighboring town of Zwickau benefited from their connection with the industry, as miners or workers in a variety of service industries. In the 20th century, strong radioactive springs were discovered at Oberschlema, contributing to a flourishing spa enterprise.

There existed in the region the famous "Markus Semmler Stolln," a water conduit for the Schneeberg and Schlema pits. It was laid down in 1503, reaching the length of 44 kilometers in the 19th century. The danger of flooding increased as excavations went deeper. Flooding affected the Schneeberg pits on a large scale at least three times, in 1470, 1491 and 1511. Every time, the miners succeeded in bringing production back on stream, but the center of mining moved elsewhere, to Annaberg and Sankt Joachimsthal, or Jáchymov in Czech.

Prosperity spread from the Schneeberg district to the village Aue, which became a town in 1626. Aue specialized in the mining and processing of tin, a product that enjoyed a high reputation at the time. In two settlements nearby, Auerhammer and Niederpfannenstiel, sheet metal and blue dyes were made, and the locality became a well-known metal-processing center. After 1945, the town became the administrative center of the uranium industry. Annaberg became a free mining town in 1497 and, until the middle of the 16th century, it grew faster than any other town in Saxony. Home to about 12,000 inhabitants in 1540, it was the second largest town in the Erzgebirge after Freiberg.[3]

Silver mining was the first business in the Erzgebirge that paid wages, giving the miners a freedom rarely enjoyed by other workers. Hard work and its dangers created strong bonds in the mining community, leading to the establishment of its first benevolent organizations. Apart from the welfare function miners' guilds performed—supporting those members who had fallen on hard times—they played an important role in improving the living and working conditions of the miners. Guilds helped with projects such as the establishment of health commissions and the building of accommodation for the miners.

The silver boom ended about 1570, when silver imports from the new world diminished the value of the Erzgebirge silver. The towns lost their prosperity and many of their inhabitants. The decline of the region reached its nadir between 1618 and 1648, during the Thirty Years War. Silver mining in the Erzgebirge revived about 1660, when new finds were made in the neighborhood of Johanngeorgenstadt. The town was founded by Protestant refugees from Bohemia who had come under the rule of the Catholic Habsburgs. Protestant families from the mining towns of Platten, Gottesgab and St. Joachimsthal crossed the border to Saxony. In February 1654, the ruler of Saxony, Kurfürst Johann Georg I, gave them permission to found a town and give it his name.[4] In addition to silver, cobalt, iron, tin and sulphur were also mined there.

The town had a special connection with cobalt, which was used for the production of blue dyes (Blaufarben); the skill first came to Saxony from Platten and Joachimsthal in Bohemia at the beginning of the 17th century. Workshops were established at Johanngeorgenstadt, Schneeberg and later at Oberschlema. The rulers of Saxony were aware of the importance of the dyes for export. Under the threat of capital punishment, they forbid the export of cobalt ore in 1683.[5] By the middle of the 18th century, mining at Johanngeorgenstadt was in a severe crisis, and had to be supported by state subsidies. In 1838, seven profitable pits formed a union under the name "Vereinigt Feld am Fastenberg." They produced silver, bismuth and uranium.

As metal prices declined in the 19th century, the profitability of Freiberg's mining was affected: nationalized in 1886, they were kept alive by subsidies. After 1908 two state mines, "Himmelfahrt" and "Himmelfürst," as well as a trade union mine, were still operating. The state mines were closed down shortly before the outbreak of the First World War. At the beginning of 20th century, some cobalt, wolfram, tin and nickel were still mined in the area.

Sankt Joachimsthal was an important part of the mining region on the border between Saxony and Bohemia which shared a common past. The valley of St. Joachim is situated in a fold of the Erzgebirge mountains, or Krušné hory in Czech. The German name of the mountain range refers to its abundance of ores; the Czech to the harshness of the region's climate. It formed the border between Saxony, in the East Franconian march, and the Kingdom of Bohemia. A trade route led through the forest from Annaberg to a valley on the Bohemian side, where a settlement called Konradsgrün came into being. Emperor Sigismund gave the Konradsgrün estate as a fief to his chancellor, Caspar Count Schlick, in 1437.

After the flooding of the mines at nearby Schneeberg in Saxony, the search for silver was carried over to the Bohemian side of Erzgebirge in 1511. In the following year, two experienced miners began to sink a shaft at Konradsgrün but soon ran out of money. Further excavations were financed by Stephan Count Schlick, together with a few neighboring nobles. They founded a mining company and hired miners, who struck rich veins of silver. The news spread fast and attracted more workers, most of them from Saxony. The stream of migrants was described in verse: "Ins Thal, ins Thal, mit Muttern, mit All." (Into the valley, the valley, with mothers and all.")

Konradsgrün was home to 400 houses in 1516. The Schlicks completed the Freudenstein fortress, built to protect the mining town from plunder and remind the miners who their master was, the following year. Unrest broke out in 1517. On 2 August 1518, a mining statute based on the statute of Annaberg and containing 106 articles, was published. On 6 January 1520, Konradsgrün became a royal town by decree. Its name was changed to Sankt Joachimsthal, or the valley of St. Joachim, the father of the Virgin Mary. Most of the immigrants were miners from neighboring Saxony, coming from Annaberg, Marienberg and Josefsstadt: Joachimsthal was added to keep the holy family together. The town was awarded a coat of arms and the right to choose its own local government.

Despite the royal ban on the export of silver (dating from the year 1300, early in the first silver rush in Bohemia), a part of Joachimsthal silver was sold to the great merchant house in Germany, the Fuggers in Augsburg and the Welsers in Nürnberg. Count Schlick, concerned about the loss of revenue, successfully negotiated with the Bohemian Diet the right to mint coins. In 1519, the first Joachimsthaler was issued from the cellars of the Freudenstein fortress. With a silver content of 931 out of 1000, the coin became common currency throughout Europe. It was also well known in the New World, as the derivation of the word "dollar" from "thaler" shows. Over 3 million thalers were minted between 1520 and 1528.

With 18,200 inhabitants, St. Joachimsthal had become, after Prague, the most populous city in the kingdom by 1532. Silver production had reached its peak: 134 silver veins were mined by some 8,000 miners and 800 supervisors. Some of the veins were uncommonly rich: a single mine in a quarter of the year 1524, for instance, yielded as much as 1.6 tons of silver. Total production between the years of 1516 and 1600 was put at 330–350 tons.[6] The ore was processed in 13 large smelters, which filled the valley with poisonous fumes. The mining town became unbearably crowded, with 10–15 people living in each house. The miners brought the ore to the surface in sacks, from long, unventilated galleries (stretching 11.5 kilometers in Barbora in 1589) and shafts with climbing ladders. Floods, famine and plague were common in Joachimsthal.

Though the silver boom was comparatively short—by 1574, the number of inhabitants in the valley of St Joachim had fallen to some 4,000 individuals—the town acquired several handsome buildings. The chapel of All Souls, the oldest surviving building near the cemetery, was completed in 1516, the brewery in 1518 and the old town hall in 1520.

The Bread Market formed the centre of Joachimsthal before the business of the town moved a few hundred meters further up the valley. The new town hall was completed, in the renaissance style, around the year 1531. The royal mint was opened in 1536, the church of St. Joachim close by was built between the years 1534 and 1540, and a Latin school with an outstanding library came into existence. Johannes Agricola (Bauer), who wrote a treatise on mining and was fascinated by the healing powers of minerals, was the local physician.

Traces of the original mines in the forests around the town still exist, as do mounds of soil running through the woods, left in the wake of the miners as they followed the veins of silver. The seams were gradually exhausted, and the mines abandoned. The Thirty Years War began to ravage central Europe, confirming the dividing line between Catholic Bohemia and Protestant Saxony. Joachimsthal suffered the final blow when the royal mint was established in Prague in 1670. Intermittent and unsuccessful attempts, usually sponsored by the state, were made to revive the mining of silver during the 18th century.

The great mining towns of the Erzgebirge never regained their former prosperity. In 19th century Joachimsthal, the endeavor to expand traditional crochet work was overtaken by the glove industry, a factory for the production of bottle corks was constructed, and then a paper mill. The new enterprises were joined by a tobacco factory in the 1860s, and then by a puppet factory, run by the Samuel brothers. About the same time, glassmakers and potters in Bohemia started using uranium-based dyes, which added special luster to yellow, green and orange colors. The price of uranium went up, and some of the old pits were saved as mining activities were resumed. Between 1854 and 1860 some 7.4 tons of uranium were produced in Joachimsthal. In the following decades leading up to 1910, production increased from 38.1 to 39.9 tons, and then 44.5, 53.7 and 104.4 tons. It declined to 50.1 tons between 1911 and 1920, picking up again in 1921 and 1930, reaching 154.2 tons before declining again to 130.4 tons between 1930 and 1938. More than 621 tons were produced overall.[7] Uranium-related trade became so brisk that a railway branch line was built between Joachimsthal and Ostrov nad Ohří in 1896. Uranium was used for the production of dyes at Schneeberg and Johanngeorgenstadt as well. Uranium paint factories were founded in Johanngeorgenstadt and later at Oberschlema.

On the Bohemian and Saxon side of the Erzgebirge region, there also existed local spa industries based on the belief in the healing property of radioactive waters. On the German side, the springs at Ronnenburg in

Thuringia were the first to acquire reputation for their healing properties. Duke Frederick III of Saxe-Gotha, who suffered from gout, spent a considerable part of the year there. Ronnenburg lost its position as the leading resort early in the 20th century, and its spa service was suspended.

Jáchymov, where the healing properties of the local waters had long been established, acquired an international reputation. Initially, the silver miners at Joachimsthal had drawn the waters to cure their rheumatic ailments, and Emperor Mathias II (1612–1619) had the waters carted to Vienna. Following the discovery of radium, Professor Neusser of the General Hospital in Vienna, asked the Joachimsthal mines for 500 kilograms of uranium waste in 1904. Neusser used the waste for experiments in healing. Heinrich Mache and Stefan Meyer, physicists at the University of Vienna, came to Joachimsthal on 7 January 1907. They compared the radioactivity of the local waters with other spas. Water drawn on the second floor of Werner yielded 2500 Becquerel (bq) per liter, the highest known radioactivity in water anywhere in the world.

In 1906, two cabins were opened in the house of Josef Kuhn, the local baker. A miner provided them with large, forty-liter wooden pails of water. Kuhn's establishment prospered: one bath cost three crowns—the price of two good dinners—and the cost increased if the medical services of Dr. Gottlieb were used. In 1908, the ministry of public works in Vienna started taking interest in Jáchymov. A pipeline was laid, and a spa building was opened in October 1911, with a celebration in which the local miners' brass band took part.

The year before, in 1910, a private company, financed from Vienna, had started building the Radium Palace Hotel. Set deep in the valley, off the main road before it enters the town, the luxury hotel contained 250 rooms and 40 cabins. Radioactive baths were available, there were reading and gambling rooms, a concert hall, and electric lighting and telephones were provided in every room. The hotel however kept losing money, and the war did not help its financial situation. After 1918 it was sold to an English company, and the Czechoslovak state bought it in 1922.

On the Saxon side of the border, the success of Joachimsthal in attracting visitors caused the ministry of finance in 1908 to ask Professor Carl Schiffner, of the mining academy in Freiberg, to examine connections between uranium deposits and the ground water.[8] Germany was known to have the greatest amount of radioactive springs in Europe, with some 85 accounted for. Among them, Oberschlema, Bad Brambach, Bad Kreuznach and Baden-Baden became the best known.

Dr. Max Weidig and Richard Friedrich, a builder at Schlema, took part in Schiffner's researches. Their work helped establish the spas at Oberschlema and Brambach. The springs in the vicinity of Oberschlema were strongly radioactive, and the town became Jáchymov's leading competitor. The radium spa, opened in 1918, became very popular.

After the First World War, Jáchymov returned to the spa business with renewed vigor. There was brisk speculation in local property and new spa buildings, hotels and pensions were built. A spa pharmacy selling popular radioactive ointments opened in May 1925. The pharmacist benefited from interest in radioactive products, of which Jáchymov was the leading maker.

An Early Atomic Age in the Erzgebirge

The Erzgebirge, both on the Bohemian and the German side, went through two comparatively short periods of economic prosperity. The first era was marked by the abundance of silver, the second by its absence. Both the German and the Czech languages reflect the facts of geology. When silver miners came to the end of the seam, they often struck a shiny, resin-like substance, which they called *Pechblende*—pitchblende, or *smolinec* in Czech. The German and the Czech words, *Pech* and *smula*, from which the names of the ore derive, have the same double meaning: they are the words for resin and for bad luck. *Pechblende*, or *smolinec*, the ore containing uranium, is not only shiny and resin-like; it also meant bad luck for the miners. When they struck it, the miners knew that there would be no more silver, and they had no use for pitchblende.

The element uranium was first described in 1789 by Heinrich Martin Klaproth, who used Joachimsthal pitchblende for his experiments. In 1895, Wilhelm Conrad Roentgen discovered X-rays in his laboratory in Goettingen. They were invisible, and penetrated paper, flesh or even metal. The physicist put various materials under the rays and took pictures of them. Roentgen's discovery was well received in scientific circles. Doctors and physicists soon realized the practical application of X-rays and began to experiment with them.

In Paris, Henri Becquerel, professor of physics at the Ecole Polytechnique, tried to find out whether luminous materials could replace the new rays. He put various minerals on photographic plates wrapped in transparent paper and left them in the sun. When he used a piece of uranium ore, Becquerel was astounded to discover that the metal gave out a new kind of radiation. His February 1896 discovery was the first step into the nuclear age.

Marie Curie-Sklodowska based her graduate research work on Becquerel's findings. She described the rays spontaneously created by uranium as "radioactivity," and in December 1898, together with her husband, Pierre Curie, she identified a new element, radium, with much higher radioactivity than uranium. Her tireless research work in the instability of matter has become legendary, and the Curies received the 1903 Nobel Prize for physics. Ernest Rutherford later discovered that uranium gave out several kinds of rays, and described the structure of the atom.

Radium excited the imagination of scientists, medical doctors and quacks. It was available only in small quantities and was thus extremely expensive. For the production of one gram of radium, about two tons of uranium were needed. The processing plant at Joachimsthal produced 1–2 grams of the precious element a year. In 1902, the French paid 15,000 francs for one gram of radium, before its price sharply increased. In the United States before the First World War, one gram of radium cost about $100,000; in Czechoslovakia after the war, it could be had for 10–12 million crowns. The price collapsed to 1.7 million crowns for a gram in 1933.

Because of the long half-life of radium (1,590 years), it was usually leased rather than sold, sealed into glass or platinum tubes. Trade in radium also took the form of the sale of radium salts. The new element was enthusiastically taken up in medicine. At the end of the 19th century, experimental treatment of diabetes, cancer or various cysts with radioactive materials was introduced. Before the First World War, radium preparations were applied to skin disorders. A variety of home remedies came on the market and radiation therapy became popular. The boom in radium-related products led to the 1927 foundation of a Czechoslovak company called Radiumchema, which made radium-based soaps, "Radisapon" and "Sulforadon," as well as hair shampoo. It also produced a luminous souvenir, called "spinthariscope," and a device called the "pocket spa," which produced radioactive water.

This device retailed for about $100 in 1932, and caused internal radiation, which sometimes proved to be fatal. This "Pocket Jáchymov" was banned in America, but Radiumchema continued to sell the product throughout the world, maintaining that it corresponded to the highest medical standards. The popularity of radioactivity affected the local brewery as well, which launched its Radiumbier in 1930. The reputation of Jáchymov as the prime source of uranium and of radium in Europe had been established prior to World War I.

Trade in luminous dyes also found a new outlet at that time: the dyes started being used in the aircraft industry, and watches with luminous dials became popular. In America alone in 1920, four million luminous wrist-watches were produced. Workshops using luminous paints employed more than 2,000 workers, and the US Radium Corporation was one of the best-known companies in the field. The women workers who painted the dials had the habit of licking the paint brushes used in their work, exposing themselves to a variety of health risks. The company was sued, and the trial attracted much public attention.[9] Towards the end of the 1920's, the US public health authority tackled the problem of radium poisoning.

Several German companies also made radium products. They included Buchler of Braunschweig, which employed a leading radium expert, Dr. Friedrich Giese, until his death in 1924; Auergesellschaft and Allgemeine Radium Gesellschaft in Berlin and Radium Chemie AG in Frankfurt.[10] It processed uranium which largely came from Jáchymov.

Early in the 1920s, Union Miniere du Haut Katanga, located in the Belgian Congo, was the most prominent business dealing with radium and its derivates. During the economic crisis at the end of the decade, the demand for radioactive materials dwindled. At that time, rich sources of uranium were discovered in Canada by Great Bear Lake. Through its plant at Port Hope in Ontario, the Eldorado Gold Mine company became the world's largest producer of uranium. In the year 1937, Czechoslovakia ranked third in world radium production, with 11% of the total, while the Belgian Congo produced 15% and Canada 66%. The remaining 8% was produced in other countries, and total world production amounted to 39 grams.[11] About half of this radium was used for medical purposes, particularly for the treatment of cancer, more than a quarter was used for the production of luminous paints, and the rest for scientific purposes.

There was a marked increase in the production of radium before the Second World War. The Americans, the British and the Germans wanted to lay down strategic reserves of the material,[12] and military demand for luminous paints also increased sharply. Because of its link with radium, uranium was considered to be an important war material. For the time being, it was only needed in small amounts.

National Tensions in Jáchymov

The population of the St. Joachimsthal district was predominantly German when it passed from Austrian to Czechoslovak administration in 1918, shortly after the end of the First World War. Mining remained the main occupation of the Germans, and the district continued to be administered by former Habsburg officials. The Czechs started arriving in larger numbers in the mid-l930s, many of them to fill vacant posts in the district administration. Economic crisis weighed heavily on the Erzgebirge industries.

Most Jáchymov Germans voted for the Social Democrat Party, which ran a separate, German organization in Czechoslovakia. The voting habits changed during the general elections in 1935, and many Germans in the Sudetenland, including the Jáchymov district, turned to Konrad Henlein's Sudetendeutsche Partei (SdP). With open anti-Czech propaganda and secret financial support from Berlin, the SdP gained more votes than any other political party in Czechoslovakia.

The slump affected the Sudeten Germans more severely than the Czechs, and Henlein's SdP exploited discontent in the Jáchymov mining community. The miners complained that they were paid short hours and that they had to buy their own tools. Safety measures were inadequate and mortality rates kept on rising. In the 1920s, the miners derived meager comfort from the belief that the deaths were a part of the aftermath of the First World War, before other explanations were offered in the 1930s.

After the onset of the world economic crisis, miners' wages fell below the level earned by workers in the local glove factory. Charitable organizations supplied school children with milk and bread—often the only meal they had during the day. The miners went on strike in February 1938, and sent a delegation to Prague to complain to president Beneš. The delegation waited twelve days for an audience with Beneš, but was apparently told that the president would not talk to the strikers.[13]

Tensions between the Sudeten Germans and the Czechoslovak government were about to culminate in 1938. The issue between Czechoslovakia and Hitler's Reich was no longer whether the German minority would be granted—and accept—autonomy. In March, Hitler annexed Austria. In September in Munich, the Reich was awarded the border territories of Bohemia and Moravia, including Jáchymov. Politics sped up in Central Europe, and the Czechs twice mobilized and twice sent their troops home. After Munich and the occupation of the Sudetenland, many

Czech mining officials left Jáchymov: out of 42 Czech employees, only seven stayed behind.

It so happened that, in the Kaiser-Wilhelm-Institute at Berlin-Dahlem at the end of 1938, Otto Hahn succeeded in splitting the atom. He had used radioactive materials from Jáchymov for his experiments for some twenty years, and in 1938 he received five tons of radioactive waste from Jáchymov. Inside Hitler's Reich, the former Czechoslovak Sudeten territory became Sudetengau and Konrad Henlein, the leader of the former Sudetendeutsche Partei, was its political master. On 7 February 1939, the Reich ministries of economy and of finance recommended to the cartel of German radium processing companies that they establish, as soon as possible, a limited company in Joachimsthal. A leasing contract was to be made between the state and the company. Because of the high costs involved in running the uranium mines, there was no hope they would return a profit. The difference between production costs and the price of raw materials was to be made up by the Reich treasury.

The St. Joachimsthaler Bergbaugesellschaft was nevertheless established on 7 March 1939, eight days before Hitler occupied Bohemia and Moravia. Its capitalization was to be 300,000 RM, which was equally divided between the Auer Gesellschaft AG in Berlin, the quinine factory Buchler & Co of Braunschweig and Treibacher Chemische Werke AG. The company had three mines employing 234 workers, and new pits were to be opened in case the demand for uranium or radium increased.

Between 3 and 4 grams of radium were to be produced in 1940,[14] requiring the production of uranium to double. The company did not meet its target, producing only 1.75 grams of radium from of 12.5 tons of uranium. Later increases in radium production never managed to achieve the original target.

The Joachimsthaler Bergbaugesellschaft GmbH was an essential war industry and was given priority in the allocation of labor. After the invasion of the Soviet Union, problems with the labor supply became acute, though the "Göring program" promised increased employment of prisoners of war and of foreign labor. At the end of 1941, Göring sanctioned the use of Soviet POWs on a massive scale. The forced labor program was fully implemented under Fritz Sauckel, who became the Reich plenipotentiary for labor in February 1942.

Himmler's SS had tried and failed to use prisoners in Jáchymov as early as 1939. The following year, after the first German campaign in the West, French POWs started arriving in Jáchymov. There were no miners among them, and many were peasants from Brittany. (After pressure from the International Red Cross, they were moved to other jobs. According to the Geneva Conventions of 1929, which were ratified by the German government in 1934, prisoners of war could not be used for unsafe jobs.) They were replaced by Soviet POWs, who were not so protected. They would remain in Jáchymov until the end of the war.

After the medical officer in the Sudetengau administration proposed that foreign labor should be employed in Jáchymov, replacing German miners,[15] the mining office (Bergamt) in Karlsbad opposed the proposal on the grounds that it devalued the miners' work. The mining official argued that the miners should remain at their posts, as did the soldiers. The Karlsbad office received support from a higher instance, the Oberbergamt at Freiberg.[16] The miners stayed in their jobs, but no more Germans were recruited for work in Jáchymov during the war.

At the beginning of 1942, there were 196 workers employed in Jáchymov mines, together with 21 technical and administrative employees. There were 37 French POWs still working at Werner, where the management was planning to extend mining operations, as well as at the mine of the Saxon Nobles. On 27 January, in the business plan for the year 1942, Kurt Patzschke, the manager of the Joachimsthaler Bergbaugesellschaft, requested the allocation of 25 foreign workers or Soviet POWs. They were carefully selected, replacing the Frenchmen in June 1942. The manager reported an increase in productivity.[17]

In 1943, some 100 Soviet POWs worked in Jáchymov, about matching the number of German workers there. Production targets were not high: in June 1944, the Oberbergamt in Freiberg set monthly production at 0.9 tons of uranium oxide. The management argued that, at the request of the Ministry of the Economy (Reichswirtschaftministerium), 26 miners had been released. It asked the Karlsbad office to replace them with Soviet POWs. There were shortages of surface workers and of craftsmen such as locksmiths and electricians. The management suggested that the production norm should therefore be lowered to 0.8 tons of uranium oxide. Allied intelligence agencies noted that the production of uranium at Jáchymov was not significantly increased during the war.[18]

Sickness and Nationality

Man-made disasters, including the virtually continual political regime changes of the 20th century, as well as the demographic decline effected by the expulsion of the Germans from Czechoslovakia in the aftermath of the Second World War, tended to overshadow the natural calamities from which the Erzgebirge region suffered.

There existed a well-established tradition among the Erzgebirge miners that their work was somehow linked with disease. The "Schneeberg mining sickness" (Schneeberger Bergkrankheit, Bergsucht) was accompanied by a cough and breathing difficulties, and it killed many young miners. The sickness was identified by modern medicine as lung or bronchial cancer, arising from radioactive conditions underground. The first recorded references to the miners' disease were made in the 1520s. Georg Bauer, known as Agricola, who worked as a physician in Joachimsthal, pointed to dusty, contaminated air as the cause of the disease, and recommended better ventilation. His successor, Magnus Hund, described the symptoms as "difficult breathing, anxiety and pressure in the chest," and recommended change of place and of occupation (hard advice for the miners to follow). Debate on the disease was initiated by Theophrast Bombast von Hohenheim, known as Paracelsus, in a 1567 book entitled *Von der Bergsucht oder Bergkrankheit drey Bucher*.

Apart from medical reports, the texts of several sermons offer information on the conditions in the mines. Clerics at the time of the reformation tried to link the religious message of their sermons to the working lives of their flock. Johannes Mathesius (1504–1565) was the most influential among them. His sermons, published in Nurnberg in 1562 under the title *Sarepta* (or Bergpostill) contain references to the condition of air underground, to the spread of fires and to avoidable accidents.

Connections between the conditions in the mines and the miners' complaints continued to be made in the 17th and 18th centuries, particularly in a 1728 study of mining and other occupational diseases by Johann Friedrich Henckel, the town physician at Freiberg. Carl Leberecht Scheffler, the Annaberg physician, devoted his 1770 book to the dangers of the mining profession. The "Schneeberg sickness" came into prominence again in mid-19th century, when systematic uranium mining began and cases of lung cancer became more frequent. Two doctors, Harting at Schneeberg and Hesse at Schwarzenberg, studied the disease and reported on their findings in 1889. They identified cancer in miners who stopped work between the ages of 30 and 35, and who died of illnesses of the respiratory organs within ten years. Bergsucht was identified as lung cancer, and in 1881 "Schneeberg lung cancer" was entered as an occupational disease in a reference book on public health (Handbuch des offentlichen Gesundheitwesens).

Radon gas, a product of natural radioactivity and as the source of alpha rays, was discovered in 1900 by Rutherford. Radon-222 is of special importance in mining. With half-life of only 3.8 days, it originates from crystalline minerals and it can be released from water and dispersed by air streams. The link between the "Schneeberg sickness" and exposure to radon was traced by Professor Carl Schiffner at the mining academy in Freiberg. Schniffner was commissioned in 1908 by Saxony's ministry of finance in to explore the effect of uranium deposits on waters underground. Schiffner detected a high content of radon in the air in the shafts as well. A few years earlier, it had been established that contact with X-rays may cause cancer, and the discovery helped to indicate the radon link with cancer. When H. Müller, director of the Zwickau mines, gave reference for one of the miners, he wrote "I therefore regard the Schneeberg lung cancer as a particular occupational disease of the mines, where the minerals contain radium and where the air is contaminated by a strong emanation."[19] The theory was disputed: was the sickness cancerous, tubercular, or was it simply a lung condition caused by the dust?

Between 1922 and 1926, health officials in Saxony carried out numerous examinations of the miners. They worked in pits reaching depths of 300 meters which had no hoists. Some were very damp, with temperatures fluctuating between 9 and 16°C. Pickaxes were used in soft soil, while the rock was worked with pneumatic drills; they were available for two hours every day and they created a lot of dust. Dust masks were available but were hardly ever used. The working day was eight hours.

The examinations involved I54 miners from Schneeberg, as well as a control group from the Oberschlema paint factory and 186 other inhabitants, making 516 people altogether.[20] The results of the enquiry were published in 1926 and created a considerable stir in medical and mining circles. 21 of the miners died, with lung cancer given as the cause of death for 13 of them. Though the cause of the Schneeberg lung cancer was not yet identified with absolute certainty, it was included in the list of occupational diseases in the regulations concerning such diseases of 12 May 1925. The regulations coupled silicosis with lung cancer and caused disputes with insurance companies.

Geological conditions in Joachimsthal were similar to those in Schneeberg, and the question arose whether the miners were exposed to similar dangers. The investigations at Joachimsthal took place about the same time as those in Saxony. The medical faculty at the university in Prague asked the local spa doctor, Max Heiner, to carry out the investigation.

Two years later, Heiner tried to convince the Prague medical association to work out health regulations for the protection of the miners. In the years 1924 and 1925, František Běhounek of the Radiological Institute in Prague took systematic measurements of radioactivity in the local waters, the open air and in the mines. They showed high levels of radiation in the mines and in their immediate neighborhood.

Early in 1929, Professor Julius Löwy of the university clinic in Prague published the first results of examinations of two Jáchymov miners who died of lung cancer. "Conditions in Joachimsthal and Schneeberg mines are similar" he wrote, "insofar at Schneeberg, the emanation content in the air is 50 and in Joachimsthal 40 Mache units (Macheeinheiten) and the dust in the mines contains arsenic. We have therefore the same causes, the same symptoms of the sickness and we can make the same diagnosis." Löwy regarded the existence of occupational lung cancer in Jáchymov as highly probable. Post-mortem examinations of a large number of Jáchymov miners carried out between 1929 and 1938 gave grounds for alarm. The findings were disputed and little was done to act on them.

For a long time, concern with occupational health hazards remained confined to the Erzgebirge region. The medical research was internationally acknowledged much later, when doctors in America after the Second World War discovered the older studies, including the work by Harting and Hesse. In the US, both the medical profession and the mine operators tended to argue that the special conditions obtaining in the Schneeberg region, such as its poverty and malnourishment, caused the disease. This was an error, which once again sent the debate in the wrong direction. In the 1950s, American research finally established the connection between the inhalation of radon and its byproducts with lung cancer.

Löwy's research in Prague encouraged further studies on the subject of radiation-related diseases, which later acquired a curious political significance. The world economic crisis diminished the living standards of the mining community, and the Nazi regime would misuse the crisis to their own ends. Investigations, started in 1936 by Saxon physician Artur Brandt, led to Berlin placing increased pressure on the Czechoslovak government to improve the working conditions in the mines. Henlein's supporters complained that regular examinations of the miners did not take place, and that a Czechoslovakian law passed on 1 June 1932, which classified lung cancer as an occupational disease, was being disregarded. After protest meetings by the miners, supported by Henlein's party, the Czechoslovakian Ministry of Health decided, in the autumn of 1937, to establish a medical counseling and research service in Jáchymov.

In the meanwhile, Dr. Max Heiner had assembled considerable documentation on the miners' sickness, combining his own research with information from a long series of observations dating back to 1907. He wrote a memorandum and several studies, including a list of deceased miners between 1913 and 1938. Heiner put the average age of the miners at 45.5 years. According to his calculations, the causes of early death amongst the miners were cancer of the respiratory organs (46.6%), acute silicosis (*Staublunge*) (22.4%) and tuberculosis (20.6%). He described the cases of lung cancer as damage caused by radiation and the miners' poor physical condition, a consequence of undernourishment.

The damage caused by dust began being reduced at Jáchymov early in the 1930s, when American water-jet drills were introduced to the mines. Ventilation in the Einigkeit and Werner pits was deemed insufficient because of the inadequate provision of air shafts, and unprofessional mining techniques there were criticized.[21] Heiner's complaints had political undertones: Czech managers in the state mines supervised predominantly German workers. Personnel changes occurred in the 1930s, and the old Austro-Hungarian officials were replaced by Czechs from the local administration and the nationalized radium and spa industry. Heiner's criticism of the conditions in Jáchymov was in line with Berlin's increased pressure on Prague prior to the Munich agreement.

Heiner proposed medical examinations of all the newly employed workers. Those whose health was in doubt would immediately have the right to up to eight weeks paid leave. In cases of proven radiation-related disease, the worker would be retired early. Working time should have been reduced to seven hours and the miners were to be entitled to six weeks holiday a year. The causes of excess dust were to be reduced, and radioactivity in the air was to be measured once a fortnight. The miners were also to have the opportunity to change their clothes before and after the shift.

The living standard of the 300 Jáchymov miners and their families deteriorated in the 1930s, and this decline made the miners receptive to Henlein's party propaganda. The SdP positioned itself as the natural protector of the miners, praising and flattering them while making much of their low pay and insufficient social welfare. They linked the short life expectancy of the miners with the Czechs' poor management. It was all grist to the Nazi propaganda mill, helping Henlein and Goebbels to pillory Czechoslovakia. In the centrally-directed press of Hitler's Reich, the German miners in Jáchymov were described as living in appalling conditions and dying early.

The agitation by the SdP created very bad working conditions in the mines, which nothing but Hitler and National Socialism could cure. Early in 1938, well before the Munich agreement, the SdP had asked the Czechoslovakian parliament to recognize the Jáchymov miners as a separate occupational group and prepare a special law for the protection of persons working with radium and related materials. The failure of the miners' February 1938 visit to Prague to discuss their grievances with president was exploited by SdP propaganda.

The Munich agreement was signed on 29 September 1938, and war was postponed at Czechoslovakia's expense. Konrad Henlein became Hitler's proconsul in the Sudetengau, and the Joachimsthal mines and radium processing plant passed under control of the German state. Kurt Patzschke, who became the secretary of the new company, Joachimsthaler Bergwerke GmbH, complained about the low productivity of the miners. For generations, the miners had been forced by inadequate pay to have another occupation in agriculture, cattle breeding or forestry. They were, in Patzschke's words, "half-baked miners." In recent months they had started receiving an adequate basic wage, Patzschke maintained, and they should be able to maintain a higher standard of living, afford a better diet and take more pleasure in their occupation. He referred to the psychosis linked with the diffusion of radon, expressing the hope that better ventilation and the reduction of dust would remove the obsession with the effects of radiation. The obsolete processing plant, which became a danger for its employees as well as for the visitors of the popular spa, was to be razed to the ground.[22]

In November 1938, the Nazi Deutsche Arbeiterfront (DAF) promised the Joachimsthal miners improvements in their pensions and standard of living. In July 1939, Dr. Artur Brandt became the head of a research institute in Dresden, from where he started to build an occupational health service in Sudetengau. When the medical examinations of the miners in Joachimsthal were completed, they were rewarded with a fortnights' holiday at a miners' rest house at Hartenstein. Martin Mutschmann, the Gauleiter of Saxony, arranged the treat for them.

Brandt ordered investigations that, in addition to lung cancer, also traced the incidence of silicosis. It was found that 80% of the miners who had worked more than twenty years suffered from various degrees of silicosis. Brandt's report pointed to three severe health hazards: silicosis, lung cancer and diseases of the blood due to radioactivity. It also emerged that from the insurance point of view, the miners would have been worse off if German Reich laws were applied in the Sudeten territory, than they had been in Czechoslovakia. Dr. Brandt therefore asked for a new definition of "miners' sickness," as well as a review of the pension provisions.

He proposed to supplement the existing Reich regulations on occupational diseases of 12 December 1936 with a clause on the "Schneeberg miners' sickness." He argued that the miners expected an improvement in their living conditions after the incorporation of the district into the Reich, and they had been promised such an improvement "in the heat of political struggle."[23] After the political struggle against the Czechs had been won and there was no longer any need to blame the Czech authorities, it was admitted that ventilation in the mines had been improved in the 1930s and that jet-cooled drilling machines had been introduced.[24] These measures led to the decrease of radioactivity in Jáchymov mines, which fell below the levels recorded at Schneeberg. The fear of the "Joachimsthal miners' sickness," which had been promoted by Nazi propaganda, now threatened to rebound against the new political masters.

Brandt was aware of this situation, and a promise was made to limit the miners' employment in the industry to fifteen years, as well as the immediate introduction of new regulations regarding compensation. However, the Reich insurance agency and labor ministry took exception to the hurried proceedings, and the issue was shelved.

During the debate in the spring 1940, doubt was once again cast on the connection between radiation and silicosis. The health office of the DAF organized a conference on the Joachimsthal sickness at the end of March 1940. Brandt, Patzschke and Dr. Boris Rajewsky, the head of the Frankfurt Institute for Biophysics, reported on the latest findings. Examinations of 389 miners revealed put life expectancy of the miners working underground at only 39 years. Rajewsky presented the results of some 2,000 emanation measurements, according to which the air underground in Joachimsthal contained between 3 and 5 Mache units, with occasional measurements of 18 units. These were considerably lower values than those found in Schneeberg, but the waters in Joachimsthal had a much higher radioactive content. Experts at the conference regarded any concentration of radon up to 3 Mache units as harmless to health.

It was confirmed that lung cancer, silicosis and blood diseases were directly linked with the miners' work, whereas TB was only aggravated by radiation.[25] Insurance matters were raised, and it was understood that miners' fears about their health had to be allayed. Representatives of the mining authority in Freiberg suggested that the description "Schneeberg sickness" should be substituted with "radium sickness," so as not to alarm the Schneeberg miners.

Difficulties with definitions persisted. On 6 November 1940 the Reich Ministry of Labor wrote to the minister of economy and labor in Saxony that "The Schneeberg lung sickness is in fact lung cancer, which seems to be linked with silicosis. The degree of silicosis is however not decisive in the development of lung cancer. It can be assumed that the effect of radioactive rays or emanation has an adverse influence on the origin and development of silicosis and that silicosis may be regarded as the Schneeberg lung sickness..."[26] The Ministry of Labor was at the time aware that silicosis in the uranium industry had a different character than in other mining sectors, and that the development of the disease, in the case of the uranium miners, was harsher.

On 21 November, the Karlsbad mining office issued a provisional order on "preventive measures for the employees of St. Joachimsthaler Bergbau GmbH." It included regulations on ventilation, dust reduction, rooms and baths for the employees, medical examinations and working hours. The maximum permissible level of radiation was put at 5 Mache units, and places with over 22 units were to be reported to the mining authority.[27]

The Karlsbad regulations were the first ever attempt to achieve protection against radiation in the uranium industry. They were formulated in an area under double pressure: of genuine medical necessity on the one hand, and of Nazi propaganda on the other. The Nazis had been hoisted by their own petard in the Sudetenland and had to deal with the panic they helped create. In addition to the Karlsbad Bergamt measures, the DAF "office for health and public protection" made a "proposal for compensation in sickness in the Joachimsthal uranium mines" in September 1940.[28]

Medical evidence was carefully assembled, but it was not followed by improvements in the living conditions of the miners. The wartime economy made uncompromising demands on the population of the Reich, shifting the debate to the question of whether foreign labor should not be forced to work in the mines. Brandt's report of 10 October 1942 had shown that the diagnosis of Schneeberg lung cancer equaled a death sentence, with a life expectancy of only three to six months. Once again, Brandt played the national card. He argued that measures for the protection of the miners would no longer help those who had worked "under

Czech conditions," and proposed the "final solution of the Joachimsthal problem."[29]

The president of the Reich health office, Hans Reiter, used sentiments such as "principles of chivalry" and "heroism of labor," casting doubts on the special position of the Joachimsthal miners. In the mines of Schneeberg, he pointed out, the risk of lung cancer was not lower than in Joachimsthal, and he objected to special regulations for uranium mines anywhere. In the end, Rajewsky's views on the matter prevailed. He summed them up in a memorandum on 15 March 1943. The decisive criteria for him were the improvement in working conditions and the miners' length of service. He proposed alternative surface and underground employment, and turned down the substitution of German miners with foreign labor.[30]

The final decisions were made on 24 March 1943 at the Ministry of the Economy. The miners were to be retired after fifteen years' work underground, after which they could be replaced by workers from the East. Special regulations for the Joachimsthal miners were to be retroactive, effective 1 January 1943. Every miner who had worked for more than 180 months was entitled to a pension of at least 80 marks a month. The ruling had fatal consequences in wartime—the miners had to join the army straightaway. "Disabled for the mines, but not for the army," a miner commented.[31]

Soon after the end of the Second World War, early in July 1945, uranium mining at Jáchymov was about to be resumed (after a short break) under Czechoslovak management. The enterprise employed 122 people, and a lorry and two horse-drawn wagons were at its disposal. The town was in a state of chaos as the Germans left and the new Czech settlers moved in. "Wild" expulsions of the Germans were taking place, not yet sanctioned by the conference at Potsdam. That summer, seventeen people, including whole families, chose voluntary death. German houses were ordered to fly white flags, and the Germans wore white armbands. The Germans feared the incoming Czechs, who in turn were frightened of the Germans returning at night to their former houses, so as to retrieve as many of their possessions as possible.

The Czechoslovaking "retribution decree" of 19 June 1945 provided for the punishment of "Nazi criminals, traitors and their assistants," and for the establishment of a "national court" with jurisdiction in Czech lands (i.e. not in Slovakia). It was followed, on 27 October 1945, by the so-called "little retribution decree," concerning "offences against national honor." It proved to be even more insidious—and more flexible in its application—than the original decree.* Many innocent people, Czechs and Germans alike, suffered gross injustices against which they had no recourse.

Several hundred Germans from the Joachimsthal district were tried under the special legislation and some of them were sentenced to forced labor in the uranium mines, where they worked alongside the German POWs.[32] Most of the POWs left the mines by 1950, but some 800 Germans, sentenced under the retribution laws, worked for a time in the uranium industry. A few German miners' families stayed behind and later applied for Czechoslovak citizenship. After the release of the prisoners and their departure for Germany, they founded an association called Kameradschaft der ehemaligen Gefangenen des Uranbergbau von Joachimsthal und anderer Lager und Gefängnisse (The Friendly Society of Former Prisoners in the Uranium Industry and in Other Camps and Prisons) and published their own periodical, *Der Joachimsthaler*.

The association devoted itself to searching for men who had disappeared, and it looked after the returnees' pension claims. It combined charitable activities with rather aggressive propaganda. In the memories of many Germans who had worked in Jáchymov, as well as in *Der Joachimsthaler*, German concentration camps and the Jáchymov camps were considered to have been equivalent. A miner asked in *Der Joachimsthaler* whether it was "really so much worse in the concentration camps than it was for us, who had the misfortune to become Russian prisoners of war?"[33] To attract public attention, the functionaries of the Joachimsthaler and of the Sudeten German associations (Landsmannschaften) made far-fetched comparisons, describing the Jáchymov camps as a "gigantic Katyn" which claimed more victims than did the mass murder of Polish officers by the NKVD.[34] This was an absurd claim, as some 11,000 Polish officers were murdered by the Soviet secret service at Katyn. Among some 800 German civilians who worked in the Czechoslovak uranium industry, most were Sudeten Germans by origin, born between 1893 and 1923. Most of them went through several prisons or camps, spending between 3 and 12 years at Joachimsthal. They were

---

* The "peoples' law courts" handled, until the middle of 1947, some 132,000 cases. More than 38,000 cases were tried. Of the 713 death sentences, 475 were handed down to Germans, and at least 20,000 people were sentenced to long-term imprisonment. Between 1945 and 1948, an estimated 250,000 people, including Czechs and Slovaks suspected of collaboration, as well as Germans, were tried under the retribution decrees.

released in 1955, 1956, 1962, or as late as 1968. Many of them arrived in the German Federal Republic a long time after their release because, the Czechoslovakians maintained, they were in the possession of state secrets and there had to be a cooling off period. The idea of Czech "uranium concentration camps" was intended to balance the guilt of the Nazi Germans. There were no references in *Der Joachimsthaler* either to Nazi war crimes, or to the behavior of members of occupying forces in Bohemia and Moravia during the war.

Many of the former miners arrived in Germany with their health seriously damaged. Two doctors, Dr. Wolfgang von Nathusius and Dr. Grischek, who had both worked in the uranium mines in Johanngeorgenstadt and Joachimsthal, examined them in the 1950s. They came to the conclusion that damage caused to the miners' health in Czechoslovakia was more serious than that found in the miners who had worked for Wismut AG in East Germany, the largest producer of uranium in Europe. They blamed the disparity on the poor nourishment that had been proffered to the Czech prisoners.[35] Their findings were based on a small sample and the research was not pursued further. The report did provoke some public interest, for a while reviving fading memories of the conflict between the Czechs and the Germans.

It was natural that, in the German Federal Republic, the Joachimsthaler association was sharply critical of Chancellor Brandt's Ostpolitik. At the beginning of the 1980s, Der *Joachimsthaler* ceased publication, most of the association's members having died in the meantime.

Notes

1 Eberhard Wächtler and Ottfried Wagenbreth, *Bergbau im Erzgebirge: Technische Denkmale und Geschichte.*
2 Walter Bogsch, Die Marienberger Bergbau seit der zweiten Halfte des 16. Jahrhunderts, Koln 1966.
3 W. Roch, "Annaberg 1496–1946," *Chronik*, Annaberg 1946.
4 Siegfried Sieber, *Um Aue, Schwarzenberg und Johanngeorgenstadt*, Berlin 1972.
5 Hellmuth Semming, *Die wirtschaftliche Entwicklung der Exulantensiedlung "Johanngeorgenstadt" von der Gründung 1654 bis zum Stadtbrand 1867*, Dresden 1931.
6 Vladimír Horský, Paměti královského horního města Jáchymova a jeho stříbrných a uranových dolů, MSS dated January 1993.
7 Private archive Aurand, file Joachimsthal, State Uranium and Radium-Works Joachimsthal.
8 Frieder Jentsch, Begegnungen mit einem Giganten, in: Ans Licht gebracht. Begegnungen und Erinnerungen, Chemnitz 1998, 19.
9 Catherine Caufield, *Das strahlender Zeitalter*, 46.
10 Walter Buchler (ed.), *Dreihundert Jahre Buchler: Die Unternehmen einer Familie, 1651–1958*, 115.
11 Weltmontanstatistik, vol. IV 1939; Statistisches Jahrbuch fur das Deutsche Reich 1939–1940, 70.
12 Vladimir Picugin, Aus der Geschichte des sowjetischen Atomprojektes, in Rainer Karlsch, Harm Schröter, Strahlende Vergangenheit, 50.
13 Gine Elsner and Karl-Heinz Karbe, *Von Jachymov nach Haigerloch*, 25–26.
14 GARF Moscow, Sonderarchiv Nr 1458-10-308, 53. Gaukreditausschusssitzung 4 September 1941.
15 cf. Artur Brandt's memorandum of 10 October 1942, Elsner and Karbe, op cit., 61 et seq.
16 *Idem.*, 69–70.
17 *Idem.*, 89 et seq.
18 cf. F. H. Hinsley et al, *British Intelligence in the Second World War*, London, HMSO 1979–1990, volume 3, Appendix 29, 931–954.
19 Werner Schüttmann, "Die Geschichte Schneeberger Lungenkrebses," *Der Anschnitt*, no. 1/2, 1998.
20 Ursula Jehles, *Der Bergbau im Erzgebirge und die Arbeitsmedizin.*
21 Private archive Aurand, Joachimsthal folder, Max Heiner, Denkschrift über die Bergkrankheiten in St. Joachimsthal, deren Ursachen und Bekämpfung, November 1938.
22 GARF, Fonds 1458, Liste 10, 308, 14. Kreditausschusssitzung on 7 June 1939.

23 Bundesarchiv Koblenz (BAK), R 89, Nr 15 138, Bl 55.
24 GARF, Sonderachiv, Nr 1458-10-308, memorandum by Dr. Kurt Patzschke for Reichswirtschaftsministerium.
25 Privatarchiv Aurand, Handakte Prof. Boris Rajewsky, Bericht über die Besprechung über die Joachimsthaler Bergkrankheit am 28. und 29. März 1940.
26 BAK, R 89, Nr 15 138, Bl 258f.
27 Sächsisches Hauptstaatsarchiv (SÄHStA) Dresden, LRS, MAS, Nr 2174/1, Schreiben des Bergamtes Karlsbad and die Direktion der Joachimsthaler Bergbau GmbH, 21 November 1940.
28 Aktennotiz Prof. Rajewsky, Privatarchiv Karl Aurand, Handakte Prof. Boris Rajewsky.
29 Elsner and Karbe, *Von Jachymov nach Haigerloch: Der Weg des Urans für die Bombe*, 63.
30 Privatarchiv Karl Aurand, Handakte Rajewski.
31 Elsner and Karbe, op cit, 86.
32 Sudetendeutsches Archiv, Munich, item "Joachimsthaler" (box 8).
33 A letter from Adolf Appel of 1 February 1970 to *Der Joachimsthaler*.
34 Der Sudetendeutsche, 7 December 1957.
35 Sudetendeutsches Archiv, Munich, Joachimsthal box 9, Wolfgang von Nathusius' letter to Hans-Peter Ullmann, 9 January 1970; E G Schenk and W von Nathusius, Extreme Lebensverhältnisse und ihre Folgen, Schriftenreihe der ärtzlich-wissenschaftlichen Beirates des Verbandes der Heimkehrer Deutschlands, Cologne 1958.

Part 3

The Politics of Czechoslovak Uranium

A Fatal Friendship: Beneš and Fierlinger

For the second time in the 20th century, Czechoslovakia emerged in 1945 among the victors of a great war. This was despite the fact that Slovakia had taken an active part in Hitler's military campaign against the Soviet Union. Stalin had assisted President Edvard Beneš's government-in-exile in London restore pre-war Czechoslovakia. Beneš and his country thus remained in Stalin's debt.

The politician who played the foremost role in developing ties with Russia was Zdeněk Fierlinger, a close associate of president Beneš since the First World War. Seven years younger than Beneš, Fierlinger was born in 1891 at Olomouc, the third child of a teacher of French and English at the local high school. He attended the German commercial academy in Olomouc. An indifferent student in most subjects, he was good at languages. At the age of nineteen, Fierlinger left his native Moravia for Russia. He worked at Rostov-on-Don for MacCormick International Harvester Company.[1]

It was only the beginning of Fierlinger's long association with Russia. He was among some 70,000 Czechs who lived in the tsarist empire when the war broke out. He joined a group of Czech volunteers organized in Rostov, which later fought alongside the Russian army on the Eastern front. The Česká družina became the Czechoslovak Legion (Československá legie), which came into armed conflict with Lenin's regime in the spring 1918.

Fierlinger was remembered by one of his contemporaries as a brave and selfless soldier.[2] He was among the Czech officers who came to France after the Bolshevik revolution in November 1917, thus missing the civil war and Allied intervention in Russia.

In Paris towards the end of the war, Fierlinger worked with Edvard Beneš, the secretary of the Czechoslovak national council. Fierlinger had the rank of full colonel, and the two young men got on well together. Neither had any close friends, and both were industrious and ambitious. There was a trace of arrogance in Fierlinger, who was extremely practical and less academically inclined than Beneš. They complemented each other and similarities in their characters helped to draw them together. Their long association left a strong mark on the history of Czechoslovakia.

At the close of the First World War, Fierlinger became the head of the economic section in the foreign ministry of the new Czechoslovak state, where Beneš was the minister. Fierlinger was self-confident, blunt in expressing his views and ready to stand up to experts.[3] He was sent as the minister to The Hague in 1921, when he was thirty years old. Fierlinger spent a short time in Romania and then three years in Washington, from 1925 to 1928. He developed a kidney ailment there and was sent to Berne, where suitable medical help was available. His second job in Switzerland often took him to Geneva, as a permanent delegate at the League of Nations.

Fierlinger remained in Switzerland until 1932, where Beneš was a frequent visitor to Geneva and the League of Nations. He was probably the only person in the foreign ministry who could afford to stand up to Beneš in a political argument. The debates sometimes took place in front of the other members of the Czechoslovak delegation and usually concerned problems before the League of Nations, such as disarmament. Beneš resented arguing with Fierlinger, and yet he made excuses for him on the grounds of his youth and poor education.[4]

As the economic crisis deepened in the 1930s, interest in the Soviet experiment in the West increased. Fierlinger knew Russia well, though he had left it before Lenin's regime got going. In the last two years of his diplomatic stint in Switzerland, Fierlinger addressed himself to the advantages of planned economy. He used one of the vacations to visit Moscow, where Stalin had routed Trotsky and emerged as an uncontested ruler. The outcome of Fierlinger's journey was a book published in Prague 1932, called *The Soviet Union on a New Track* (Sovětský svaz na nové dráze).

His book was similar to many others published at that time. European intellectuals were turning to the Soviet Union, which looked to them a more attractive place than Hitler's Germany. There were so many admirers of the Soviet experiment in New York that it was described as the liveliest part of the Soviet Union; in London, Sidney and Beatrice Webb were about to start visiting Russia to collect material for their comprehensive study; André Gide and H. G. Wells also made their journeys of discovery, and took part in the debates on the nature of the Bolshevik experiment.

During the Geneva disarmament conference, Beneš moved Fierlinger to Vienna, where he stayed until 1936. Fierlinger kept up an interest in Russia, and wrote several articles for the Czech press praising the achievements of Stalin and of Soviet economy. He believed that there was a general European trend towards socialism, and that its eastern and western parts would draw closer together. The theory of convergence between capitalism and socialism, as it became known later, was well-received in Prague. It provided the justification for Czech diplomacy towards the Soviet Union. In 1935, the French-Soviet treaty of mutual assistance suited Beneš so well that he decided to add Czechoslovakia to the alliance.

Beneš, who insisted that employees of the foreign ministry be personally loyal to him and have no ties to political parties, continued to shelter Fierlinger. He was free to express his views on the Soviet Union in public, and criticize Czechoslovak economic policies for their conservatism. In the spring of 1936, after being recalled from Vienna, Fierlinger became the head of the political section of the foreign ministry. Beneš had been elected president at the end of 1935, and he wanted to have someone in a top position at the foreign ministry that he could rely on.

While he was in charge of the key department of the ministry, Fierlinger stopped a shipment of Czechoslovak arms to Portugal, on its way to Franco's rebel army. Portugal broke off diplomatic relations with Czechoslovakia and scrapped the trade agreement. Fierlinger was sharply criticized for his action, for which he had had no authorization. He nevertheless kept his post and, in the summer of 1937, was sent to Moscow as the minister. There was a question mark over the importance of Fierlinger's new post. The Soviet Union had played a subdued role in European diplomacy until joining the League of Nations and negotiating the treaty with France. Fierlinger made no secret of his opinion that many Czech politicians would rather turn down offers of Soviet friendship than offend the Western powers. Long before Munich, Fierlinger believed it to be his duty to bring Czechoslovakia closer to Russia.

He witnessed the last round of the Great Purges in Moscow and expressed his admiration for Stalin's courage in dealing with the enemy within. He had no doubt that the conspiracy against Stalin existed and that it was assisted by the enemy outside; that the opposition in Moscow worked with the Germans to overthrow the Soviet regime. Fierlinger believed that there were young members in the communist party of the Soviet Union who, "free of old revolutionary theories,"[5] followed Stalin with enthusiasm. According to Fierlinger, after overcoming the current crisis, the party would return to its former tolerant methods and to cooperation with democratic states. Like Fierlinger, Beneš was convinced that Stalin was under threat, and not only by conspiracy at home.

In August 1938, the Czechoslovak Ministry of Foreign Affairs was notified by a Western embassy that Fierlinger was passing copies of confidential material to the Soviets. The foreign ministry took the "necessary precautionary steps,"[6] while Beneš had other things on his mind. As the Munich crisis was about to culminate, Beneš and Fierlinger met at Sezimovo Ústí, where they had neighboring country houses. Fierlinger assured the president that Czechoslovakia could fully rely on the Soviet Union, though he avoided directly answering the question whether it would come to the assistance of the Czechs even if France did not. Beneš, deeply disappointed by the attitudes of the French government, was depressed. To cheer him up, Fierlinger played Beneš a gramophone record of Red Army marching songs.[7]

After the conference in Munich and Beneš's resignation as president, Fierlinger returned from Moscow for a brief visit to Prague in January 1939. He believed that the new president, Hacha, would adhere to democratic principles and to the policy of cooperation with the Soviet Union. Fierlinger decided to remain in his post in Moscow. After the occupation of Bohemia and Moravia by Hitler, and the establishment of the Slovak state in March 1939, Fierlinger came under double pressure. While the government in Prague asked him to resign, German diplomats demanded that he hand over the Czechoslovak legation in Moscow. Fierlinger took a vacation in April 1939 and went to Paris and London in order to meet the Czechoslovak leaders in exile. He did not see Beneš, who was lecturing at the University of Chicago.

When Fierlinger returned to Moscow late in June, Molotov had replaced Litvinov as the commissar for foreign affairs. When he reported to Beneš on the Nazi-Soviet pact on 26 August, he admitted that he had not expected such a drastic reaction to the West's disinterest in establishing an alliance with the Soviets, and pointed out that Germany did not make the mistake of underestimating Soviet strength. "I admit that they are playing for high stakes, but I think that vacillation of the West was the only reason for the sensational turnaround of Soviet foreign policy."[8]

Fierlinger was shocked when the Soviets instructed him to leave his post on 14 December. Though he was offered to stay on in Moscow as a private citizen, he left for Paris on Christmas Day, 1939. There he sought out Jan Šrámek, the nominal chairman of the Czechoslovak national committee, and volunteered to help with its work. Fierlinger expected that he would soon return to Moscow, and said he was not seeking an official position. Critical of Czechoslovak exiles because of their negative attitude to the Soviet Union, Fierlinger wrote a detailed memorandum for Beneš on 20 April 1940. He stressed the difficulties arising for Russia from its isolated position, and tried to persuade Beneš that there was little significance in Soviet diplomatic moves such as the recognition of Slovakia, or the abolition of the Czechoslovak legation in Moscow. Fierlinger's arguments helped Beneš to console himself with the thought that Moscow was doing no more than trying to make the period of neutrality last as long as possible.[9]

When Fierlinger found himself on the margins of exile politics, he set out to write a study entitled "The Present War as a Social Crisis."[10] It was a peoples' war for Fierlinger, with a strong social and economic undertow. He believed individual enterprise needed to be replaced by collective action if the war was to be won. Economic liberalism would become a thing of the past, and the socialist economy would, for the time being, use the instruments of capitalism to achieve its own ends. This was not an original view, and it closely mirrored Beneš's thinking. Fierlinger kept in touch with Maisky, the Soviet Ambassador to London. According to his own recollections, Fierlinger passed information from Prague, sent to the Czechoslovak government in exile, on to Maisky. This information included Hitler's plans to invade Soviet Union.[11]

Fierlinger returned to the post in Moscow in August 1941. Initially the minister, he later became the Czechoslovak ambassador. Never criticizing the Soviet Union, Fierlinger found fault with the Western Powers for not understanding the difficulties it faced. He expected the Czechs and the Slovaks in London to trim their policies to suit Soviet needs, and was critical of Beneš for trying too hard to establish continuity with the first republic. Fierlinger wrote that the fears of Bolshevism deeply influenced the peace negotiations after the end of the First World War, and that only "Lenin's genius could...grasp and explain...the meaning of historical development in the new era."[12] In his mind at least, Fierlinger started parting company with Beneš.

Fierlinger was on friendly terms with the Czechoslovak communist exiles in Moscow. He agreed with them that the government in London was too conservative, and he blamed it for keeping him badly informed. Fierlinger opposed the plan for the Czechoslovak–Polish confederation, which Beneš developed with the encouragement of the Foreign Office. Soviet diplomacy was also unenthusiastic about the plans for the confederation, which Fierlinger believed aimed to exclude the Soviets from Central Europe. He spared no effort, on the other hand, in achieving the Russian-Czechoslovak treaty. With Fierlinger's help, Beneš switched from the Polish negotiations to the plan for an alliance with the Soviet Union. It was a crucial break in Beneš's wartime diplomacy: "The idea for the Russian-Czech treaty rose as a phoenix from the ashes of the proposed Polish-Czech confederation."[13]

Fierlinger, who regarded himself as the true father of the Czechoslovak–Soviet Treaty of Friendship, Mutual Aid and Post-War Cooperation, had to wait for it until December 1943. Beneš went to sign it in Moscow, despite advice from Eden and the Foreign Office. Negotiated during the summer and autumn of 1943, Beneš considered the treaty essential for the future of Czechoslovakia. Fierlinger provided the necessary arguments for Beneš's resolution.

Instead of acting as a mediator between the Eastern and Western partners in the alliance, Beneš became increasingly inclined to side with the East. He was caught in communist scissors, with the Soviet Union on one side and the Communist Party of Czechoslovakia on the other. It was hard for Beneš to act decisively when Fierlinger congratulated Osubka-Morawski, the leader of the Moscow-backed Poles, on the entry of the Polish army on Polish soil. The British government protested, as did Jan Masaryk, the foreign minister. Fierlinger answered his chief in a robust letter. The tension between the government in London and its emissary in Moscow developed so far that the government unanimously voted, on 28 July 1944, to have Fierlinger recalled. Beneš refused to accept the recommendation.

In Moscow, Fierlinger had been in touch with the Czechoslovak communist leaders since his return as the head of mission in 1941. He helped them keep in touch, through diplomatic channels, with their comrades in London. In a memorandum for Gottwald from November 1944, Fierlinger referred to large-scale confiscation of all enemy and collaborator property. He insisted that all financial institutions come under the control of the state, and that the Soviet Union should become Czechoslovakia's main trading partner. The memorandum was written after consultations with the communist leaders, when it was agreed that the mild introductory phase of "peoples' democracy" would be followed by "fundamental socializing measures."[14] In Moscow at the end of 1944, Fierlinger trailed his coat as a possible candidate for a high political office.

Fierlinger went to Teheran to meet Beneš in March 1945, and told him that the communists wanted him to become the prime minister. Beneš was unprepared for the news and tried to talk him out of accepting the offer. Fierlinger apparently became cross with Beneš and ignored his advice.[15] When Gottwald told Jan Masaryk that the communists planned to put Fierlinger forward as their candidate for premiership, Masaryk refused to discuss the matter. He confined himself to saying that the government in London had been critical of Fierlinger's activities.[16]

Jan Masaryk did not hide his dislike of Fierlinger from Beneš. It was difficult for Beneš to review his long association with Feirlinger. The two men shared similar views on the development of European societies, and were bound by a relationship that was as close to friendship as their cold characters permitted. They bought neighboring plots for their houses in the country and in Prague, and lived on the same street between the wars. Beneš used Fierlinger as an important piece on the diplomatic chessboard, one that guarded the flank of Czech diplomacy. Instead of defending the flank, Fierlinger briskly moved to the leading formation. By the end of his life, Beneš had developed a hysterical distaste for his former associate.[17]

The East-West fault line that developed at the end of the war began breaking the unity of Czechoslovakian exile politics much in the same way that it divided the eastern and western wings of the Polish exiles. For the time being, Beneš succeeded in concealing the growing rift. He believed that Stalin would remain on good terms with the Western Allies, and that the two principles of social organization, communism and capitalism, would draw closer together.

A Proof of Friendship

Moscow could not have wished for a more dedicated supporter than Fierlinger. His loyalty to the Soviet Union was beyond doubt, and there is little need to speculate whether he was under contract as an agent of the Soviet secret service. He manipulated Beneš's disillusion with the West after Munich—he was Beneš's bad conscience.

The Czechoslovakian-Soviet treaty on uranium, concluded on 23 November 1945, was the most significant foreign policy act of Fierlinger's government. It was Fierlinger himself who, virtually singlehandedly, succeeded in bringing it into existence. He provided cover for the initial Soviet approaches to the Czech side and prepared the way for the conclusion of the secret treaty.

Three weeks after the destruction of Hiroshima, General Mikhailov and Colonel Alexandrov visited Jáchymov. Mikhailov was the commander of the Red Army district Karlovy Vary; S. P. Alexandrov was a geologist in the service of the NKVD. On 26 and 27 of August 1945 they inspected Svornost, the mine in the centre of the town, with a processing plant and a store of concentrates. They asked about the known reserves of uranium and discussed, with the Czech management, the cost of running of the mines. They enquired whether there had been any other foreign visits to the mines, and took samples of pitchblende with them.

According to General Boček, Chief of Staff of the Czechoslovak army, "From their questions, it became apparent that their interest in our uranium mines was strong."[18] Boček added that, on a different occasion, a Russian civilian visited one of the mines, asking why it was not working two shifts a day, and why labor was not being recruited from amongst the local Germans. The Russians had come to Jáchymov unannounced, and General Boček wanted to know whether such visits should be authorized in the future.

On 13 September, the management of the mines informed the General Staff that three detachments of the Red Army, consisting of about twenty men each, had occupied the three pits. A week later, they were visited by a Red Army general, who ordered the guard to be strengthened. Admission to the mines required special passes, issued by the Red Army command at Karlovy Vary. The miners had to undergo bodily searches before and after the shift. No ore was allowed to leave the premises. When an official of the Ministry of Defense inquired, in September 1945, what should be done about the visits to Jáchymov by Red Army officers, Fierlinger told him to "do nothing."[19]

The issue of the Jáchymov mines concerned the Czechoslovak military as well as the ministry of industry. The military made another enquiry, on 2 October, about the status of the Red Army guards posted at the uranium mines. On 23 October, the ministry of industry, led by Social Democrat minister Bohumil Laušman, noted that the presence of the Red Army, while in no way interfering with the operation of the mines, gave the Russians access to vital information on the uranium industry. The ministry was unable to judge whether Czech interests were thus adversely affected. It merely requested that the issue of passes to the mines be reviewed, as Red Army procedures were rather slow.

The ministry of industry was responsible for mining in Czechoslovakia, as well as the production of radium. Fierlinger told Laušman nothing about the early stages of the negotiations with the Soviets, and Laušman initially assumed that production at Jáchymov could be restarted as soon as the Soviet troops left. In a report for the minister dated 10 October 1945,[20] the start of mining activities was mentioned without any reference to the Soviet interest in the industry. The report referred to individual veins that could be exploited, and to the depletion of reserves during the German occupation. It complained of a labor shortage and of low production amounting to only one tenth of the Jáchymov mines' former capacity. By the time Laušman read the report, he had already been aware of the negotiations of the secret treaty for about a fortnight.

During the preparations for the treaty, Fierlinger skillfully exploited the authority of president Beneš, whom he informed of the progress of the negotiations when he thought it necessary. The Soviet Ambassador to Prague, Valerian Zorin, received the Czech draft of the treaty from Fierlinger on 12 September 1945. The draft contained provisions agreed upon by Beneš and Fierlinger. These provisions concerned Czechoslovakian property rights, the fixed annual amounts of deliveries of the ore and cooperation in mining and research activities.[21]

On 26 September, Fierlinger chaired a meeting in which the ministers of the interior, industry and defense, as well as the chief of the general staff and several high officials, took part. The Soviet delegation, at a lower level, consisted of only four members. It included Ivan Bakulin, of the ministry of foreign trade (He is referred to as the trade attaché in Prague in Czech documents, and there exists a reference to Bakulin as deputy minister of foreign trade), and Colonel S. P. Alexandrov, who had taken part in the first inspection of Jáchymov.

When Bakulin read out the Soviet draft treaty to the meeting, it contained a proposal for the formation of a joint stock Soviet-Czechoslovak company. Fierlinger said that technical cooperation with the Soviets would suit the Czechoslovaks better, as they were in the middle of nationalizing their own industries. Laušman remarked that the Czechoslovak government would be charged with "denying the Western allies economic rights and positions, while conceding such rights to the eastern ally; the Czechoslovak cause would be damaged politically." Laušman made his point and pursued it no further. For Fierlinger, there was no question of political or economic justification of the treaty. He confined himself to asking that the draft be reconsidered, adding that he was certain that his side would be able to meet the Soviet wishes.

Members of the Soviet delegation were inclined to regard German investment in Jáchymov as a reparations item, and they indicated that they were ready to make further investments. They proposed that the whole production go to the Soviet Union, and that they should provide the wherewithal for geological exploration on Czechoslovak territory. Fierlinger replied that the uranium question concerned Soviet as well as Czechoslovak military interests; and Alexandrov left the Czechs in no doubt about the Soviet resolution to break the American nuclear arms monopoly.

Towards the end of the meeting, Alexandrov said that the mines had been for six years in the hands of the "German fascists, who wanted to mine 50 tons a year." His remark sounded like a complaint that the Czechs allowed the situation to arise, so he went on in a more conciliatory tone. He said that the Czechoslovak proposal was personally sympathetic to him, and that he would take it back to Moscow. "It was not the Soviet Union who declared the atom bomb to be the weapon of the future," Alexandrov added. "Now we know where we stand," Fierlinger concluded the discussion.[22]

On 6 October, when Bakulin visited Fierlinger, the Russian urgently pressed him with questions. Can the Soviets be certain of receiving all the mined ore and concentrates? Will they be able to influence production and exploration according to the needs of the Soviet Union? Will they participate in the day-to-day control of the production plan? Could Soviet specialists and workers be employed by the enterprise?

Table 1. Production and Export of Uranium from Czechoslovakia to the Soviet Union 1945–1989

Year	Uranium Metal	Chemical Concentrate	Total (in tons)
1945	30.8		30.8
1946	18.0		18.0
1947	49.1		49.1
1948	102.7		102.7
1949	147.2		147.2
1950	241.4		241.4
1951	524.2		524.2
1952	807.8		807.8
1953	1104.2	48.9	1153.1
1954	1439.9	115.6	1555.5
1955	1896.6	164.6	2061.2
1956	2138.7	197.8	2381.5
1957	2544.7	199.8	2744.5
1958	2734.4	189.9	2924.3
1959	2805.1	181.6	2986.7
1960	2812.2	225.1	3037.3
1970	500.0	2100.0	2600.0
1980	0	2467.2	2467.2
1989	0	2400.9	2400.9
1945–1989	38745.7	57925.9	96660.6

Source: Oskar Puskal, *Surovinové zdroje uranu ČSR*, Manuscript, Prague, 1993.

On the following day, a Sunday, Fierlinger left for his country house at Sezimovo Ústi, in south Bohemia. He walked through the garden to the neighboring house, which belonged to president Beneš. He explained the Soviets' interest in the rapid development uranium mining to the president, and told him that the Soviet government regarded the matter as being of great political and military importance.

Beneš objected to the proposal that the whole production should go to the Soviet Union. Fierlinger however insisted that if the Czechoslovaks met Soviet demands, they would receive compensation from the Soviets. Fierlinger explained to Beneš that the Soviets would help them with the oil fields in Slovakia and Moravia, and with the extension of Czechoslovakian territory at the expense of Austria. Soviet goodwill was also necessary for the rectification of the border with Poland, especially in the Kladsko district. Fierlinger had convinced Beneš of this, although Beneš expressed the wish that the Soviets be told exactly what was expected of them.

While the uranium treaty was being considered, Bakulin reminded Fierlinger that he had agreed to hand over 38,516 kilograms of radioactive material available on Czechoslovakian territory to the Soviets. Fierlinger immediately gave orders for the transfer of the materials to the Soviet foreign trade organization, Torgpredstvo, and requested that the ministries of industry and foreign trade take care of the transfer. On 14 October, 37,012 kilograms of uranium paints, (containing up to 58.4% uranium oxide [U_3O_8]), were taken away from the Příbram smelting works. On 29 and 30 October, barrels and boxes containing 9,725 kg of 58% uranium concentrate followed. Additional pitchblende from the slagheaps was collected, and the amount of uranium the Soviet Union received from these sources has been calculated at 30.838 tons; only 0.919 tons of uranium was actually mined in 1945.[23]

In the meantime, in Moscow, the Czechoslovakian diplomatic mission came under pressure from Molotov in connection with the uranium treaty. He told the Czechs that the treaty should correspond to the spirit of alliance between the two countries: the Soviets must be represented on the board of the company, and the treaty should be concluded as soon as possible. Instead of comments on their proposals, the Czechs received a reiteration of the Soviet position. Fierlinger assured Molotov that Czechoslovakia was a faithful ally of the Soviet Union, and that it would try to meet Soviet wishes. He nevertheless stood firm on the matter of the joint stock company. On 4 November, Molotov summoned the Czech Ambassador, Jiří Horák, to see him. Molotov avoided any reference to the joint stock company, speaking instead of a national enterprise which would conclude a special agreement with the Soviets. He insisted on the need to increase production and conclude the treaty as soon as possible.

On 14 November, Bakulin had a new draft of the treaty ready. The Jáchymov mines were to remain a solely Czechoslovakian enterprise, and was to be run by a joint commission of four members, two of them Soviet. The Czechs made no objection to the idea of shared management, but they regarded thirty years as being an unnecessarily long contract and wanted to keep a part of the production. Bakulin insisted that the Soviet Union was to be the sole buyer, and Fierlinger succeeded in convincing Beneš that the joint commission should decide the amount of uranium to be retained in Czechoslovakia. Beneš confined himself to saying that the treaty should run for twenty years only.

As late as 17 November, six days before the treaty was signed, Fierlinger reported to the presidium of the government (the prime minister and his five deputies, representing their political parties). He stated that "the very urgent matter of the extraction of uranium ore in Jáchymov" had been conducted since August, and that the Soviets were "examining especially the possibilities of economic exploitation of atomic energy." One of the deputy prime ministers, an elderly Catholic cleric and leader of the Peoples' (Lidová) Party, expressed lively interest in the subject. He had been in charge of the ministry of health before the war, and told his colleagues that doctors assured him that the use of radium for healing was a crudely empirical method. He insisted that the Czechs should get an adequate price for the ore. When the matter of Soviet technical assistance was discussed, Július Ďuriš, the communist minister of agriculture, was convinced it would surely be magnificent. Jan Šrámek, on the other hand, feared it might go too far. Jaroslav Stránský, deputy prime minister in the national socialist interest, assumed that the uranium business with the Soviet Union could be profitable, though the treaty ought to be designed so that "we should keep control over the enterprise."[24]

The presidium agreed that the treaty was to run for twenty years, and that an unspecified part of the production should be retained in Czechoslovakia. Fierlinger countered objections by enumerating the benefits for Czechoslovakia. He was asked to carry on the negotiations in the spirit of the discussion just concluded. The deputy prime ministers were well acquainted with Fierlinger, and they knew he would do his best to meet Soviet wishes.

On 21 November, Bakulin and Alexandrov informed Fierlinger that Moscow was satisfied with the new draft, providing that the Czechs would retain no more than 10% of the production. Fierlinger then asked Beneš to approve the treaty, which the president granted. It was put before a secret meeting of the full cabinet on 23 November. It was the first time most of the ministers, including Jan Masaryk, the foreign minister, heard of the treaty. An official read out the draft and the supplementary protocol. Fierlinger told the cabinet that the negotiations had gone on since August with the president's knowledge, and that the narrow cabinet had been unanimous in recommending the acceptance of the treaty. He said that it was a "singular agreement," and offered the opinion that the treaty would "make no demands on the financial resources of the state, on the contrary, it means financial gain."[25]

About three hours after the conclusion of the secret meeting, Alexandrov and Bakulin came to sign the treaty. They were in a hurry, and brought with them their own text. It was identical to the draft treaty read out to the ministers in the afternoon, but the supplementary protocol differed from the original Czech version. The clauses concerning secrecy and continuity of production had been strengthened.[26] On 27 November, Fierlinger requested that Laušman, the Minister of Industry, get in touch with Bakulin and establish the new national enterprise at Jáchymov.

The Russians did not forget the kindness extended to them by the Czechs in 1945. In 1980, on the occasion of the thirty-fifth anniversary of the secret uranium treaty between Czechoslovakia and the Soviet Union, N. I. Chesnokov, the deputy minister of "medium" industry—that is, the entire Soviet military-industrial complex—spoke as a representative of the sole buyer of Czech uranium. He reminded the Czechoslovakians that it was the first treaty between the two countries concluded after the original Czech-Soviet treaty in December 1943.[27]

A Unique Treaty

The secret treaty between the Union of Soviet Socialist Republics and Czechoslovakia was signed in Prague on 23 November 1945. It was an unusual document in that it contained not a single reference to its subject—that is, uranium. It was a political rather than a business treaty, and it was regarded as such by both the partners. It was negotiated and signed by the highest functionaries of the Czechoslovakian state, and by a comparatively low-level delegation on the Soviet side.

As a result of the treaty, a small industry was transformed into a large enterprise. The Czechoslovaks helped the Soviets solve the uranium problem without even being aware of its existence. They did not know that the nuclear program came under the control of Lavrentii Beria and his security organ, the NKVD. They were also unaware of Stalin's methods of mobilizing resources and their place in his plans. After the conclusion of the Second World War, Stalin concentrated on a single goal: making the Soviet Union a leading world power. In its absolute single-mindedness, Stalin's system seemed to be effective in achieving specific goals. For the Czechoslovaks, the uranium industry became the gate through which they entered Stalin's empire.

At the time of the negotiations with the Czechoslovaks, the Russians had a model agreement ready for the exploitation of uranium deposits outside the Soviet Union. Its central requirement was the establishment of a joint stock company with the country concerned, which would become the monopoly producer of uranium and supplier to the Soviet Union. Similar agreements were concluded with other countries in the Soviet empire, such as Bulgaria, on 17 October 1945. Wismut, the Soviet company in East Germany, was also eventually converted into a joint stock enterprise.

Table 2. Uranium Production in the Soviet Block, 1946–1950

Year	GDR	CSSR	USSR	Bulgaria	Poland	Total (in tons)
1946	15.7	18.0	50.0	26.6	0	110.3
1947	150.0	49.1	129.3	7.6	2.3	338.2
1948	321.2	103.2	182.5	13.2	9.3	629.4
1949	767.8	147.3	278.6	20.1	43.3	1257.1
1950	1224.1	281.4	416.9	54.1	63.6	2040.1

Source: W. I. Wetrow, "Bildung von Betrieben für die Gewinnung und Aufbereitung von Uranerz," in Ministerium für Atomindustrie der Russischen Föderation (ed.), *Die Entwicklung der ersten sowjetischen Atombombe* (German working translation), Moscow, 1995, 22.

The agreement with Czechoslovakia was unusual in that it did not provide for the establishment of such an enterprise. In offering the Soviets the next best thing, a Czechoslovak national enterprise run by a joint Czechoslovak-Soviet commission, the Czechs won a Pyrrhic victory. In any case, the Czechoslovakian uranium industry came under Soviet control. Soviet managers initially tried to revive the plan for a joint stock company. They argued that production was not being increased quickly enough, and that the NPJ (Národní Podnik Jáchymov, or National Enterprise Jáchymov) was uneconomical and wasted valuable resources.

The provision for the joint permanent Czechoslovak-Soviet commission in Part 4 of the treaty laid the foundations for Soviet control of the industry. Part 6, concerning technical posts for Russian experts, further strengthened the Soviet position. The geological service was not provided for in the treaty. Run by the Soviets alone, it became another means of exercising their influence.

The cost of uranium quickly became a matter of dispute between the partners. The treaty gave the accountants little guidance; the Soviet definition of cost (*svestojne naklady*) was unfamiliar to the Czechs. It related to parts of production, and made no provision for investment in the necessary infrastructure, including housing, policing and doctoring for the workforce. No allowance was made for the depletion of the reserves. The conflict between the Soviet and Czech members of the permanent commission came to a head in 1949, when they were unable to agree on the previous year's accounts.

There were striking similarities between the Czechoslovak-Soviet treaty and the earlier American-British-Belgian Memorandum of Tripartite Agreement of 26 September 1944. Like the Czech-Soviet treaty, the Agreement was a political act. It was a long-term contract providing for the shipment, until 1956, of the Belgian Congo's entire uranium production to Britain and America for military purposes. The text of the agreement referred, in virtually the same terms, to production costs plus "reasonable profit" as the basis for the calculation of the price of uranium; though in the case of the Tripartite Agreement, provision was made for Belgium to share in any profit should the two powers sell the ore on the commercial market.

In Czechoslovakia, the Soviet commitment to cover production costs was intended to secure the highest possible output of uranium. The industry disposed of inadequate technical equipment and patchy geological information. In 1946, it was calculated that 1,330 crowns was spent to produce 1 kg of natural uranium. In 1949, the amount was 1,834 crowns. It wasn't until 1953 that the cost came down to 724 crowns, dropping to 636 crowns in 1955.[28] At a time when the world market price for uranium was not yet established, Czechoslovakian uranium was more expensive than either North American or Congolese uranium. The costs decreased later, as the richer deposits at Příbram came on stream. (The original contract between Sengier [the head of the Union

Minière], the Belgian operating company in the Congo, and the US Army had provided for the shipment of 3,400 pounds of uranium oxide for as little as $1.71 per pound. The Congolese ore was much cheaper than the uranium produced in Canada and America. After the war, the exchange rate stood at 50 crowns to $1.)

There are two contrasting ways of calculating the cost of Czechoslovak uranium. On one hand, the Soviets were buying the metal at comparatively high prices, especially in the first four years after the war. Production costs were high, and thus the Czech myth of cheap uranium, handed to the Soviets by the communists, hardly corresponds to reality. A former director of NPJ stated that "Though we lost nothing on the extraction of uranium, the Russians paid all their own expenses plus ten per cent profit, but the advantage of world demand and of world market did not exist for us."[29] This may have been true during the first decade or so, while uranium was regarded everywhere as a rare metal. As the Russians put their own resources of the metal on-stream, the Czechoslovaks developed a uranium lobby in the interest of continuing production of the metal at any cost. Production became more expensive as they started to pay their own costs. It has been calculated that, between 1967 and 1989, the state subsidized the Czechoslovak uranium industry at the rate of 23.9 billion crowns, with investment amounting to additional 7.4 billion and geological exploration 4.4 billion.[30]

The real cost of the uranium industry to Czechoslovakia, on the other hand, is a different matter. The necessary industrial and social infrastructure had to be created in Jáchymov, and then in the Příbram districts. Living quarters for the fast growing workforce had to be constructed, and schools and medical services provided. After the 1949 decision to employ convict labor, the labor camps had to be built where no former POW camps were available. None of these costs were reflected in the calculation of the production costs of uranium. Soviet payments were irregular in the early years of the uranium industry, and the Czechs had to provide investment costs themselves.

Most significantly, the uranium industry was developed hastily and in without regard to the needs of the post-war economy. NPJ achieved a privileged position in the economic life of the country and made high demands on its scarce resources. It came into being in an ambience of secrecy and fear, and helped establish the devious and secret methods of Stalin's state in Czechoslovakia. The human and social cost of the uranium industry is impossible to calculate, as is its detrimental effect on the political conduct of the state.

During the treaty negotiations, the cabinet gave the uranium problem scant attention and there was no parliamentary debate. The media was silent. A didactic article on the uses of atomic energy appeared in August 1945, a few days after the Soviet officers visited Jáchymov.[31] In the summer of 1946 the silence was broken by a singular article entitled "Atomic Energy and Jáchymov" in a Prague weekly.[32] The author, Jan Kolář, went straight to the heart of the matter. He wrote that the Czechs were gratified by Jan Masaryk's pronouncement, made several months earlier, (see in the chapter "A Serious Diplomatic Embarassment") that Jáchymov uranium would only be used for peaceful purposes. Kolář then added: "In this connection one cannot but wonder that many Germans are still employed in the Jáchymov mines, so that security needs require strong military presence, by our own units and by the Red Army as well. It will therefore be necessary to employ only Czech miners in the Jáchymov mines in the shortest possible time, as it is necessary to resolve the problem of the Jáchymov spa, renowned worldwide. Its close proximity to uranium mines of military significance has caused the refusal, for security reasons, of visitors from abroad; it has happened this year to hundreds of patients."

The article caused alarm in political circles in Prague. The author was interrogated and copies of the article made rounds of the ministries; an official made a marginal remark that "According to the opinion of the general staff of the Ministry of Defense the article reveals the presence of Russian workers at Jáchymov and it may again cause unpleasant [one illegible word] by the western press."[33]

On 23 November 1945, Hubert Ripka, the minister of foreign trade, signed the uranium treaty with the Russians. When Ripka was in exile in London three years later, he wrote a book on the communist takeover of power in Czechoslovakia.[34] He made no mention in the book of the secret treaty. The matter of the uranium treaty did not go away, and it played a rather dire role in the last years of Ripka's life. As a political exile in New York, Ripka taught at the New School of Social Research and attended meetings of the Council of Free Czechoslovakia. He was accused of collaboration with the communists and, in February 1954, gave evidence before a Senate sub-committee. Dr. Kurt Glaser, a German American educated at Harvard, who had been employed by the American forces of occupation in Germany, was present at the hearing. Glaser had published "The Iron Curtain and the American Policy," a denunciation of Czech non-communist politicians, and a pamphlet which was specifically aimed against Ripka.

The uranium treaty was among the charges leveled by Glaser against Ripka. Glaser wrote that the treaty had been initiated in London and was on Ripka's agenda when he returned to Prague. The Jáchymov mines had been handed to the Soviets free of charge, Glaser argued, and from there "the Soviets now acquire atomic materials for the bombing of the United States."[35]

In a draft reply to Glaser, Ripka stated that the uranium treaty had not been initiated in London, but in Moscow. The negotiations were conducted, according to Ripka, by Fierlinger and by the Czechoslovak communists, exiled in Moscow. Ripka maintained that "I had heard of the agreement for the first time at a secret meeting of the government in exile. I telephoned president Beneš at once, so as to ask for his opinion and I was told that it had to be accepted as an inevitable consequence of our alliance with the Soviet Union."[36] Ripka further argued that, in the concluding stages of the war, with the whole of eastern Central Europe- and East Germany under Soviet influence, and because the West was unwilling to risk a conflict with the Soviet Union, there existed no other option than the policy pursued by Beneš. Ripka added that, when the Czech leaders later discovered that uranium was used for the production of atomic bombs—"a fact we had previously ignored" —they were unable to grasp why the western Allies, during the partition of Germany in 1943 and later, did not keep Jáchymov and the neighboring region in Saxony for themselves.

A draft of Ripka's reply to Glaser's charges contains several inaccuracies in the passage dealing with Jáchymov. No evidence has been found that the government in exile ever discussed an agreement on uranium. Immediately after the war, when Beria's experts began visiting Prague and Jáchymov, they left Czechoslovak officials in no doubt why they needed uranium. The passage on Jáchymov, as written by Ripka, confirms how marginal the uranium problem was for Ripka in the months immediately following the war.

The first public reference to the treaty with the Soviet Union was made by another exile from Czechoslovakia, Otto Friedman, in *The Break-up of Czech Democracy*.[37] Friedman gave October 1945 as the month of its origin. According to Friedman, the treaty guaranteed the Russians the right to mine uranium in Jáchymov. Neither the government nor the parliament were party to the decision to conclude the treaty and were not informed of it. Only Fierlinger, Jan Masaryk and Hubert Ripka knew of its existence, according to Friedman. Czechoslovakia made all the necessary investments. The Russians employed prisoners, and the mines were out of bounds to Czech officials.

Inside Czechoslovakia, the treaty could not be kept secret either. The postmaster at Jáchymov, who was a member of the People's Party, the only Czechoslovakian party at the time that could be described as non-socialist, knew about the treaty. He wrote two letters to party headquarters in Prague in February 1946. In one of them he mentioned a treaty with the Soviets, though he was not quite certain whether Fierlinger had signed it with or without the knowledge of the government. He took exception to the officially published news that there were no more Red Army units stationed at Jáchymov. (Most of them, both Russian and American, had left the territory of Czechoslovakia at the end of 1945; but the Red Army unit guarding Jáchymov stayed on.) In the second letter, dated 13 February, the postmaster reported to Prague that two Soviet officers, accompanied by two Czech officials—one of whom was Svatopluk Rada, a member of the Czechoslovak–Soviet uranium commission—visited Jáchymov, and that in the next few days 100 German POWs and 40 members of the Czech militia were about to arrive.

Militia representatives treated the "...manager of the enterprise in such manner that he cannot stay here. They behave arrogantly...The militia is exclusively communist; it may happen that I and other members of the peoples' party will have to leave, or they will arrest us all!"[38] The postmaster did not know if an agent in his own party passed his letters on to the Communist Party headquarters. They ended up in the personal archive of Klement Gottwald, the communist leader who was then the prime minister.

A Serious Diplomatic Embarrassment

When the representatives of fifty countries met in San Francisco on 25 April 1945, to draft the Charter of the United Nations, Jan Masaryk, the Czechoslovak foreign minister, was among them. The UN Charter was unanimously adopted on 24 June and came into force four months later. In January 1946, Masaryk headed the Czechoslovak delegation at the first meeting of the UN General Assembly in London.

Despite the signal failure of the League of Nations to keep the peace between the wars, hopes for the new international organization were high. It was expected to settle disputes between its members by peaceful means, to ensure adherence to a high standard of human rights and to assist in the social and economic advancement of mankind.

Most of the people who met in San Francisco knew nothing of the development of the atomic weapon. (It was first successfully tested on 16 July 1945 in New Mexico, soon after the adoption of the UN Charter.) The explosions in Japan in August moved the control of nuclear energy into the public domain and to the top of the diplomatic agenda. In a statement on 6 August, the day the bomb was dropped on Hiroshima, Truman expressed the hope that atomic power could become a "forceful influence towards the maintenance of world peace." (Truman's statement announcing the bombing of Hiroshima.) Washington regarded UN regulation of the world's uranium supply as being central to international management of nuclear energy. At that time, Czechoslovakia was the only known producer of uranium in Europe. It was not under the control—as the Canadian or Congolese resources were—of America or Britain.

At the first meeting of the UN General Assembly in January 1946, Jan Masaryk suffered a deep humiliation. He was in his sixtieth year, the son of the revered founder of Czechoslovakia, Thomas Masaryk. His mother was American, and Jan spoke English with an American accent and played the piano rather well. He succeeded in presenting himself to the public as a person at peace with the world.

Masaryk's public personality concealed a divided and complicated character, and there were questions regarding his private stability and public staying power. He had failed to establish himself in America before the First World War, and spent the war years in the Austro-Hungarian army. Masaryk became close to his father after the war, joined the foreign ministry, and later became the minister to London. It was said in Prague that, between the wars, only few people knew Jan and those who knew him did not take him seriously.[39]

Jan Masaryk's wartime broadcasts from London made his reputation among the Czechs and helped to keep up their spirits. He was a close associate of Beneš, and there was a chance that Masaryk would become the president's successor. But Masaryk did not have Beneš's toughness and single-mindedness of purpose, nor his dedication to politics. Masaryk was an intelligent man with an aversion to theorizing, unable to work out conceptually coherent political attitudes. He discovered his talent as a public speaker late in life, although he showed an interest in propaganda and in publicizing the cause of the new state of Czechoslovakia since the beginning of his employment in the foreign ministry.

Masaryk needed somebody to direct him—after his father it was Beneš—and to shelter him. He never acquired the habit of methodical work, nor the skills necessary for running an organization. He tended to absent himself when he was most needed, and he often relied on his gift of gab to get him out of trouble. Beneš, who retained his hold on Czechoslovak foreign policy, relied on Masaryk's support. Towards the end of the war, an overworked Beneš became increasingly frail. On 9 March 1945, he suffered a mild stroke the night before he was to fly from London to Moscow. It seemed that Beneš's Russian policy was leaving Masaryk behind. Masaryk did not accompany Beneš to Moscow at the end of 1943, when the Czechoslovak-Soviet treaty was signed. Masaryk did not oppose the treaty outright, and criticized the British for questioning the purpose of Beneš's visit to Moscow. In June 1943, he nevertheless expressed his skepticism to Philipp Nicols, the British representative to the Czechoslovak government in exile, about the proposed treaty and his fear that it would isolate the London Poles.[40] He was unable to challenge Beneš's optimism about the changing nature of Soviet communism, or cast doubt on the president's conviction that Stalin was a reliable partner.

The treaty of December 1943 with the Soviets was designed by Beneš and inspired by Fierlinger. In a speech broadcast on 16 February 1944, Masaryk said that, on his return from America, his first visit was to Beneš. He talked about Moscow, while Masaryk told Beneš of his extended trip to the United States, which lasted from 17 October to 10 February. He said that Beneš's visit to Moscow was "approved of in America," though there were "a few reactionaries who hide their own selfish interests behind the pretence of fear of bolshevism." But he added that "...we should get used to calling it the Soviet empire. Because it will be the Soviet empire which will play the most important role on the continent." Masaryk added: "At the conference at Atlantic City, our cooperation with the Soviet delegates was ideal. Working with those young, energetic, educated and nice lads was very pleasant."[41]

Jan Masaryk drifted in the wake of Beneš's eastward course. He loved comfort and was at home in London or New York, in contrast to many of his colleagues in the Czechoslovakian government in exile, for whom London was an alien and temporary place of residence. He could not avoid traveling to Moscow in March 1945 for negotiations regarding the composition of the new government. He was disappointed by Moscow, and the Russians were disappointed by him. Stalin's Moscow was not a place where he could display his talents to their best advantage. In the background, there was the malign presence of Zdeněk Fierlinger, the Czechoslovakian Ambassador and communist nominee for the post of prime minister. Masaryk could not abide him.

He did not wait to travel to Prague from Moscow with the rest of the government. Masaryk flew instead to a meeting in San Francisco, where the United Nations was being organized. He had felt uncomfortable in Moscow, yet was unable to translate his feelings into a clear-cut intellectual position or any kind of political action. His views were unsteady, and there is evidence of his private anti-communist attitudes from that period.[42]

He kept on repeating Beneš's view that Stalin would not interfere in the internal politics of Czechoslovakia. He said that the country enjoyed full intellectual freedom and that he could see no signs of a divided Europe, or of an iron curtain coming down. He strayed beyond the limits of good taste when addressing a meeting of the Slavonic committee in March 1947: "We, the Slavs, have found each other. And, as the beautiful Czech song goes—we have found each other and we shall never leave each other." He deceived the public as much as he deceived himself and, in his confusion, he lost sight of the shape of the emerging post-war world. Beneš no longer served as a reliable guide to him.

Even after the unconditional surrender of Germany, Beneš remained obsessed by the German threat. He believed that the Eastern and Western allies would remain united, and that Czechoslovakia would create a bridge between them. Beneš continued to believe in the possibility of convergence between the socialist and the capitalist worlds, and failed to grasp the reasons for the emerging hostility between Russia and America. Neither Beneš nor Masaryk considered the possibility that Stalin feared the advantage the Americans gained by possessing the nuclear weapon.

Though Masaryk had taken no part in the preparation of the secret uranium treaty between the Soviet Union and Czechoslovakia, he was present when its text was read out at a meeting of the full cabinet on 23 November 1945. On his arrival in London in January 1946, he should have been forewarned of the importance of the subject. When the US delegation landed in London, journalists immediately questioned Secretary of State James Byrnes was about American attitudes regarding the control of atomic energy.

Masaryk's own political future hung in the balance at the time of the meeting in London. The pressure of Czechoslovakian politics weighed heavily on him, and there was the chance of being offered a high post at the United Nations. After they arrived at their hotel in London, members of the US delegation had a private discussion about who should be appointed to the post of the Secretary General of the United Nations. When the name of Jan Masaryk was raised, one of the delegates remarked that he would not take the post. Another delegate objected, saying that Masaryk was now ready to take a job at the UN, including the post of the Assistant Secretary General.[43] The post went to Masaryk's wartime colleague Trygve Lie, the Norwegian minister of foreign affairs in exile. Lie became the first secretary general of the UN on 1 February 1946.

Jan Masaryk led a delegation to London, composed of representatives of all political parties, who helped him prepare his speech. He preferred speaking off the cuff, and found the found the collective effort hard to take.[44] Masaryk knew that the passages concerning uranium were especially sensitive. The delegate Jan Bělehrádek was a biologist, rector of the University of Prague and a Social Democrat. He drafted the passages concerning the uranium problem, although he knew nothing of the secret treaty with the Soviet Union.

Masaryk explained to Bělehrádek that he thought the uranium problem was important, and that it must be included in the speech. The question was, Masaryk said, how it should be done. He knew that every socialist would be against continuous arming, and in favor of controlling the arms industry. He told Bělehrádek that he had discussed matters of disarmament during the war, and that he favored international control. Bělehrádek was delighted that Masaryk turned to him, and agreed to draft a few notes. The minister asked him whether the new situation would require control of the atom, saying that he was not a scientist and wanted to hear the views of an expert.

Bělehrádek began explaining the problem to Masaryk at length, making free use of scientific terms. Masaryk let him carry on, asking him an occasional question. Bělehrádek said that he would include the question of controlling the atom and of "our Jáchymov." Masaryk doubted whether Bělehrádek was better informed on the nuclear problem than he was himself. Bělehrádek's aide's memoir became too involved and moralistic, but Masaryk did his best to adapt it for his audience. "Does it not approach manipulation?" Lumír Soukup, Masaryk's private secretary, asked later. He replied that he was not Machiavellian. "I never used the word Jáchymov myself" Masaryk insisted, but Bělehrádek did. "The initiative came from a party comrade of the prime minister." He turned over the pages of the draft of the speech, saying he was happy with it and would not change a single word. He spoke on behalf of the great majority of people all over the world, who had experienced great suffering in the war. Masaryk said that he wanted to avoid empty phrases, which created indifference in the listener. When drafting the speech, he told Soukup, he saw the eyes of the people in the streets and in the villages, and he spoke for them and to them.[45]

Masaryk opened the speech with a reference to the League of Nations, in which his father, as well as Edvard Beneš, had placed so much faith. There had been a tremendous outburst of idealism after the First World War, Masaryk said, adding "I was one of those whose head was rather in the clouds and whose feet were not too definitely on the ground."[46] This time, Masaryk wanted to voice "calm, realistic optimism." The speech had peaceful undertones: humanity must be assured that war is not inevitable, and that peace is indivisible. The Czechoslovakian cabinet would challenge part of Masaryk's speech, which easily fit into its general tone:

> Wars should be stopped by controlling all the means of war, whether they are physical, chemical, biological, psychological or sociological. Within the framework of our Organization there should be an international protection of science against the abuse of its progress for political or militaristic schemes; humanity should be safeguarded against the result of abuse of scientific inventions. The armament industry, together with its latest devastating invention, should be put under the control of the United Nation. [Applause] I speak with a certain amount of knowledge on this subject, because our radium mines [sic] in Jáchymov were among the first to serve humanity by supplying radium for medical purposes before new mines were discovered. And may I here, in all humility but with profound conviction, express the hope, which I know you all share, that not one particle of uranium produced in Czechoslovakia will ever be used for wholesale destruction or annihilation. [Applause] We in Czechoslovakia want our uranium to do exactly the opposite—to build, to safeguard, to raise the standard of living, to make our lives more secure and efficient. To this purpose we wish to dedicate our radium mines. Please do help us."[47]

After the speech, in an interview with Reuters' correspondent, Masaryk indicated that the Jáchymov mines were open to international inspection. In Prague, the reaction to Masaryk's speech was less than enthusiastic. It was the first item on the agenda for the 18 January meeting of the narrow cabinet, and created collective hysteria. Everybody present was of the opinion that Masaryk should have never made the speech. Stránský described it as a "terrible faux pas," while Gottwald thought that it would threaten good relations with the Soviet Union. Sramek believed that "after Masaryk's speech, we look very odd in the eyes of the Soviet Union." Masaryk's recall to Prague was discussed, and Prime Minister Fierlinger did not want to take responsibility for the drastic step. He left the meeting to consult the president, who advised that Masaryk should stay in London and be asked to explain his statement. Deputy Foreign Minister Vlado Clementis was sent to see Valerian Zorin, the Soviet ambassador, to tell him that the government had not authorized the speech, and that the delegation in London had received no instructions that would justify making it.[48]

Fierlinger immediately cabled Masaryk in London that the presidium of the government dealt with his speech on the control of the armament industry and the Jáchymov mines, even making a declaration to Reuters. "We beg you to report on this in detail and straightaway. We fear very unpleasant political consequences, because it is inadmissible that we should be dragged into such discussions, which require our utmost reserve, not least because of our treaty obligations to the Soviet Union, which must remain secret."[49]

The message profoundly shocked Masaryk. It was delivered to him at the UN plenum at Westminster Hall. Masaryk left hurriedly, and was very agitated on the short car trip to his flat at Westminster Gardens. Soukup noted that "...I had never seen him so upset..."[50] Masaryk sent the Ambassador his apologies and said he would get in touch later. He enquired whether the Ambassador was free in the evening, but then he changed his mind, asking Soukup to collect the reply later. Soukup found him cross rather than upset. Masaryk held two pages of his reply in his unsteady hands.

Masaryk's reply showed how deeply upset he was. He again confused uranium ore and radium, and wrote that the Czechoslovaks had never made any secrets about the production of radium. He insisted that the Soviet–Czechoslovak treaty must be kept secret. In a confused sentence, he expressed the hope that there would be no war, and that neither "the Soviets nor we" would use the atomic bomb.*

---

* The full text of Masaryk's letter can be found in Appendix 2.

In the morning on 19 January, Soukup dictated the cable to a secretary at the Embassy. The cable was encrypted, and Soukup kept the original. Masaryk came to the embassy later and brought with him a new version of the cable. Soukup told him that the first one had been sent, and offered to rewrite the second version. Masaryk hesitated for a while, and then said that it should be sent as it was. "They will have to read them carefully and have something to compare in their dumb heads..." Masaryk left the sentence unfinished.[51]

He looked more ill than the previous evening, explaining that he couldn't sleep and had pains. He said that he would not go to the General Assembly meeting, and he asked Ripka to lead the delegation. Masaryk emerged from the Ambassador's room later and showed Soukup the drafts. The minister's handwriting was illegible, with many corrections. The envelopes were sealed and put into the diplomatic bag. Masaryk sat in silence, breathing in short gasps, with sweat on his forehead. After a while he began to smoke and the color came back into his cheeks. He said that they want to get rid of him, without mentioning any names.

It took two days before Masaryk recovered and the members of the delegation began to consider the aftermath of his speech to the UN. Hubert Ripka had himself signed the secret treaty with the Soviets. He took time off on 21 January to write from London to the chairman of his party, Jaroslav Stránský. Ripka pointed out that Masaryk "was especially depressed by the fact that the explanation [to the Russians] was offered before he himself could inform the government. He had a stroke yesterday and he wanted to give up the leadership of the delegation, until the whole matter was cleared up. We talked him out of it, because it could cause here diverse speculations. He is really ill now, but it does not make a strange impression, because other delegates have also fallen ill."[52]

Ripka contrasted the reception of Masaryk's speech in Prague with the positive reaction to it in London, including its reception by the Soviet delegation. The reference to uranium was occasioned, according to Ripka, by two factors. From the time of their arrival in London, the Czechs were questioned from every side about their uranium, and whether it was true that the Russians had exclusive claims to it. "We declared categorically that it is not true, and that the deposits of uranium ore and its mining are in our hands."[53] Ripka knew that Czechoslovak uranium would come under discussion in a special commission, which was established on the day he wrote the letter. Ripka explained that Masaryk wanted to preempt any doubts that may have existed. He was convinced that the Soviet delegates understood the uranium reference in Masaryk's speech, and that they had no objections.

Masaryk merely expressed, according to Ripka, his agreement with the principle of international armaments control. It was generally acceptable, and Czechoslovakia would adhere to it only if all the other states did so. "It is self-evident that we would not accept control unilaterally applied only to us, or relating specifically to our uranium only," he wrote. There was no doubt in Ripka's mind that the Czechoslovak declaration was in accordance with the Soviet policy. He argued that, in the political commission on that very same day, a Polish delegate proposed a solemn declaration binding all states to never use atomic energy for warlike and destructive purposes. According to Ripka, the Soviets agreed with Polish delegation's proposal, which was identical with Masaryk's declaration. Everybody who carefully considered the Czechoslovak and Polish proposals would come to the conclusion that they were more embarrassing for the Americans than for the Soviets.

Ripka was younger and more resilient than Masaryk. He felt no need to refer to the uranium treaty in London, even to his colleagues on the delegation. He was aware of the existence of the problem of the international control of atomic energy. In contrast with the timid politicians in Prague, constrained as they were by secrecy and guilty knowledge of the treaty's implications, Ripka felt no qualms about covering up its existence. He knew that, in the short term at any rate, the Czechoslovaks were constrained to follow Soviet moves in connection with the international control of the sources of uranium.

On the same day, 21 January 1946, Professor Jan Bělehrádek sent a letter to his party comrade Fierlinger.[54] He explained how Masaryk's speech was drafted and added that it was well received—it was considered to have been one of the best made at the London meeting . This was apparent from the people who came to congratulate him, including the Soviet delegates. Some of them wished him well in connection with the candidacy for the post of the Secretary General of the UN. Bělehrádek referred to the positive press response in Prague, especially the reference in *Rudé právo* (the communist news organ) to the possibility of misusing science and the proposal for arms control.

Bělehrádek further explained to the prime minister that the telegram from Prague created embarrassment amongst the members of the delegation. They all tried to find a reason for it: perhaps the slanted Reuters interview created an adverse impression in Prague, either in the case of some members of the cabinet, or at the Soviet Embassy. According to Bělehrádek:

I should like to note here...we support the Russians even when we know that we may lose. The weakness of the Soviet delegation means that cooperation between the great powers suffers and that the respect of the small nations for the new organization is acquiring looser and more democratic forms. In any case the endeavor of the small nations not to be pushed into passivity and subservience to the great powers is apparent here...It was hinted to us several times here that we are in fact subservient to the Soviet Union against our conviction, and that we have given up our international political freedom without intending to do so. In my view Masaryk's speech therefore strengthened the impression, even at the cost of appearing to go against the great powers, that we have preserved a great part of independence of a small nation.[55]

Bělehrádek's optimism was made possible by his ignorance of the uranium treaty. In London, Soukup had a chance to ask Masaryk about it. He did not know who had initiated the treaty, but he had his suspicions. There was a "super comrade" who was not a member of the Communist Party and who had lost Jáchymov for Czechoslovakia, Masaryk said that he wanted to please our friends, and he was generous in his offers. "The ground had been prepared and when the bombs hit Hiroshima and Nagasaki, Zorin at once claimed our ore. He soon discovered that it would not look well on the international scene if they brought their own miners, and so they left the mining in our hands, and they will buy the ore from us. You know that on the international market we would get hard currency and our economy would be greatly helped, but the pennies we get from them..." Soukup remembered the conversation many years later, when he found out that his friend, another member of the Jan Masaryk secretariat, Dr. Antonín Sum, worked in the Jáchymov mines as a slave laborer.

It may be that Masaryk wanted to bring the matter of Czechoslovak uranium to the attention of the international public. If this were the case, he chose an oblique way of doing so. If he wanted to express his own views on the question of international uranium control, he would have had to convince his government and the president of their correctness. Jan Masaryk did not attempt to do so.

It soon became apparent that the secret treaty was becoming a very sensitive and crucial point of Czechoslovak diplomacy. In a personal letter to Beneš from Moscow on 20 February 1946, Jiří Horák, the Czechoslovak ambassador, complained that he was badly informed from Prague: "...we were short of any information concerning the agreement about the mining of uranium in Jáchymov, though I intervened in that matter with V. M. Molotov." In the case of Jan Masaryk's speech in London, Horák thought that "The hasty approach to the Soviet Ambassador to Prague definitely harmed the reputation of the government as far as the local authorities were concerned, who are impressed by calm dignity rather than by too much anxiety, which may be regarded as evidence of weakness."[56]

The uranium incident in London cast a sharp light on the state of politics in Czechoslovakia early in 1946. At the top of the political hierarchy, the president gave little guidance during the negotiation of the secret treaty with the Soviets. He was not even aware of its implications. He was sick, and out of his depth in the emerging system of international relations.

The Czechoslovak government regarded the treaty with the Soviets as a guilty secret while its members, regardless of party affiliation, were united in condemnation of Jan Masaryk's speech to the UN plenum. Masaryk, who was Beneš's closest associate in the government, suspected Fierlinger of plotting against himself as well as against Beneš. At the very least, Fierlinger used the incident to cut Masaryk's reputation down to size.

When Masaryk recovered from the first shock, he fell back—if the memory of his young private secretary is to be trusted—on stressing his guile in inviting Bělehrádek, a highly placed member of Fierlinger's party, to draft the uranium passage in the speech. He was however unable to hide his deep nervousness, and the fact that he was lost in the maze of the developing international system. The atomic era, and atomic diplomacy, had not yet been so named.

Industry with a Future: A National Asset or a National Disaster?

The secret uranium treaty was a political act with far-reaching consequences. The uranium industry soon became the flagship of Russian enterprise in Czechoslovakia, where Soviet methods of doing and making things were first introduced. They included the Soviet planning methods, production undertaken regardless of human and economic cost, and eventually, the use of slave labor.

According to the official communist version, the central committee of the Czech party (Ústřední výbor KSČ) created a successful industry by providing suitable conditions for its development.[57] One of the most experienced communist bosses, the state uranium company and its long-serving head, Antonín Schindler, wrote that "the Soviet Union sent us experienced specialists and gave us every form of material assistance."[58]

The alternative, post-communist view of the development of the uranium industry also assigns the Soviet Union a place of prominence: "The conduct of the Czechoslovak uranium industry was from the moment of the signature [of the secret treaty in November 1945] fully subordinated to Soviet interests, without any regard to the letter of the treaty and the existence of the inter-governmental commission."[59] The Soviets were no doubt influential in the uranium industry from the very beginning of the NPJ, but did not always control its operation. There remained areas of friction. It took at least three years before differences on the interpretation of the treaty, especially with regard to the cost of uranium, were settled to the Soviets' satisfaction.

The Czechs concluded the secret treaty without knowing the position of the nuclear program in Soviet political thinking. They did not know that it came under the control of Lavrentii Beria and his security organs, and that the scarcity of uranium in Russia would put the Czechoslovak government under intense pressure. The struggle for the control of the uranium industry continued beyond February 1948, when the communists took power in the state, and involved communist and non-communist political players in Czechoslovakia itself.

In the old silver-mining town of Jáchymov, the church spire and the tower hoist over the centrally-located Svornost mine still dominate the skyline. In 1945 there existed two other uranium mines in addition to Svornost in the district. Bratrství (Brotherhood), which had been formerly called the Mine of the Saxon Nobility, and Rovnost (Equality), formerly Werner. Bratrství was tucked away in a valley that had developed into a small holiday center between the wars, with tennis courts, a swimming pool and a popular dance hall. Rovnost had a good view of the surrounding countryside and harsh climatic conditions.

The ways in which a small mining business was transformed into a huge state company in the years after the Second World War were singular. It was partly a privileged and secret enclosure, and partly a disorderly Eldorado, which the state, the Communist Party and the Soviets all tried to get under control. Slave workers added later and housed in concentration camps.

The number of employees for 2 July 1945 was given as 122.[60] The State Mining Board [Statní báňské ředitelství] had the existing pits in its charge, and mining operations were restarted with largely German workforce. There were eight Czechs on the managerial side. In April 1946, a few months after the Czechoslovak-Soviet treaty was signed, the new company, Národní podnik Jáchymov (NPJ) had 320 miners in its employment, and 100 German POWs. In its early years, NPJ relied on a German workforce. In addition to the local, Jáchymov district Germans—24 of which worked for NPJ a year after the liberation—there was a growing number of POWs, who were sentenced for war crimes, and Germans who were resettled in Jáchymov. Czech officials did not approve of strengthening the German ethnic element in the district, but labor was scarce and the pressure from the Russians strong.[61]

There were 69 Red Army guards on duty at the pits, commanded by four officers. Behavior of the Red Army troops was reported to have recently improved, and drunkenness, brawls and thefts became less common. The NPJ was in the process of organizing its own security unit, under the command of Lt. Koryma, a young Czechoslovak officer who had returned from the Eastern front.[62]

In Czechoslovakia immediately after the Second World War, priority supplies of mining and other technical equipment went to the coal industry.[63] Skilled labor as well was channeled into coal mining, before the Soviet interest started asserting itself and the development of the NPJ became a top priority for the Czechs as well. The uranium industry therefore developed, in a critical sense, alongside, rather than with, the mainstream of the Czech economy and without regard for its overall requirements.

NPJ was formally established on 1 January 1946. The enterprise reported to the Central Board of Czechoslovak Mines [*Ústřední ředitelství československých dolů*]. The managing director of the Central Board, Svatopluk Rada, became one of the two Czechoslovak members of the joint Czechoslovak–Soviet permanent commission [*Stálá československo-sovětská komise*], which became an important conduit for Soviet control over the uranium industry. The commission was responsible for developing plans and methods to increasing production, for material, technical and financial aspects, and for pricing the final product. In 1946, the economic counselor at the Soviet embassy in Prague, Dashkievich, became one of the two Soviet members of the Commission, and K. Volokhov was the other. The latter was replaced by S. N. Voloshchuk (he later became the managing director of Wismut AG) in 1950, and Dashkievich's place was taken by A. P. Morozov a year later. D. N. Sukhanov replaced Voloshchuk in 1954.

On the Czech side, J. Kovář was paired with Rada until 1948, when Kovář was replaced by Dr. O. Pohl. After Rada's suicide in 1952, Antonín Schindler joined the Commission, and remained in the industry until the beginning of the 1970s. He was a mining engineer from the Silesian coal district, who spoke German better than Czech and survived the industry's many political and organizational changes.

Among the Soviet commissioners, Volokhov had the most clear-cut views on the management of the NPJ. He resented the fact that the joint Soviet–Czechoslovak company had not come into existence, and tended to ascribe the flaws in the running of the mines to incompetent Czech managers. As the Soviet experts took up their posts during 1946, the Czechs sometimes found it hard to work with them. The Russians were ignorant of Czech mining regulations. The Czechs complained that they were unable to grasp the fact that pit managers were responsible for the safety of the miners, regardless of their origin. Nor did the Russians take much notice of the works councils, a socialist innovation achieved by a recent presidential decree. They sometimes seemed to be so irritated by the Czechs that their displeasure was "virtually tangible."[64]

Most of the pit managers were Russian and they kept Czech assistants by their side "in case of trouble."[65] The disputes did not only run along national divides. The Russian team was not always united, and there were open tensions, especially between Krivonosov, the production manager, and Pavlenko, one of the chief mining engineers. In one of the earliest reports on the encounter between Soviet and Czech management practices, the sympathy of a member of the Czech team was on Pavlenko's side, as he came close to "our idea of a good mining engineer." He was able, knowledgeable and experienced. His calm self-confidence contrasted with Krivonosov's explosive and touchy nature. Krivonosov was an authoritarian boss who wanted to have everything under his control. Both Krivonosov and Pavlenko nevertheless were, for Czech taste, "too magnanimous in financial matters."[66] The Czechs knew that Alexander Pavlov, employed in the planning office, was responsible for security matters.

Many of the difficulties encountered by the management were connected with the ill-defined position of the NPJ in the post-war economy of Czechoslovakia. Delays occurred in supplying the mines with machinery and transport equipment, and there was a chronic labor shortage. Government ministries in Prague, whose officials were aware neither of the existence of the secret agreement nor of the privileged position of the uranium industry in the Soviet scheme of things, often refused to meet the management's wishes. The Czech members of the commission kept on turning for help to Klement Gottwald, who became prime minister after the general elections in May 1946.

The Secret Treaty Becomes a Subject for Debate

In the early days of the NPJ, the Czechs held diverse views on the future of the uranium industry, which did not remain hidden from the Soviets. Shortly before the treaty was concluded, radium production was discussed at an inter-departmental meeting on 8 November 1945. The ministries of health, national defense, foreign trade and finance were represented at the meeting. It was agreed that the old technologies should be abandoned. Koblic was in charge of the spa and of the production of radium in Jáchymov, and was the most outspoken advocate of modernization. He was referred to the Ministry of Defense.[67] In March 1946, Koblic turned instead to the presidium of the cabinet, expressing concern with the possible effect of intensive extraction of uranium on the radioactive springs. He recommended that only processed uranium should be exported to the Soviet Union and made recommendations on its price.[68]

Anxieties that the reserves of uranium would soon be exhausted were a reaction to the secret treaty—or to the rumors about it—and awoke anxieties among Czech geologists and other experts as to the exhaustion of the uranium reserves. A member of the commission, Václav Kovář, was among them. Most of the criticism came from the ranks of the Social Democrat Party and were communicated to Bohumil Laušman, the Minister of Industry.*

Fears as to the scarcity of uranium were first expressed in a brief memorandum drafted on 22 June 1946. It argued that, while the Czechs could not compete in the development of nuclear weapons, uranium would become as important as coal "in a few decades," and that one kilogram of uranium would create as much energy as 3,000 tons of coal. The report mentioned only one uranium deposit in Czechoslovakia, e.g. at Jáchymov, and that this unique resource should be carefully husbanded for the future.

"Today, we are exporting the ore in considerable quantities. Experts fear that the sale of the raw material, which will have an enormous value

---

* Laušman was minister until November 1947, when he replaced Fierlinger as party chairman. He served, for a few months, in the communist government after February 1948. He emigrated to Yugoslavia in 1949 and was kidnapped by the Czechoslovak secret service in Austria in 1955. Laušman was sentenced to seventeen years imprisonment, and died in unexplained circumstances in prison.

for future generations, will harm us and future generations." The memorandum further stated that if Jáchymov was the only deposit within reach of the Soviets, then it "would of course be our moral duty to our great liberator and ally" to put the ore at their disposal. The Soviet Union, however, with its vast natural resources, would probably find uranium in Siberia, the Urals and Central Asia. The memorandum further recommended a review of the sale of uranium, because Czechoslovakia was short of other sources of energy and the possession of uranium could have a beneficial influence on the whole industrial life of the country.[69]

The four member permanent commission became the scene of hard bargaining. The price of uranium remained to be agreed on. Soviet payments were made at sporadic intervals: before 16 September 1946, the Russians had paid 5 million crowns and delivered equipment for about half that sum. Czechoslovakian credit was used to finance the NPJ, as investments needed for the level of production required by the Soviets were high: for the two year plan in 1947 and 1948 they were put by the Czechs at 32 million crowns (160 million before the currency reform in 1953, when the rate between the old and the new crown was set at 5 to 1). It became apparent in 1947 that a much higher sum would be needed, at least 230 million; the Soviets offered the Czechoslovaks a loan of 200 million crowns.[70]

At the end of 1946, the two Czech members of the permanent commission, Kovář and Rada, drafted a note for the minister of industry, Bohumil Laušman. It contained a reference to the shortage of labor in the NPJ, and concentrated on calculating of the price for uranium exports. The "production costs" referred to in the treaty were understood differently by the Russians and the Czechs, while "reasonable profit" remained undefined.

The Czech experts, concerned as they were with the future shortage of uranium, proposed to include an item in the final price representing the cost of depletion of natural resources. It was put at 6,200 crowns for one kilogram of uranium oxide. The argument for the inclusion of the depletion cost of the proven reserves (*dolová podstata*) was anchored in the old mining laws of the kingdom of Bohemia. The Soviets asked for a written proposal concerning the price, along with a justification of the depletion costs.[71]

In his calculations, Kovář achieved some bizarre results. He compared the potential of coal and of uranium to generate energy, working on the assumption that the reserves of uranium at Jáchymov amounted to no more than 400 tons. The price reached astronomic heights: a broad price band between 30,000 and 75,000 crowns (at old currency values—the price was between 6,000 and 15,000 after the reform) for 1 kg of uranium oxide. According to rumors circulating in Prague, the Americans were paying $300 for 1 kilogram for uranium from the Belgian Congo, when the price in fact was closer to $35. A portion of the US supplies had been secured for less than $5, when uranium was regarded as a by-product of the process of making radium. The Russians knew what the Americans paid, though they made no mention of the US purchase price to the Czechs. Their reticence was prompted by their desire not to put the Czechs off developing their industry as fast as they could.

Kovář discussed the problem with Dashkievich on 5 February 1947.[72] The price of Czechoslovak uranium was to include, according to Kovář, a depletion element in addition to the production expenses. It was also to include a reasonable profit, calculated at 18%. Production expenses were to include wages, salaries, social and administrative costs. Kovář did not include the costs of building the infrastructure—schools, medical services, housing etc.[73]

Dashkievich argued that the mining of uranium and its deliveries to the Soviet Union did not deplete reserves. On the contrary: assistance from Soviet experts led to the discoveries of new deposits, thereby enhancing the mineral wealth of Czechoslovakia. He added that, at the time of the conclusion of the uranium treaty, the Czechs were aware that it was not a business contract. However, if they were convinced that prices on the world markets were higher, and that it was necessary to change the agreement, the commission did not have the power to act. In any case, Dashkevich said, the question of the depletion costs had been raised when Gottwald, the prime minister, visited Jáchymov. Gottwald agreed that it should not be included as did the other Czech member of the Permanent Commission, Svatopluk Rada.

The other controversial part of the uranium agreement proved to be the 10% of production the Czechs were entitled to retain for their own use. Dashkevich was convinced that the Czechs could do so only with the agreement of the Soviet government because, as production increased, the portion set aside for the Czechs would be too large for their requirements.

Kovář's letter to Fierlinger on 19 February contained a request for the elucidation of the share of production the Czechs were entitled to keep. He expressed concern with the swift exhaustion of reserves in Jáchymov, and hinted that his colleague, Rada, held different views. Rada would not insist on the Czechs getting their agreed share of the production of uranium. Kovář used a news item from the *Neue Zürcher Zeitung* of 5 February 1947, concerning the project of a nuclear power station in Britain, in support of his argument. He asked Fierlinger as well as Laušman, the minister of industry, to support his views.[74]

The dispute over the price of Czech uranium continued until Svatopluk Rada visited Moscow, accompanied by J. Kašpárek of the Foreign Ministry, at the end of 1947. They did not receive a warm welcome. The Russians argued that uranium from the stores and slag heaps had cost the Czechs nothing to produce, and that its value was no higher than the cost of transport. Rada returned to Prague as a resolute defender of Soviet positions; Kašpárek defected to the West after February 1948.[73] Kovář was replaced on the permanent commission by O. Pohl in the latter half of 1948.

By the end of 1947, the Soviets had made most of the necessary points on how the uranium industry should be run and had learned to use the permanent commission as the main venue for controlling the industry. It took some four years or so before the conditions of sale of Czech uranium started to improve. Its cost reached the highest level during 1949–1950, sharply declining thereafter. The prices reflected the uranium hunger of the Soviets: for about four years after the war, pitchblende was mined in Czechoslovakia with little regard to the cost of production.

The "Geological Service"

In the uranium industry's hierarchy, the "geological service" ranked lower than the permanent commission. It nevertheless exercised crucial influence on the NPJ's day-to-day operations. The Czech communists paid high tribute to the work of the Soviet geologists, attributing the fast growth of production to their assistance.[76]

A. E. Vorontsov was among the first group of experts who came to Czechoslovakia, and his task was to establish the service. It was divided into two sections; one dealt with production, the other with exploration. (*rudnichnaia and perspektivnaia geologia*) At the end of 1946, the geological service employed 26 men: 14 Soviet experts and their 12 Czech assistants.[77] Vorontsov was well-liked by his team. Several of its members later achieved international reputations in their fields, such as Getseva in mineralogy, and Krasnikov in geology.

The first brushes between the geological service and the management of the NPJ were reported in 1946. They concerned the fulfillment of production targets as well as the geological justification for some of the mining work. Vorontsov had differences with Krivonosov, whom he blamed for ordering unnecessary excavation work.[78] The service employed Czech assistants, who were often denounced by their colleagues: an extensive security operation involved assistants who took company papers home with them.[79]

The survey of the Jáchymov district started in a rather chaotic way. The Soviets had little geological documentation at their disposal and did not know where to turn for advice. They talked to local miners and could not understand why, despite the long history of mining in Jáchymov, no comprehensive geological, geophysical or hydrogeological documentation was available. They were surprised to discover that experienced pit managers were left in charge of the works, without the benefit of expert advice from the geologists.

The Soviet service used an old geological map—its scale was 1:14000—and two German treatises. One of the treatises dated from 1916, the other was prepared recently by a geologist by the name of Elstner during Colonel Alexandrov's visit to the Soviet zone of occupation in Germany in the spring 1945. It seems that Vorontsov's team made little or no use of the work by professor J. Kratochvíl, published between the wars,[80] which mapped the dislocation of minerals on the territory of Bohemia. It also provided a historical account of silver mining in the Jáchymov region.

The archive of the State Geological Institute (Státní Geologický Ústav) in Prague proved helpful to the Soviets, though close cooperation between Czech deposit geologists and the Russian geological service was not established. On 28 August 1945, the day after General Mikhailov and Colonel Alexandrov left Jáchymov, two members of the institute, Dr. V. Zoubek and Dr. J. Koutek, carried out a survey of the existing Jáchymov mines.

Zoubek had worked in the Jáchymov district before the war,[81] and both he and Koutek were skeptical about the uranium reserves on the lowest floors that were then accessible in the Jáchymov mines. Their findings contradicted the discoveries of Kurt Patzschke, the manager in Jáchymov during the war. Patzschke's findings were, incidentally, known to the Russians. In their report, the Czech geologists pointed to the rule which applied to the usual grouping of the ores, (the "five element" theory: Ag-Bi-Co-Ni-U) and recommended that exploration be extended to the more distant parts of the Jáchymov district, including the sites of the former silver mines. They mentioned Boží Dar on the Saxon border, close to Johanngeorgenstadt, and promised that the Geological Institute would keep an eye on the uranium problem.[82]

It is likely that the initiative for the visit by the two geologists to Jáchymov in August 1945 came from the Ministry of Industry. Dr. Zoubek later became a member of the Czechoslovakia Academy of Sciences and Koutek the professor of deposit geology at Charles University in Prague. They were regarded as top experts in their fields. After August 1945, they never returned to the Jáchymov district, though they did not give up work on uranium deposits.

The Czechs went on looking for uranium independently of the Soviet geological service. In 1946 and 1947, Koutek and Zoubek continued to explore districts other than Jáchymov. They prepared several reports for the Institute of Geology. One of report, dated June 1947, pointed to Krkonoše, Falknov and Cheb regions in northern and western Bohemia as being promising, and a rather diffident recommendation was made in favor of the Příbram district. While Příbram in South Bohemia was to become the richest source of uranium in the country, small deposits in the Krkonoše region were located on the Polish, rather than Czech, side of the border.[83]

Soviet geologists relied on the mining experiences of different deposits lying close together and on the archaeology of disused silver mines, including old slag heaps. The geologists carried Geiger counters in the shape of short hockey sticks. On one occasion, a party of Soviet geologists and their Czech assistants rested their Geiger counters against the wall of one of the houses in a deserted German village. They were disturbed by high-pitched noise from the instruments. All the houses in the village were subsequently dismantled and the radioactive building materials were sent to the Soviet Union.[84]

For the time being, Soviet attention remained focused on the Jáchymov and the Horní Slavkov districts in the Erzgebirge. The continuity of mining operations proved to be of considerable help to the geological service: 16th century silver miners had had a good understanding of the metal-bearing seams, dividing them into "midnight" and "morning" seams. The former ran northwest to north-northeast, the latter west to east.[85] It was found that uranium seams in the district predominantly followed the "midnight" line, and often replace other mineralization as the seam plunged deeper.

The neighborhood of Rovnost, in the mountains high above Jáchymov, looked promising; in 1946, considerable amounts of pitchblende were found on the slag heaps of Eliáš, a disused mine nearby. In the following year, another site was opened up further west, and the new mines were named Barbora and Eva. Abertramy was the last part of the Jáchymov district to come on stream, which occurred as late as 1959–1961. During this time, mining operations were winding down elsewhere in the district. Production in two of the original mines, Bratrství and Svornost, never met expectations.

The sites at Horní Slavkov, in the neighboring district, started being opened up at the end of 1947.[86] This was largely accomplished by the method of slanting shafts, following downward the seam with surface exit, a technique that helped speed up excavations. The two districts had several features in common, including the morphology of the seams: there existed many shorter seams, with only a few longer than several hundred meters. The seams were also comparatively thin, some only several centimeters in diameter. Uranium was found in the form of irregular nests or lenses, which made mining operations difficult.

The NPJ suffered shortages of technical and material equipment, as well as shortages of skilled labor—especially locksmiths and carpenters. Uranium was mined regardless of cost and without sufficient investment in the necessary technology. High numbers of workers made up for the low level of mechanization. This was writing on the wall; the first hint of the progressive Sovietization of production methods.

The equipment of the NPJ, at the beginning of its existence, was modest. It disposed of two mine locomotives (*důlní lokomotivy*) four mining machines (*těžní stroj*), 7 compressors, 26 pumps, 32 ventilators, and a uranium processing plant (*zařízení na úpravu rud*).[87] The secret Czechoslovak–Soviet uranium treaty combined the provision for increased production with the promise of technical help from the Soviet Union. In the first years after the end of the war, Moscow supplied the Czechs with whatever it could spare from its own severely limited resources.

By the end of 1945, 237 miners had produced less than a ton of uranium. In the following year the figure for the amount of mined ore was stated to have been 14.5 tons, while 3.5 tons came from other sources, slag heaps, stores, etc. In 1947, 49.1 tons were sent to the Soviet Union, of which 44.7 tons resulted from mining operations. In that year, 3,742 people were employed by the NPJ, including 1956 German POWs and 107 Soviet experts. In 1948, when exports to the Soviet Union rose to 102 tons, of which 84.1 tons were mined, 3,563 German POWs worked among 7,966 employees.[88] The low content of pure metal bore witness to the hurried nature of the mining operations.[89]

Preliminary excavation works were often based on inadequate assessment of the deposits, and the cost of production soared. The extraordinary extent of excavations was confirmed by data from the Jáchymov district, where 25 pits were sunk between the years 1946 and 1960, with passages branching out of them on 162 different levels, some of them many kilometers long. For instance, a 3,000 meter-long passage, started in 1946 at Rovnost, was extended to 7,000 meters in 1949. It would reach more than 20,000 meters in 1952. Excavated soil amounted to between 1.3 to 1.5 million cubic meters a year between 1953 and 1957, the largest amount on record for the Jáchymov district.[90] The shape of the countryside around Jáchymov was consequently changed, the slagheaps threatening to overwhelm the town itself. The local chronicler compared the countryside to a dead animal,[91] with its inner organs spilling out.

There is evidence that information on the volume of the mining operations became significant in its own right, indicating the achievement of the NPJ workforce. In addition to monthly production figures, the two communist heads of state, Gottwald and, after his death in 1953, Zápotocký, were sent information on the volume of mining operations. They proved to be costly; a calculation of production expenses invested before 1957 showed that some 1.8 billion crowns remained unaccounted for.[92] The economic regime imposed on the districts under the control of the NPJ was incapable of adjusting production according to information yielded by the cost of the mining operations.

The Czech uranium industry was given new lease on life, and the terms of sale for the Russians were improved by the opening up of the reserves in the Příbram district in South Bohemia. It had a longer mining history behind it than Jáchymov, a better climate, and more favorable geological conditions. After Niederschlema on the German side of the Erzgebirge, Příbram had the second largest uranium deposits in Europe. While Jáchymov needed a decade of intensive development before it reached its peak in 1955, Příbram overtook the production in Jáchymov in 1956 and reached the maximum of 2,087 tons a year in 1962.

Příbram was first approached by the NPJ as a possible source of skilled labor. On 15 March 1946, Dr. Kovář and a trade union representative came to Březové Hory to negotiate the transfer of 400 miners to Jáchymov. They were promised food and board, a special allowance of 65 crowns a day for living apart from their families, and the same pay with better premiums than in Příbram. According to the entries in the

Příbram town chronicle—kept in a public document despite the secrecy of the uranium operation—150 volunteers had shown interest by 1 April and only 69 of them left. On 12 April another, smaller group travelled to Jáchymov, so that altogether some 100 volunteers from Příbram worked for the NPJ.

At the start of the uranium mining operations, Příbram had, like Jáchymov, a declining population.* The development of the uranium industry gave the town a new lease of life. The attention of Soviet geologists was drawn to the district by archival materials: Příbram became a mining center in the late middle ages, and the available *bergbuchy*, or chronicles of mining, indicated that ores belonging to the uranium group (nasturan, uranium ochre, gummit and eliasit) were found among the local minerals. Nasturan or pitchblende—uranium oxide (UO_2)—was the most plentiful, and was first sold to the dyes manufacturers in Jáchymov in 1857.[93]

A group of geologists led by A. I. Zubov arrived in Příbram in the summer 1947. Several young assistants were equipped with Geiger counters and their task was to take surface measurements of radiation. One of them, who had just left the local high school and attended a short course at Jáchymov, remembered the early days of the uranium rush in the district:

> We were so impatient that, before we arrived at Hotel Horymír, the place where we were meant to stay in Příbram, we left for the slagheap at the old pit named Lill, assembled the radiometers and went to look for the first catch. It did not take long before there were shouts from every side 'I have got it!' And around Zubov, who was visibly excited, there grew a collection of bigger and smaller stones with clear elements of uranium mineralization. It was a beautiful feeling especially for me, a native of Příbram, a real confirmation of preliminary information.
>
> We returned to Příbram triumphantly. And we at once divided into search parties. Some of us went down the Anna pit at Březové Hory on the same day, where there was apparently uranium in the Janska vein. The other group was to examine in the first place the material on the slag heaps of the old pits—the relics of the old mining glory of the Příbram district.[94]

---

* Together with Březové Hory and Zdabor, Příbram accounted for 18,743 people in 1890, but its population was reduced to 12,445 by 1950.

The survey started by Zubov was carried on by the first independent unit of the NPJ located in Příbram, named K2, with the Soviet geologist A. G. Stepanov at its head. Uranium followed the old silver seams, usually at a considerable distance (about five km) to the southeast. The K2 unit included a mining division, which could do in the district whatever it pleased. The park of the castle at Kamenná was bulldozed during the search for uranium minerals, while open-cast mining took place in a field nearby, at Třebsko. The ore was immediately transported to Prague airport at Ruzyň and flown to the Soviet Union. Members of the geological service as well had privileged access to any object they decided to inspect. Příbram was a less remote region of Czechoslovakia than Jáchymov, and Soviet disregard for agricultural cultivation as well as property rights was all the more obvious.

The only registered employees of K2 in 1948 were 65 German POWs. They were moved to the Příbram district from Jáchymov, and started excavation works on the Vojna hill, where they built a camp of the same name. The Czech civilian employees, working alongside the POWs, remained on the books of the local state mining company, [*Rudné a tuhové doly*] (RTD), Ore and Graphite Mines. It is possible that the arrangement was made so as to keep the employment of Czech miners in uranium mines secret; or that the management of the RTD wanted to keep them on their books.[95] There exists another explanation: the early stages of mining for uranium in Příbram had a makeshift, temporary quality about them, and Soviet geologists did not know that they had come upon one of the richest concentrations of uranium in the world. They may have been misled at first that the common grouping of five minerals, familiar in Jáchymov (and including uranium), were not obtained in the Příbram district.

More than sixty sites were examined when uranium prospecting began there, including the polymetallic mines in the Březové Hory-Bohutín district and the slag heaps and the neighborhood of the pits, both working and closed. Zubov's group established that a small amount of uranium, less than a ton, was among the 86 minerals that had been mined in the district. Old mining books (bergbuchy) showed the presence of uranium in the pit called Anna, on the slagheap of the closed pit Lill, and in another location [*Svatomatějská štola*]. Altogether five promising sites were staked out: the report for the end of 1947 concluded that "according to our findings, the Příbram district emerges as a new, independent uranium province, possibly with industrial perspectives."[96] The first important discovery was made on the slag heaps of a disused mine near Vojna. In 1948 it was followed by the opening of several promising exits (*vychozy*) of uranium seams in the vicinity.

As the Czechs were trying to learn Soviet methods of planned economy, the situation in the uranium industry became acute. In 1949, a new institution was created, the Main Organization for Research and Extraction of Radioactive Materials [*Ústřední správa výzkumu a těžby radioaktivních surovin*, or USVTRS], with its own operational, supply, planning and personnel departments. In theory, it shared control over the NPJ with the permanent commission. Svatopluk Rada, one of the two Czech commissioners, became the head of the new office. His position obliged him to bear responsibility for an enterprise almost entirely under Russian control.

While the head office of the NPJ remained in Jáchymov, the running of the Příbram branch of NPJ was assigned, as of 1 May 1949, to the NPJ's mining Inspectorate no. VII. The employees seconded by the RTD to the uranium industry were moved to the new unit, and Petr Polák became its first head. He was replaced by Vladimír Liška a year later. They had both come from the RTD, and while both were experienced mining engineers their job description did not include anything regarding the technical side of mining operations. They dealt with wages, accountancy and personnel matters, whereas the production of uranium, including its processing, was supervised by the Soviets of the K2 unit. The terms of the Czechoslovak-Soviet treaty were applied in the Příbram district even more rigorously than in Jáchymov. The Czech involvement at the highest management level was lower, and the Russians were better represented at lower levels as well, including the planning and fiscal departments.[97]

No reliable data exists on the production of the Příbram district before the end of 1949. In the years between 1950 and 1955, the district yielded 1,297 tons of uranium for export to the Soviet Union. The production in 1955 alone amounted to 629 tons.[98] The growing importance of the Příbram district occasioned a meeting in the office of the managing director of the NPJ on 2 August 1956. Eighteen experts took part in the meeting, including comrade Růžička, who represented the central committee of the Czechoslovak Communist Party, together with five Czech and twelve Soviet mining engineers. The Russians, who proposed going to depths beyond 1,000 meters, assumed that the reserves in the district would provide sufficient pitchblende for ten years or so. The district was divided into a number of mining units, with Vojna (which was subsequently divided into two parts: Střed and Kamenná) being the most prominent among them. Vojna was joined by the Sever and Bytíz units, and Svatá Hora together with Východ and Jih.

The secret uranium treaty had, at the time of the 1956 meeting in Jáchymov, less than ten years to run. The uranium rush had peaked, and the metal was no longer regarded as a rare commodity as it had been in the years after the war. The Soviets were still interested in Czechoslovak uranium and hoped that the conditions of sale would be improved. High production costs in the Jáchymov and Horní Slavkov districts were partly set off by the comparatively inexpensive production in Příbram. The rise of the Příbram district coincided with the decline of Jáchymov, where mining was winding down by the second half of the 1950s.

In the mid-1950s, the management of the uranium industry gradually returned to Czech hands, and the uranium lobby came into existence. The so-called "placemen" (*umístěnkáři*) began filling the lower level posts. Most of them were young workers, loyal to the party, who had passed through either a specialist high school or a technical university. The intention was to make the personnel more professional and give the young workers sufficient experience to eventually to replace the departing Soviet production managers.[99] Employment in the uranium industry was still attractive: in 1955, differences between the average basic wage in industry and in the NPJ were considerable. While industrial workers earned 1,298 crowns a month, uranium miners received a basic wage of 2,065 crowns.[100]

The Uranium Factor

In contrast with the earlier searches for precious metals, the worldwide uranium rush between 1944 and 1948 was more concealed from public view. It was different in other regards as well: ownership of uranium did not promise wealth for individuals or companies, but rather unlimited power for the state. For Churchill, the wartime nuclear project was pursued in the background in deep shadow. Only three sources of uranium were known at that time: in Canada, the Belgian Congo and Czechoslovakia. Countries where uranium deposits existed were likely to acquire new political and diplomatic alignments. For instance, the relationship between Belgium and the Congo changed and became closer, while Canada was drawn into a tripartite relationship, its diplomacy largely concerned with the Anglo-American nuclear effort.

In the case of Czechoslovakia, the political impact of uranium ownership was likely to be more serious. In London and Washington, it was generally assumed in government circles that Czech uranium mines were one of the key reasons why Stalin was determined to bind Czechoslovakia to Moscow. In the numerous studies of the communist takeover of power in Czechoslovakia in February 1948, the uranium factor has received no attention whatsoever.

Well in advance of the "victorious February" of communist propaganda, the existence of uranium deposits and the secret Czechoslovak-Soviet treaty of 23 November 1945 helped create a climate of fear and nervousness in Czechoslovakia. which facilitated the transfer of control of the uranium industry to Moscow. The Russians had a close-up view of the way the country was poised between its former allegiance to the West and the alliance with the Soviet Union. In addition, beginning early in 1946, the fate of the uranium industry was one of the most important indicators of the direction Czechoslovakian foreign policy was taking for Western powers.

In London there was no doubt as to the impression the atomic explosions made in Moscow. A British diplomat reported that the Russians feared their victory in the second World War would be annulled. In a memorandum for Ernest Bevin dated 11 September 1945, the Under Secretary in charge of the North and South American Department in the Foreign Office, expressed grave fears concerning the future. The wartime allies were all aware of the possibility of a rift among the Big Three, caused by the atomic weapon. "Russia will make every effort to acquire the weapon also, and until she has done so, she will be more suspicious and resentful of the pressure that the Americans and we will have to put on her over many other issues. She will also become more doubtful of the allegiance of her satellite States…" The memorandum went on to say that it was not "…solely a question of the English-speaking combination possessing exclusive processes and plants. Raw materials add a potentially serious complication. Our measures to acquire to ourselves exclusive supplies and control of these raw materials have in fact brought into the English-speaking world orbit the countries with whom we have negotiated agreements. If and when these become known to the Soviet Govt they will increase their suspicion; and that Govt's probable desire to make similar agreements with e.g. Czechoslovakia, will increase the drift towards those rival spheres of influence which we desire to avoid."[101] The Foreign Office assumed that Russia would be less sensitive if it had ample supplies of uranium on its own territory. The Soviets had attempted to purchase large quantities of uranium from Canada, but their offer was rebuffed.

The "absolute weapon" in the hands of the Americans threatened to upset the post-war equilibrium and forced Stalin to review not only the foreign, but also the domestic policy of the Soviet Union. Stalin was perfectly informed by his intelligence services about the British and the American nuclear projects, and he knew that Churchill and Roosevelt had concealed vital information from him for a long time. After the heavy losses inflicted on it by the war, Soviet society was forced to make a great technological leap. Hundreds of thousands of tons of uranium had to be mined, and laboratories and factories had to be built. Billions of rubles were invested. It was the beginning of the development of the Soviet military-industrial complex and the arms race with America.[102] Czechoslovakia had a special place in Stalin's reassessment of Soviet policies. Less than three months after fears were expressed in London that Czechoslovakia would conclude a uranium treaty with the Soviet Union, such treaty was indeed concluded.

Soon after the end of the war, the American, British, French and Swedish governments all showed interest in the Jáchymov district. Many diplomats, as well as professional and amateur spies, visited the district. The British consul in Karlovy Vary (Karlsbad), O. Bamborough, reported to the embassy in Prague that he visited Jáchymov on 4 January 1946, where the "mines are in Russian hands and the alleged number of the garrison is about forty men." He added that after his return he met M. Dejean, the French ambassador, who was about to leave Karlovy Vary for Jáchymov.[103]

On 6 February, the day after the Consul's visit to Jáchymov, the report was forwarded to London. The report mentioned that the Czechoslovakian news agency had announced that the Czech authorities took full control over the Jáchymov mines after the departure of the Red Army, and that mining operations were renewed for the sake of the extraction of radium for medical purposes. The report also announced that Mr. Higgs, the Director of Continental Mines, would arrive in Prague in the next few days to find out more about the matter. In London on 11 February, a young diplomat drafted a top secret memorandum on his conversation with Robert Luc, a friend of his employed at the French embassy in London. Luc informed him that Stalin had recently asked Fierlinger to hand over the whole output of the Jáchymov mines to the Soviet Union.

The British diplomats continued to keep an eye on the events in Jáchymov. In the middle of March, Sir Charles Hambro came to Prague on a business trip in the interest of the family bank.* Hambro asked the

Czech authorities whether it was possible to export Jáchymov uranium. He discovered that the Czechs were not interested in the business—claiming that the volume of the ore would make heavy demands on transport. Hambro was informed that they decided to process the ore locally and produce radium. It was a clumsy excuse, and Hambro asked whether the Russians had shown interest in Czech uranium. He discovered that they were interested, but that the ore was not exported anywhere.[104]

There were other straws in the wind. A few days after the secret treaty with the Soviets was signed, on 26 November 1945, the Czechoslovak Ministry of Foreign Affairs asked the British authorities for the return of 9.223 kilograms of "uran ore" which was allegedly taken from Jáchymov to Treibach on 26 February 1945.[105] As the note from Prague gave such an exact weight, the British assumed that it could not refer to raw material. The request provided further proof that the Czechs, as well as their Russian allies, were not interested in radium for medical purposes, but rather in uranium concentrate. An expert explained to the Foreign Office that nine kilos of the concentrate would contain only about one milligram of radium.[106]

For the Americans, control of the uranium sources remained one of the key demands in the negotiations on nuclear energy in the UN. Although the US was by no means short on uranium, General Groves never lost interest in its sources in other countries. As early as the end of January, American newspapers (alongside a reference to Masaryk's recent speech in London) published the news that the Russians controlled the only uranium mines in Europe.[107] A year later, Washington knew of the Czech–Soviet uranium treaty, about the details of the security measures as well as the mining technology available in the Jáchymov district.[108]

Politicians in Prague, on the other hand, failed to assess the international implications of their decisions. In a personal letter to Beneš from Moscow on 20 February 1946, Jiří Horák, the Czechoslovak ambassador, recommended that the cabinet to deal with the uranium problem with calm dignity. It was difficult to remain dignified in a situation

---

* Sir Charles Hambro, a member of the family which founded Hambro Bank in the City of London, was the head of Special Operations Executive (SOE)—during the war. After the war ended he supervised the transportation of uranium from Germany to America.

where the imperial interests of the Soviet Union crossed the interests of the Czechoslovakian state, as well as the policy of the Czech communists.

When Horák wrote to Beneš, the political situation in the Jáchymov district was still fluid. The district council at the time was doing its best to complete the expulsion of the remaining Germans. At the end of 1945, some 3,000 Czechs lived in the district. In the elections to the national council on 28 May 1946, the Communists received 1,122 votes, the Social Democrats 473, the National Socialists 593 and the People's Party 159.

Although the security section of the Jáchymov national council did its best in the matter of the NPJ, it could not satisfy eager Communist Party members. They resented that they were not in full control of the uranium industry. Its manager was Hegner, a member of the people's party; his deputy, Čmelák, was a newly recruited Communist Party member; the second deputy was Zalud, a Social Democrat. The communists used the political composition of the management to explain the low efficiency of the Jáchymov mines.

In a comprehensive report to Gottwald from the end of 1949,[109] Svatopluk Rada, Deputy Minister of Industry and one of the two Czech members of the permanent commission, complained that in the NPJ "The party simply did not exist. Party secretaries and functionaries changed frequently and one of the party secretaries went as far as shooting himself during an interrogation, when his contacts with the West were proved. I want to show how far the employment of spy cells in the national enterprise reached, and still reaches." On 20 September 1947, comrade Svoboda of the district secretariat of the Communist Party in Karlovy Vary reported to the general secretary of the party, Rudolf Slánský, that "On 18 this month I had a consultation with the leading comrades in the Jáchymov mines. It was stated that the production plan for the last month was fulfilled only at 60%." Svoboda blamed Hegner for the poor results, together with other non-communist members of the management.

The Communist Party in Karlovy Vary, as well as the district organization at Jáchymov, including the communist organization within the NPJ, became the object of Slánský's effort to strengthen the regional party organization. After the "victorious February" in 1948, Slánský and his party comrades had reasons to believe that they had won the struggle for the control of the NPJ.

Hegner left his post on the initiative of the communist action committee at the NPJ; other personnel changes followed. Slánský, in his eagerness to assert the control of the KSČ in Jáchymov, failed to notice that the Soviets were not interested in promoting a strong Czechoslovakian Communist Party organization in Jáchymov or within the NPJ. They intended to run the enterprise in their own way.

The Soviet party leaders were not great admirers of the Communist Party of Czechoslovakia, suspecting it of lacking in revolutionary zeal. The Czech communists, on the other hand, were more reliable, from the point of view of Moscow, than other parties in Czechoslovakia. Their leaders might agree to trade Czechoslovak uranium on the international market, and their members passed on news about the mining of uranium to Western intelligence agencies with a clear conscience.

At the beginning of 1947, 107 Soviet experts came to Jáchymov with the agreement of the Czech government. Rada wrote of them in his report for Gottwald that they had the "advantage of being narrowly specialized and, secondly, they are far more politically reliable than a great many of our experts." [110] Though the Czechs and Russians had a common task in increasing the output of uranium as quickly as possible, they were accustomed to different styles of work. When the Soviets tried to help their Czech colleagues, two administrative systems came into being. In 1948, it resulted in the management's inability to agree on the accounts for the year and especially to separate production from investment expenses. Rada was of the opinion that in the first two or three years of the existence of the national enterprise the Czechs "wasted the trust of their Soviet comrades."

Scarce Labor

The Soviet hunger for uranium, the shortage of labor in post-war Czechoslovakia and the methods of Beria's agencies were woven into an uncommon pattern. Jáchymov was situated in a predominantly German territory. The expulsion of the Germans after the war profoundly changed the demographic pattern in the border region. Depopulation of the German districts of Czechoslovakia, together with the losses suffered during the Second World War, caused an acute labor shortage throughout the country.

War losses among the Czech and the Slovak population were relatively low, and are usually estimated at 2.7% of the total population. According to the last pre-war census taken in 1930, the Germans of Czechoslovakia accounted for 22.3% of the population. If an adjustment is made for the loss of the population of Carpathian Rus, which became a part of the Ukrainian SSR in 1945, post-war Czechoslovakia suffered the loss of about a quarter of its population, i. e. more than 3 million people. The total loss of population in Czechoslovakia was much larger than losses suffered by those countries most devastated by the war, such as the Soviet Union or Yugoslavia.

Jáchymov acquired the reputation as a kind of Eldorado, a place of easy pickings. The company nevertheless did not attract, in the first years after the war, enough civilian workers. From the beginning of its existence, the NPJ paid higher wages than were common in other industrial sectors. A variety of premiums were introduced, including special awards for working underground or in Jáchymov. The last round of awards was agreed in November 1951, when Prime Minister Antonín Zápotocký, discussed the problem with members of the permanent commission. For increases in productivity of 31.5%, the amount set aside for the miners' wages was to be increased from 56.47 million to 81.8 million crowns, while professional salaries were to go up 2.1–2.8 times. The proposal was accepted on 12 December 1951 and its provisions remained in force in the uranium industry until the early 1960s, when a more egalitarian wages policy was introduced. The workers in addition received special food rations (the so-called "Russian rations"), clothing allowances and cheap accommodation.

Accommodation was of the most makeshift kind and there was not enough of it. A few civilian workers moved into the deserted German houses in Jáchymov and its neighborhood, which were at least neglected. In any case, they did not correspond to the NPJ management's idea of appropriate accommodations for the new socialist workforce. It took about five years to transform the small market town of Ostrov nad Ohří into a dormitory town for the uranium industry. An estate of 2,940 "living units" was built, including the necessary schools, shops and health centers. The new estate, one of the earliest examples of the socialist "panel" (*panelova*) construction in Czechoslovakia, served its original purpose for about a decade, when mining operations in the Jáchymov district were wound down. A similar estate was built in Příbram in the early 1950's. Prison labor helped build both the estates, and the houses in Ostrov still bear witness to inexpert construction, as some of the walls do not meet at right angles.

It so happened that the Communist Party came to power in Czechoslovakia in February 1948, a time of acute crisis in the uranium industry. The communist government tried to direct labor to Jáchymov and recruit the Germans still remaining in Czechoslovakia. The so-called "action J" resulted from the initiative of the Ministry of Industry, and was launched on 6 August 1948. German workers were to be recruited in eleven districts and were to be moved to Jáchymov together with their families, regardless of the wishes of their current employers. The campaign went on to require an additional 5,000 workers in 1948. "Action A," developed by the Ministry of Labor in 1949, endeavored to direct civilian workers to the uranium industry. It led to fluctuations in the numbers of civilian employees, much like it did, on a larger scale, in Wismut AG on the other side of the border.

Voluntary civilian workers did not provide enough manpower for the NPJ. From the beginning of the company's existence, prison labor was a necessary component of the workforce. German POWs had been employed in Jáchymov prior to an early 1947 agreement with the Soviet authorities to send additional POWs to Jáchymov. About the same time, in April 1947, the Moscow Foreign Ministers' conference agreed that all German POWs should be released by the end of 1948. However, their return was somewhat delayed. The total number of employees at the NPJ almost doubled from 1948 to 1949, numbering 13,653. Of this number, civilians accounted for 9,128 men, POWs for 3,390 and Czech prisoners for 1,135.[109] As late as summer 1949, the Czechs began to negotiate on the release of the German POWs in Jáchymov. The Soviets wanted them to carry on working in the uranium industry on the other side of the Erzgebirge, for Wismut AG.[110] Soviet officers at Wismut AG came to Jáchymov and tried to recruit the POWs for their own company. More than 2,000 men accepted the offer. They started leaving for the Soviet zone of occupation at the end of the year, some of them fleeing to the West on the way. (Under pressure from SED party leaders, the Soviets agreed to release the POWs, with the exception of war criminals, by the beginning of 1950. Finally, after the visit by Adenauer to Moscow in the summer 1955, and the resumption of diplomatic relations between the German Federal Republic and the Soviet Union, the last POWs, about 10,000 of them, were sent home.)

Ambitious plans for the increase of production and the shortage of labor caused the permanent commission to consider an agreement with Romania and Bulgaria on labor supply, as well as a proposal that, during the five year plan (1949–1953), more than 50,000 Soviet workers, including 43,301 miners, were to be brought to Jáchymov.[113] The proposal came to nothing—the Czechoslovakian communist government may not have wanted another powerful minority to come into existence on Czech territory.

The Czechoslovak authorities who disapproved of employing German and, it seems, also Russian workers at the NPJ, had few hesitations in turning to the Ministry of Justice to make up for the deficit of labor. In January 1949, less than a year after the communist take-over, the permanent commission began considering the idea that Czechoslovak prisons might help them out. The preliminary agreement on the employment of prison labor was made by the NPJ with the Ministry of Justice on 9 March, and finalized on 21 October 1949.

The plan of the permanent commission was facilitated by the Stalinist regime, which intended to do away with reactionary elements once and for all. Alexei Čepicka—Klement Gottwald's son-in-law and the minister of justice until April 1950—exerted pressure on the ministry to supply sufficient labor for Jáchymov.[114] The government began to plan supplies of prisoners to the uranium industry.

Table 3. Employees in the Czechoslovak Uranium Industry 1945–1990

Year	Total	Civilians	Prisoners of War	Prisoners*
1945	237	237		
1946	909	62	847	
1947	3742	1786	1956	
1948	7966	4303	3663	
1949	13653	9128	3390	1135
1950	17781	11002		6779
1951	24867	14119		10748
1952	33320	19946		13374
1953	40317	26496		13821
1954	44368	32398		11970
1955	46351	37137		9214
1956	43897	36772		7125
1957	42848	36532		6316
1958	37167	30564		6603
1959	30244	25251		4993
1960	25633	22710		2923
1970	24116	24116		
1980	30914	30914		
1990	20953	20953		

Source: Oskar Puskal, *Surovinové zdroje uranu ČSR*, Manuscript, Prague, 1993.
* Many of them were political prisoners

"Action D," later known as "action Ostrov," had a swift start. The first transport of prisoners left for Jáchymov in March. (Evidence as to the first intake of inmates from Czech prisons by the NPJ varies, but was between March and May. Internal evidence indicates that March seems to have been the more likely month.) As the administration of prisons was moved from the Ministry of Justice to the Ministry of National Security (MNB), and finally, in 1954, to the Ministry of Interior, the NPJ made contracts for the employment of prisoners with all three institutions.

On 15 June 1949, the camps came under the supervision of the Prison Warders Unit (Sbor vězeňské stráže, or SVS) at Ostrov nad Ohří. It changed name several times as it moved from the jurisdiction of one ministry to the other, until it was abolished on 1 January 1963. It acted as the liaison office between the employer and the camp commanders, and kept an archive of personal data on the prisoners, supervising their arrivals and departures. It also looked after the financial affairs of the camps. The path to establishing a system of concentration camps in Czechoslovakia had begun, and a gulag archipelago came into being on the western border of the Soviet empire.

In the summer 1949, the SVS office at Ostrov lacked both a telephone and a means of transport. The commander of the eleven-member team, the only SVS officer entitled to inspect the camps, was too busy to do so. Guard duties at the prison camps were carried out by a special unit of the police force (*Svaz narodni bezpecnosti* or SNB,) called Crane (Jerab), and the ratio of guards to prisoners stabilized at 1 to 10.[115] The jurisdiction of the guards stopped at the gates to the pit.

Some of the prisoners were sentenced on criminal charges. Those found guilty of collaboration with Nazi authorities during the war were known as "retribution" prisoners. (They had been sentenced by the special people's courts on the basis of the two "retribution decrees" signed by the president in June and October 1945.) In 1949, a year after the communist take-over of power, a new category of prisoners started arriving at the camps. They were the men who committed offences under new legislation—law number 231 in particular—for the protection of the state against political enemies. Many political prisoners working in the uranium industry were technically still in custody, not having been charged. Out of the total number of prisoners in May 1950, put at 32,638, 11,026 men and women had been charged with offences against the state. The NPJ became by far the largest employer of political prisoners: every other man sentenced for a political offence between 1949 and 1960 spent at least a part of his sentence in the camps attached to NPJ.[116]

Shortage of accommodation at the camps stopped the plan for 1949 —to supply the NPJ with 5,500 prisoners—from being fulfilled. The following year was more successful. The target of 8,500 prisoners was met and, by the end of 1950, there were 8,570 prisoners on the books of SVS Ostrov. The practice of using prison labor from local sources was established in those two years, and the system helped the management of the uranium industry make up for the lack of skilled labor and technical equipment.

The Czechs initially intended to keep the political prisoners down to 30% of the camp population, with the prisoners working on the surface. Communist officials argued that enemies of the state should not be trusted with the republic's most prestigious work, especially in a region situated close to the capitalist world. The management of the NPJ was, however, unable to keep the number of political prisoners under the required limit. Most of them worked underground, in the harshest conditions.

Discrimination against political prisoners became the norm in the camps attached to the NPJ. Stefan Rais, who became the Minister of Justice in April 1950, demanded that "class enemies" should be separated from the rest of the prisoners. The chief of prison service in the Ministry of the Interior, V. Baudyš, argued for the differentiation of the prisoners because "…the less dangerous prisoners, especially from the ranks of the criminal elements, come under the influence of the most dangerous wrongdoers against the security of the state…and they leave the prison often poisoned by hatred against the state…"[117]

The intention was to separate political and criminal prisoners into four categories according to the length of their sentence.[118] The first group was to be confined mainly in camps Vojna, Nikolaj, Vykmanov II, Rovnost and Eliáš; the second category was to be accommodated in Prokop, Svatopluk, Mariánská and Vršek (also known as Barbora). Camp Ležnice was set aside for the third group, and the fourth group included camps XII, Bratrství, Svornost, Vykmanov I and Central. "Retribution" prisoners were to be placed into the third and the fourth categories, as they were no longer considered to be a threat to the regime. In addition, camp XII had a separate section for prisoners from the ranks of the military and the police. The categories of prisoners were finally approved in April 1953.[119]

The separation of prisoners by category would have made the running of the camps extremely difficult. Camp elders and informers were usually recruited among the criminal prisoners, or from the prisoners sentenced for wartime collaboration with the Nazis. Many informers became camp or house elders. They had little sympathy for political prisoners,[120] and commonly resorted to intimidation and chicanery.

The authorities were more successful in providing labor for the uranium industry than in liquidating the class enemy. Most of the political prisoners were workers, peasants and lower officials. In May 1950, the 11,026 political prisoners consisted of the following: 3,488 workers, 1,119 peasants, 3,082 lower officials, 961 higher officials, 1,193 tradesmen and owners of small businesses, 306 businessmen and professional people and 877 others.[121]

The prison service grew along with the number of prisoners. Some 4,000 officers, excluding service departments such as building or transport, supervised the employment of prison labor. District prisons in Bohemia were instructed to send all inmates, including those capable of only light work, to a "special prison labor detachment at Ostrov near Karlovy Vary, to the camp Vykmanov,"[122] from where they were distributed to other camps. Prisons in Moravia were asked to send inmates incapable of underground work in the coal mines at Ostrava, as well as prisoners sentenced by the state court (*statni soud*), i. e. political prisoners, to Ostrov. The group of prisoners was to be codenamed "Ostrov," and no other description was allowed in official correspondence. Long-term prisoners were preferred, and collection points for the Ostrov recruits were established at prisons in Prague, Pilsen, Most, Olomouc and Brno.

Slave labor played a crucial role in the development of the NPJ, bridging the period of time when the shortage of labor was acute. It also helped to strengthen the totalitarian regime by intimidating the rest of the population. Methods first used to develop the Soviet Union's prewar industrial infrastructure were continued in the effort to build its nuclear might.

The Czech Gulag Archipelago: Introducing the System

After the March 1949 decision to use prison labor in the uranium industry, a Czech gulag archipelago, or system of concentration camps, was swiftly developed. Prisoners were put into the so-called "convicts' labor camps" (trestanecke pracovni tabory, or TPT), or in "forced labor camps" (tabory nucenych praci, or TNP). By 1953 the system had 18 camps, of which 12 were in the Jáchymov district, 4 in Horní Slavkov and 2 in Příbram.

The camps were not built to last. Half a century after their construction, even the foundations of some of the camp barracks were barely visible. Most of the buildings disappeared without a trace, and some were scavenged for building materials used in the construction of weekend houses in the neighborhood of the camps. Some of the camps were named after the villages where they were situated, such as Vykmanov and Mariánská in the Jáchymov district, Ležnice in Horní Slavkov, and Bytíz near Příbram. More often, the camps had the same name as the mine they served: Bratrství (Brotherhood) had two camps, the Bratrství and Ústřední (Central) camps, attached to it. The Eliáš pit had Eliáš I and II. Rovnost also had two camps attached to it, Rovnost I and II— that is, old and new Rovnost, the names used for the former POW camp and the new camp built nearby. Svornost had one camp, as did Barbora. The camp Nikolaj was situated about eight hundred meters from the pit named Eduard. Horní Slavkov had camps called Svatopluk, Prokop and XII, named after the local pits; and there was Vojna in Příbram, a camp which survived to serve later as army barracks and which was, in 1999, to become a museum and memorial to the Stalinist period of Czech history.[123]

In March 1949, the first 209 prisoners came to the camp at Vykmanov, and were joined by an additional 753 men the following month. They worked on the surface and helped build the camp.[124] At the beginning of June, the camp at Mariánská received the first inmates, and more than 200 prisoners moved there from Vykmanov. They worked at the pits Eva and Adam, where police sergeant (*vrchní strážmistr*) Josef Pytlík from Pilsen became the camp commander.

As the search for uranium spread from Jáchymov to the neighboring countryside, Horní Slavkov, some 35 kilometers from Vykmanov, became the site of another labor camp in July 1949. Josef Pytlík moved to the new camp with 250 inmates. The fourth camp was opened on 29 July at the Eliáš pit, where a POW camp had existed. It was disinfected and deloused, and sergeant Candra took charge of the camp's 518 prisoners. At Rovnost and Svornost former POW camps were also restored before the end of 1949. The last POWs left Rovnost in the middle of September, and Czech prisoners from Mariánská and Eliáš came to tidy up the camp. It provided accommodation for 978 prisoners, 800 of which came from the central camp at Vykmanov, 97 from Mariánská and 81 from Eliáš. They were under the command of sergeant Brouček, who had served at Vykmanov, and five assistants. The Svornost camp consisted of four wooden barracks, a separate kitchen building, a small warehouse and washhouse, as well as a clubhouse with a stage. The last German POWs left on 3 December, and fifty Czech prisoners were detailed to clean up the buildings. The camp eventually accommodated some 600 inmates.

The Ministry of Justice soon started receiving critical reports on camp conditions. After a visit to the Jáchymov district on 13 and 14 August 1949, Major František Čermák reported that NPJ management was incapable of securing decent accommodation for the prisoners. There was a shortage of water at all the camps and drinking water had to be specially brought in. at camp Eliáš, the water was assessed as being "very harmful to health." The prisoners had inadequate clothing, shoes, bed covers and food dishes. Basic first aid equipment was unavailable. The medical assessment of the prisoners' health was deemed to be unreliable, as one of the prisoners came to the camp on "two sticks."

Another inspection of the camps at Vykmanov, Horní Slavkov, Rovnost, Eliáš and Mariánská was carried out on 21 and 22 October 1949 by Karel Martinek, also of the Ministry of Justice. In his opinion, conditions were the worst at Rovnost. It was built on a very small plot of land close to the pit. It was cramped, the kitchen could feed only half of the inmates at a time, and they were short of essential clothing, including gum boots and coats. The camp at Horní Slavkov was so overcrowded that prisoners, who worked three shifts, had to take turns in the same beds.[125] In 1950, after an inspection at Vojna in the Příbram district, a Ministry of Justice report stated that "Some of the comrades who experienced German concentration camps during the occupation insisted that such bad concentration camps in Germany did not exist."[126] Political prisoners became known as *muklove*, men set aside for liquidation (the singular, *mukl*, is the acronym of *muz urceny k likvidaci*). In the prison hierarchy, they occupied the lowest rung.

The responsibility for the welfare of the prisoners was on the agenda of meetings between the representatives of the Ministry of Justice and the management of the NPJ in May 1950. The two teams were led by Lt. Col. Milan Kloss and Dr. Otakar Pohl, and the negotiations resulted in an agreement on 28 July 1950. The ministry undertook to provide the prisoners and the NPJ promised to take care of their board and lodgings. Accommodation was to cost 6 crowns per person a day, including heat-ing and lighting; board was to be charged at the same rate as canteen meals for civilian employees. Productivity premiums were to be set aside for the prisoners.

As long as the NPJ used prison labor, it had labor contracts concerning the employment of prisoners with three ministries in succession. Differences existed between the contracts in matters of health and security standards, the management of the labor camps, as well as the prisoners' wages. Neither the Soviet management nor the Czech civilian employees initially wanted the prisoners to have equal pay. It was finally agreed that the prisoners would receive the same wages as civilian employees, though the money was to be paid to the administration of the camps and not to the prisoners. "The prisoners paid their own guards, so that they would feel safe and work in peace," noted writer Karel Pecka in an autobiographical novel.[125]

In the early 1950s, total charges for bed and board at the camp amounted to about 30 crowns a day. The remainder of the wages was usually divided into three parts: one went to the prisoner's family, one was put on a deposit account, and one was given to the prisoner as pocket money in the form of coupons which could be spent only in the camp canteen.

The NPJ promised to provide accommodation for the prisoners in standard units with all the usual services, including telephones for the camp guard. Apart from wages, the company was to pay social insurance for the prisoners, which was set at 14.9% of their gross pay. The prisoners were entitled to the so-called Jáchymov premium at 30 crowns a day. As an employer of prison labor the NPJ was at least negligent. It made no payments of the prisoners' wages in 1949 at all.[128] It provided the first accounts of the Jáchymov premium at the end of 1950, and then only for the month of September. Productivity premiums due to the prisoners were distributed among the civilian workers instead. For the Ministry of Justice, on the other hand, employment of prison labor was profitable. In 1949, the ministry had a surplus of nearly 35 million crowns. SVS Ostrov returned over 51 million crowns to the state budget in 1954 and more than 43 million for the year 1955.[129]

The regime in the camps, which was found to have been wanting in basic provisions for the prisoners, soon came under fire from the newly established Ministry of National Security.* The staff of the ministry was

---

* Ministerstvo narodní bezpečnosti, MNB. The first minister was Ladislav Kopřiva, from 23 May 1950, then Karol Bacílek, from 23 January 1952 until 14 September 1953, when it was merged with the Ministry of Interior.

recruited from the younger members of the KSČ, who set about their task with zeal. They imposed a harsher order than the Ministry of Juistice had done. Officers serving in SVS Ostrov came under so tough a scrutiny that two of its members committed suicide and, in 1951, espionage and sabotage trials took place, in which SVS officers as well as members of NPJ management were involved.

SVS Ostrov was further blamed for failing to establish a reliable network of informers inside the prison camps: in April 1950, it had only twenty-eight occasional informers in the growing complex of the camps. The Ministry of National Security (MNB) sent its own men to Ostrov, who wore the uniforms of SVS officers and carried their cards. They looked after security in the camps, recruited informers and tried to collect information on the prisoners.

The Ministry of Justice remained responsible for providing prisoners for the system run by the MNB. At the beginning of July 1951, it was expected to supply a further 5,740 prisoners to reach the required number of 13,000. The new consignment included replacements for those about to be released and for disabled, escaped or dead prisoners.[130] New camps were built and Vykmanov II—code-named "L"—was one of them. It contained the "tower of death," where the ore was sorted and crushed. At Horní Slavkov, a new camp, Svatopluk, housed over 1,000 prisoners in spring 1951. Camp XII followed in the summer. At the end of 1951 it contained almost 1,500 prisoners, and became notorious for the harsh regime imposed on the inmates following the escape of twelve prisoners on 15 October 1951. By the end of 1951, the network of labor camps was in the main complete; only Bytíz in the Příbram district, which became one of the largest camps, was added to their number in 1953.

Between the spring of 1951, when the MNB established itself as the controlling security organ in the restricted Jáchymov zone, and spring 1953, when Stalin died, political prisoners suffered the harshest conditions. The communists employed by the MNB believed Gottwald's statement that the class enemy, the bourgeoisie, had been defeated in February 1948, but was not yet definitely crushed. MNB officers were determined to destroy the class enemy. Evidence from the inmates of camp Prokop indicates that conditions appreciably worsened after the MNB took over its administration, as did discrimination against political prisoners.[129]

The prisoners' self-government denied political prisoners privileged jobs, such as those of the camp elders or on the canteen staff. These jobs usually went to prisoners who had been sentenced for collaboration during the German occupation, or to seasoned criminal prisoners. The camp elders allocated work in the camps and meted out punishment in the form of extra duties. The security police, *Statni bezpecnost* (StB), had informers who tried to uncover the reasons for low productivity, watched for signs of resistance and prisoners' plans to escape. Their reports used different values for the assessment of the prisoners' conduct than evaluations by the technical management, setting standards of political correctness against the criteria of economic efficiency. For instance, one of the prisoners calculated that, if all the prisoners working underground at Barbara fulfilled their work norms, there would not be enough pit trolleys to take the ore away. The author was rewarded for this calculation by a long stay in solitary confinement in a concrete isolation bunker.[132]

Contact between free and forced labor was under constant surveillance by prison authorities, and any act of kindness towards political prisoners was punished. The prisoners' memoirs include mentions of individual acts of compassion by the civilian miners who worked with them, or by the guards and other officials who held absolute power over them. Civilian miners gave the prisoners food, smuggled letters for them, and in a few cases visited them in their free time. On Christmas 1951, the miners prepared a Christmas tree and presents for the prisoners in one of the Jáchymov pits.[133] Yet political prisoners did not encounter many kindnesses from those the regime had not yet succeeded to intimidate in their day-to-day lives. The prisoners' lives in the camps of Jáchymov were hard and many were reluctant, after their release, to remember them.

The transition from Nazi occupation to the Stalinist regime happened quickly. The camps originally built by the Nazis survived into the Communist era. In the new era, the attitude towards political prisoners of an officer of the Ministry of State Security deserves mention. Mr. Kruml was gratified by the fact that camp commanders succeeded in setting off class struggle between prisoners. He regarded collective punishment as suitable means of improving the political prisoners' attitude towards the state. He wanted them to produce as much uranium for the Russians as they could, while learning to be good citizens of the Communist state. He was an optimist among his colleagues, who commonly regarded political prisoners as men destined for liquidation.

Vykmanov II, known as camp L or "Elko," was one of the smaller Jáchymov camps, with just over 200 prisoners. The camp contained the processing and packing plants, and had a bad reputation because of its high health risks. Regular medical supervision, available to the civilians employed at the processing plant, was not on offer to the prisoners. Many of the inmates were Catholic priests and members of the Czechoslovak forces who had fought in the West during the war. Until its closure in 1956, Vykmanov II processed uranium ore from Jáchymov and Horní Slavkov districts.

The central point of the camp was the "tower of death," a four floor structure housing a large mill for crushing and processing the ore. The tower was surrounded by several buildings, as well as containers for raw and processed ore. It was first taken into the "bunker" and put on large grates, where the prisoners broke it up with mallets. It then went through the sorting hall and through mechanical crushers. Its final shape was that of small, dark gray grains with a metallic shine. The ore was then put into metal barrels and shaken down on a special machine (*trasak*). The camp consisted of two wooden barracks and a brick-detention cell, situated in a swamp. Beyond the main gate and barbed wire fence, there was a block of flats for the guards and their families. Several of the guards' wives worked as overseers in the sorting house.

Camp L was unique in the Jáchymov penal colony because it carried out no mining operations. There was no barbed wire tunnel there, connecting the camp with the pit. The uranium trains went in and out of the works area twice a day, providing a semblance of contact with the outside world. Effort was made to decorate the neighborhood of the processing plant with emblems of socialist endeavor, including five-pointed stars described in red sand.

The managers at Vykmanov II were Russians, and were assisted by a few civilian workers, including the wives of the camp guards. For them, the uranium industry was a simple business. The mines were Czech, but the "little ore" (*rudenka*) belonged to the Russians. Konoplov, the manager of the sorting hall at the time of Stalin's death, was known to the prisoners as "Parfumenko," because they could smell him before they saw him. The works foreman, Stirsky, was called "Hemeroid" for his habit of scratching his backside wherever he went. A relative of one of the camp guards, Mesiakova, was called the "Beast of Belsen." She had been interrogated in connection with the crimes committed by Belousov, a Ukrainian who lived in the Karlovy Vary district. He offered illegal border crossings to people who wanted to escape to the West, and then robbed and murdered his clients. At the trial, the defending lawyer claimed that Belousov's victims were enemies of the state.

Pibil, the Czech camp commander at Vykmanov II in 1953, was a butcher by trade. The prisoners worked three shifts, morning, afternoon and night. In the morning, the afternoon shift did extra work around the camp (the so-called *brigada*), while the morning and the night shifts carried out their camp duties in the afternoons. The prisoners worked in their spare time as cleaners in the guards' flats. They washed the floors of their own rooms every day, and the walls of the barracks twice a week. All this was done with cold water. The courtyard was swept three times every day. The most unpopular job was cleaning the carriage used by the guards on the uranium train. The most exhausting of all the prisoners' duties were the frequent, unnecessarily roll calls. Five musters took place in the courtyard every day: three ordinary roll calls and two additional musters on the arrival of the train. The prisoners were made to stand in the open in every kind of weather, often in wet or frozen working clothes.

The wash house resembled similar installations in many Jáchymov camps. It had windows without glass, a trough running through it and a few cold water taps. Washing was an unpleasant task, even when the water did not freeze up. There were wooden latrines nearby, and the toilet paper could be bought only by those prisoners who had a medical certificate. Newspapers were available at the canteen, on credit if necessary. Coal was strictly rationed and sometimes the guard, who thought that a room was overheated, ordered the prisoners to return the coal and put the stove outside the house.

The food was on the whole passable at camp L, with potatoes being the staple diet. Meat was served on Fridays, because there was a large group of Catholic clergy at the camp. Two draughtsmen edited the wall newspaper: one of them was František Voborský, who had created a popular cartoon character named Pepina Rejholcova before the war and, during the war, a striking political poster. It depicted a red claw marked with hammer and sickle hanging over a dark silhouette of Prague castle, with the text "If it seizes you, you will die." The camp library had little of interest in the prisoners, though some of them remembered improvised evenings of song and poetry with pleasure.

Family visits were infrequent, short and irregular. A permit to visit the prisoner entitled the visitor to enter the restricted Jáchymov zone. The prisoners changed into clean clothes in the morning and were taken by bus to Zdar. They waited in a room from where they were taken, one by one, to see their visitors. The visits took place at a table at first, before wire netting was put between the two parties. It was a trying occasion for both the prisoners and their families. The visitors were usually rewarded by a short and nervous chat, after a long journey to the restricted zone.

The Security Zone

The bleakness of the forced labor camps was uniform, as were their tyrannical regimes. Temporary buildings, clusters of wooden barracks, detention cells, kitchens, wash houses, parade grounds and, sometimes, assembly halls. These were surrounded by double barbed wire fences, with a strip of no-man's land between them. Guards were free to shoot prisoners on sight and without warning from watch-towers spaced at regular intervals around the perimeter of the camp. Search lights and tall wooden lampposts illuminated the dismal scene.

Variations existed between the camps. For instance, Nikolaj was known for its institution of the "Russian bus," also called the "buggers' march." The camp site lay several hundred yards from the pit,[134] and the route led across the main road. The prisoners were tightly bunched up and strapped together by steel cable secured by a padlock. The human parcel was marched off to the pit, the guns of the escort pointing at the men on the outside of the pack. One of the original camps, Bratrství, was situated in a sheltered valley with a stream, on a popular tourist route between the two world wars. Svornost was on a hill above the center of Jáchymov. It was linked to the pit with a steep and long staircase, which took the prisoners through a tunnel of barbed wire. Rovnost was located high up in the mountains, a place with a harsh climate and a good view of the surrounding countryside.

Access to the districts where uranium was mined became restricted in February 1948. A special supplement to identity cards was provided for the Jáchymov residents and for visitors and commuters. The borders of the restricted zone were guarded by the military, and security inside the zone was provided by the Jeřáb special units of the Ministry of Interior. The MNB introduced additional security measures in the form of a law on the protection of state borders passed on 11 July 1951, which contained new regulations on the establishment of a frontier zone (Ustanovení o pohraničním území krajů...) in Karlovy Vary, Pilsen, České Budějovice, Jihlava, Brno and Bratislava regions. There were also special regulations for the Jáchymov district. The rules were strengthened in the case of the "Jáchymov territory with restricted access" (*uzavrene uzemi Jáchymov*); the stricter regulations came into effect on 15 December. They affected twenty-one villages in the Jáchymov district; a similar device was used at Horní Slavkov, where the closed territory included fourteen villages, and at Příbram, in the neighborhood of the mines and the two camps.

Special passes to the restricted territories were issued only to NPJ employees, and no other civilians were allowed to live or work there. Several hundreds families were moved out (451 families in the restricted territory and 445 families in the Karlovy Vary border region were reported to have been moved out before the end of April 1952. In addition, 837 "unreliable persons" were moved out of the Karlovy Vary region).[135] The MNB further ruled that people who were so removed could not settle in any other border districts. The report on the action blandly stated that the "lists included, according to the instructions of the security organs, even people for whose transfer no concrete reasons were given as well as people for whose transfer full justification was not provided."[136]

The local government office at Karlovy Vary estimated the cost of the transfers at seven million crowns. The district council at Jáchymov was abolished in the autumn of 1951, and two trustworthy officials from Karlovy Vary carried out such administration as the former district required. All public roads leading into the restricted territory were interrupted by special barriers, which were permanently locked on minor roads. Main roads were blocked with pendulum barriers.

Between 1949 and 1951, it was difficult, but not impossible, for the prisoners to escape. 143 prisoners fled in 1949, 98 in 1950, and 156 in 1951. The number of prisoners who made escape attempts between 1949 and 1955 was put at 572. It is not known how many escapes actually succeeded: one estimate puts the number at 150, another only 54.[137] Prisoners caught escaping were either beaten to death on the spot, beaten and sentenced to death or, at best, were sentenced to additional terms of imprisonment. An official report gives indicates that 22 men were shot while trying to escape at from 1949 to 1956. 14 of these prisoners were shot in 1951.[138]

Prisoners were often moved from camp to camp because of the need to redeploy the informers among them.[139] Informers were to keep their ears open, especially in regard to prisoners' plans to escape. Spontaneous attempts to flee were doomed to failure. Escapes required careful preparation, and the longer this preparation took the higher was the risk of disclosure. The prisoners needed to have good connections with the world outside and preferably pre-arranged transport to take them as far away from the camp as possible. They needed reliable hiding places and helpers who could keep secrets. They could not rely on help from a society that was intimidated, indifferent, or hostile.

Prisoners planning to escape had to consider how best to cross the Czechoslovakian border. They faced a hard choice: the most direct route to the West led across the closely guarded border with Bavaria. During 1953, the state border was provided with a security system which was no more than a large-scale replica of the camp security arrangements. It included a double barbed wire fence—parts of which were later electrified—a plowed and mined strip between the fences, watchtowers with searchlights and armed guards.

Until 1955, the route across the border with Austria involved crossing the Soviet zone. The route across the Soviet zone in Germany, and then the GDR to Berlin was also hazardous. The choice of an escape route was usually dictated by the contacts the prisoners had along the border. Communist justice made the assumption that people who planned to escape also planned to betray uranium and other secrets in the West; prisoners captured while trying to escape were therefore charged with high treason.

The best-known escape attempt took place on 14–15 October 1951. It was organized by prisoners in camp XII who worked in pit XIV. It was a new workplace in Horní Slavkov, and the camp barracks were also new, not yet colonized by cockroaches and other vermin. A tunnel of barbed wire connected the camp with the pit, and a brisk walk through the tunnel took about twenty minutes. The moment they left the tunnel, the prisoners made their escape. Of the ten prisoners who attempted to flee, none succeeded in reaching his destination. Only two survived: five were killed in a shoot-out with the security forces near the village of Stanovice and three were captured. The mutilated bodies were brought to the camp and laid out on the grass. The three captured men stood near them. The entire camp population, numbering about 1,500 men, had to file past the bodies. Those prisoners who took off their caps were sworn at and beaten by the guards.[140]

There were 227 recorded prisoner deaths between 1949 and 1955, 92 of which were apparently political prisoners. The figures are unreliable: there were 64 men who died in underground accidents, with Horní Slavkov having the highest accident rate, at 35 casualties.[141] There exist no estimates of deaths resulting from occupational hazards, which affected both free and prison labor. It was common practice for the prisoners who struck a rich seam to be moved to another site. Civilian miners, especially those favored by the party, were put to work the seam.

They were thus rewarded by the highest premiums as well as the highest doses of radiation.

At the end of 1950, SVS Ostrov was the largest single prison unit in Czechoslovakia, with about one-third of the total prison population in its care. It provided about a half of the profits from the prison system for the Ministry of Justice. The secrecy surrounding the production of uranium, the establishment of security zones, and the use of prison labor created an extensive security apparatus in the ministries of Justice, National Security and the Interior. It is estimated that over 4,000 officials were directly concerned with 4,000 security people in 1951. The Ministry of National Security supplied the system with the most tough and determined officers, many of them plain sadists. They were shielded by the party, and by the importance of the Soviet uranium industry.

The Jáchymov district came to be referred to as "Space O." It was an Eldorado for some, but for many others it was a district of lawlessness and suffering. At the height of Stalin's power and paranoia, the whole country was treated as if it were a sizeable labor camp. At the time of the uranium crisis in the Soviet Union, Czechoslovakia and East Germany became the most important producers of the precious metal. They also facilitated the Soviet nuclear program in its early stages. Without Czech and German uranium, the construction of the Soviet bomb would possibly have been postponed, and the beginning of the arms race would certainly have been delayed.

German production soon exceeded the output at Jáchymov, and in 1950 the Germans produced 1,224 tons of uranium, while the Czechs managed only 241 tons. The Russian uranium shortage had ended by the mid-1950s, and the arms race between the two superpowers was made possible by the abundance of fissionable materials on both sides. The Russians began putting a portion of the uranium they mined or imported into storage.

The post-war era was coming to an end. After Stalin's death in 1953, the reforming regime of Khrushchev began releasing political prisoners and started to abolish the gulag system. The first summit since the end of the war took place in Geneva in July 1955, and the Austrian state treaty was signed in Vienna. Khrushchev made his speech against Stalin at the XX Party Congress in 1956, and there was unrest in Poland and revolution in Hungary. Khrushchev wanted to overhaul the Soviet system and get rid of Stalinism, and the possibility of using atomic energy for peaceful purposes was generally welcomed.

Daily Life at Vojna

Despite the changes in Moscow and the improvement in the international climate, the Czechoslovak state retained the system of concentration camps for another five years. It had been fluently transferred from Jáchymov into the Příbram mining district, where Bytíz, the last camp of the Czechoslovak gulag archipelago, was built in 1953. The larger camp, Vojna, had been in existence for some three years.[142] The worst times for political prisoners were over in the second half of the 1950s. The dire conditions of the Jáchymov camps in the early part of the decade were somewhat alleviated. Most of the political prisoners received amnesty in 1960, and the camps in Příbram were closed down in 1961.

At Vojna, as in the Jáchymov camps, the commander and the guards had their base in the guardhouse, which was situated outside the perimeter of the camp. The commander was comrade Vojíř, nicknamed Pingl, who was replaced by comrade Košulic, known as "Dry Wind" (Suchý Vítr). The change took place after 1955's"noodle affair," a strike declared by the prisoners at Vojna on 4 July, the US Independence Day. Košulic's deputy, known as "Chinese Chicken," was a coachman from eastern Slovakia. The coachman's former employer, a farmer, was among the prisoners.

Just below the guardhouse was the steel gated entrance to the camp. It opened on a sloping road leading down to the kitchen house. House K was the first on the left, and House L faced it on the other side of the road. House L contained the sick room for long-term patients. The other houses were designed A to F; the House of Culture (Kulturak) was next to House L, and then there was the house for surgeries, where the doctor and the dentist were assisted by several nurses; they were all prisoners as well.

One of the prisoners who provided the most detailed account of daily life at Vojna had suffered the first accident at Eva in the Jáchymov district in 1953, when he was buried under subsidence. The second accident took place at Lešetice, while he was at Vojna. He was trying to brake a loaded train and had his hand crushed. Dr. Razdan stitched up the hand as best he could. The doctor had a good reputation at the camp, and spent whole nights with seriously sick men. On one occasion he saved the life of a prisoner with a crushed windpipe by sticking as many injection needles as he could find into the patient's throat, under the injured place. It enabled him to breathe before he was taken to the hospital.

The prisoners changed their clothes and sheets in a warehouse behind the kitchen. There was a small garden nearby where the cook (also a prisoner) grew vegetables and looked after five bee hives. The prisoners took a shower once a week in old and the new boiler houses on the other side of the garden. There were about twenty showers in a large room, but not all of them worked and hot water was not always available.

The prisoners lived in long, low wooden barracks, with a wide corridor in the middle. The front door had glass panels and a similar door was located at the other end of the corridor, facing the peripheral fence of the camp. This door was kept permanently locked. On the left was a small washroom with a tin trough and cold water taps. On the right was a French-style stand-up lavatory, consisting of three places and lacking barriers. The prisoners never had any privacy.

The houses had ten rooms with 18 to 24 beds in each room, and two smaller rooms near the front door where the house elder and other privileged inmates lived. The big rooms had two-tiered iron bedsteads, with straw mattresses and a bolster. Each bed also had a name plate. Everybody was issued three blankets that had to be smoothly folded in the morning and placed on the bolster. Sheets were only issued to the prisoners occasionally in the summer, so that they would not be used as camouflage for winter escapes.

Each room was furnished with a heavy wooden table, benches seating about six people, a small iron stove and a coal shuttle with a shovel. A brick chimney ran up the wooden wall and a short pipe connected it with the stove. Coal should have been issued once a day, although sometimes it was not. In any case, one coal bin did not last the day. Prisoners collected firewood in the pit during work, or they stole coal at night at great risk. Prisoners who worked underground stole rubber hoses: in the winter, everything that burned went into the stove.

The wooden floors were sometimes treated with used diesel oil. Floors so treated had the advantage that they did not have to be washed, and dust did not rise when they were swept. A shelf ran around the room where the prisoners kept all their property: their bread ration, tin dish, spoon, soap, toothbrush and toothpaste. Some necessities could be bought at the camp shop with coupons issued by the camp administration. A towel was hung on the bed rail behind the bolster, as were trousers and jackets. Indoor shoes were put under the bedstead. Linen shirts and underpants were used day and night both winter and summer, and the prisoners made their own boot hose from bits of cloth, though they were later able to buy socks at the camp shop. Three vertical stripes of bright oil paint were painted on the backs of their jackets, and were repainted from time to time. The ordinary prisoners' clothes were changed once a week or less, according to the mood of the camp administration. The political prisoners' clothes, on the other hand, were used until they fell into pieces. Holes in the seat or the knees did not entitle the prisoner to a new pair of trousers. Only the men working underground were provided with work clothes.

Prisoners were allowed a weekly shave, and a record of the shaves was kept on the back of their ration card. They had haircuts at irregular intervals, but prisoners tried to avoid the barber in the winter. They were not allowed to wear scarves or neckwear of any kind, and long hair kept the back of their necks warm. Some of the prisoners worked on the surface, as well as in the kitchen, the stores, the medical room and the tailor and shoemaker shops.

The prisoners remembered the sing-alongs after Sunday lunches more fondly than the radio music broadcast on the camp loudspeakers. They spent their free time talking, walking in the camp's open spaces and playing chess with pieces made of bread. The expected amnesty was a popular subject of conversation, as were the prisoners' plans for the future. In the summer the prisoners could sometimes lie in the sun, or play football on Sundays. One of the wooden barracks (Kulturák) had a stage and benches in its auditorium. The hall was used for showing films and for concerts of the camp orchestra. The films were tendentious and dull, and were not popular with the prisoners. A TV set later appeared in a very small room, and only a few inmates were allowed to watch it. Members of the camp orchestra all worked on the same shift and were freed from voluntary work. The prisoners could subscribe, if they wished, to the party organ, *Rude Pravo*. The poor cultural life on offer was offset by the reluctance of the camp authorities to introduce courses of political education.

The monotony of camp life was interrupted by visitors, and prisoners who were expecting a visit were lent a special suit in exchange for their ration card. The visits were infrequent, between one and three a year, according to the standing of the individual prisoner. Two adults and one child were the maximum number of people allowed to see the prisoners at the same time; if the family had more members, they took turns during the visit. Every family group had its own guard in attendance, and the conversation concerned family matters only. The first breach of the regulation was followed by an admonition, after the second the visit was ended. In the early days the prisoners were taken by car from Vojna to

Milin, where there was a house near the railway station set aside for the visits. Later a visitors' house was built close to the camp. Prisoners walked there, and the visitors were delivered there from the station by bus.

A long security corridor led to the new visitors' house, situated on the edge of the forest. In the new environment, the visits settled into a different pattern. They were supposed to last an hour, but they were usually much shorter. The prisoner sat between two guards on a long bench, which accommodated up to five prisoners and six guards. Visitors sat on chairs on the other side of the table. They were all cautioned that political discussion was banned as were references to the prisoner's offence. It was forbidden to give the prisoners small gifts, including food or fruit—one prisoner could not accept three strawberries brought to him by his wife. He later wrote that "I would never have sunk as low as to beg those, who had criminal desire for other people's property and who made perverse laws for that purpose, so that I could be 'legally' robbed and then imprisoned...I despised such 'human' rabble from the bottom of my heart and their philosophy and practices were alien to me."[143]

Political prisoners at Vojna were usually allowed to write one letter a month, which had to be handed in at the guardhouse unsealed for the censor. If he disapproved of its contents, the letter ended up in the wastepaper basket. Other prisoners were allowed to write once a week to a stated address, and the letter had to be left open for the censor. Letters for the prisoners were delivered to them without envelopes, and were destroyed after the prisoners had read them. Many prisoners had nowhere to write.

The prisoners at Vojna worked on different sites. There was a security corridor above the guardhouse, used by the prisoners who worked at the Kamenná pit. The Vojna and Lešetice pits were in the opposite direction, and buses collected the prisoners near the guardhouse. Many prisoners worked in town of Příbram. A large part of the new town and some of its facilities, including the House of Culture and the water purifying station, were built by prison labor. The building site of the waterworks was fenced in and guarded in the same way as the camp. When they returned from work at about four in the afternoon, some of the prisoners were selected for a body search (*filcung*). They emptied their pockets and placed their feet far apart, before the humiliating search took place.

The reveille for the construction workers was broadcast over the loudspeakers at the comparatively late hour of 6 in the morning. They waited in front of the guardhouse for the buses for about twenty minutes, accompanied by the guard who was responsible for the prisoners while they worked and two additional armed guards. The first eight seats behind the bus driver remained empty so that the prisoners could not become a threat to the guards at close quarters. There were not enough seats for the fifty prisoners or so, and some of them sat on the floor or on other prisoners' laps. The guards stood facing the prisoners, with their machine guns ready. The works guard was responsible for them until their return to the camp, and the prisoners had to keep their heads down and their hands in their laps. On one occasion, when prisoners stood in the passage, two of them lowered the window and escaped. They were captured and paraded before the other inmates. After 1955, the minimum penalty for attempt to escape was one year's additional imprisonment.

The builders worked from 8 AM until 3:45 PM, followed by a quarter hour break. They returned to the camp at 4 in the afternoon, and there was a stampede for lunch. The main roll call took place at 5, and lasted about half an hour. The prisoners were then free until lights out at 10. During this time, latecomers finished a cold lunch while others rested, had a shower, ate their supper, wrote letters home or walked around the camp. Conversation continued after the lights went out, and then the lights were put on again so miners returning from the afternoon shift could get into bed. The miners worked three shifts, the construction workers only one.

The prisoners' free time on weekdays was often interrupted by "voluntary" work (*brigada*). They peeled potatoes in the kitchen or cleaned the camp and their rooms. There were the occasional jobs to be done, such as taking potatoes to the kitchen stores, or coal to the store or to the kitchen,. There was snow to be cleared in the winter. Prisoners made paths through the snow by treading through it four abreast. In these tight little groups, the prisoners walked between the inner and the outer fences around the perimeter of the camp. In addition to being followed by guards in watchtowers (*špačkárny*), the prisoners were accompanied by armed guards on the other side of the fence.

The prisoners wore inadequate clothing and did not spend any more time in the open in the winter than they had to. They got thoroughly chilled during the two roll calls every day. The roll call before the start of the shift was a ceremonial occasion. All members of the shift had to present themselves five abreast, in lines according to the order on the guards' list. The guard at the gate read out the name of the prisoner who then left the line, called out his number, passed through the gate and joined a similar formation on the other side. When the last prisoner changed sides, the marching order was given. The miners went through the security corridor of barbed wire which connected the camp with the pit; there they dispersed, the men working underground went to the lamp room where they handed in their tinplate numbers, collected their lamps and went to the hoist tower. The surface workers walked to their places of work.

The mornings were the busiest time at the camp. They began with the reveille broadcast on the camp network. Men on barracks duty helped in cases where the reveille failed. The sequence of dressing, making beds, washing and breakfast was spaced out so as to avoid a crush in the washrooms. After the departure of the morning shift, the afternoon and night shifts attended the roll call before they could go back to bed. The morning roll call was sometimes followed by an inspection of the living quarters, and the prisoners had to wait before they were allowed to go back to the barracks. On their return, they often found their rooms in a state of total chaos, with blankets, bolsters, mattresses scattered on the floor, covered in coal and ashes. The prisoners took a long time tidying up their rooms, and those of them who had been on the night shift snatched a brief rest before reporting for lunch.

The daily routine from Monday till Saturday was set by the needs of the workplace. On Sundays and bank holidays only the pump room operators and other prisoners with skilled jobs were on duty. During the winter snowfalls the prisoners were sent to clear the snow at the pit. Christmas was a sad time, and the prisoners waited in silence and in darkness for midnight. Sometimes they heard the bells from Svata Hora, a nearby place of pilgrimage.

On Christmas Eve, they at least had a decent meal of Wiener schnitzel and potato salad, and about a pound of Christmas bread was issued. Otherwise, the camp food was simple: one third of a loaf of bread had to last the prisoner for three days, and a ration of five spoonfuls of sugar for five days. Unrationed black coffee of appalling quality could be collected from the kitchen, and prisoners who still had sugar and a piece of bread and some milk, which had to be paid in advance for the whole month, did not go to work on an empty stomach. Most prisoners had for breakfast black coffee only.

Lunches and dinners were served in several stages at the kitchen counter. The afternoon shift came first and was followed by the prisoners who had finished the morning shift. The men who worked at night could choose to lunch with one shift or the other. Prisoners on the morning and night shifts usually took their suppers first, and the afternoon shift ate on their return from the pit. One of the cooks supervised the issue of the meals and punched the prisoners' ration cards. The dining room had long trestle tables and benches. The food was good on the whole, the usual Czech sauces with dumplings, pasta or potatoes with a piece of meat, as well as vegetarian dishes such as dumplings with egg, or pulses.

Intervals between the meals varied: the morning shift lunched between 3 and 4 o'clock in the afternoon, and had supper at 6. The afternoon shift lunched at 11:30 and supped late in the evening. During lunch on Sundays the prisoners also collected their suppers, a slice of cheese and another of salami. Men who were hungry ate the two meals at the same time. Some prisoners carried their food in tin cans to the barracks, and whoever came first sat at the table. The rest of the prisoners sat on their beds.

Prisoners who failed to fulfill the monthly norm were put on reduced rations, and were not allowed to write home or receive visitors. Purchases in the camp canteen were made in exchange for the coupons issued by the camp authorities. The guard at the counter inspected the prisoner's ration card—it served also as his identity card—and checked the list of work results for the past month. If the prisoner fulfilled the norm, he could buy salami, biscuits, soap, toothpaste, a packet of tobacco, cigarette papers, socks and toilet paper. The keeper of the shop sometimes added to the order a pair of wooden clogs, which he was unable to sell. It was unwise to refuse the offer.

There were three kinds of ration cards for the prisoners, the best one being number 3 (*trojka*), which included 150 grams of meat and 10 grams of bacon per day. This was issued to the prisoners who worked underground and met the target. The number 2 ration card was the most usual kind, with 50 grams of meat and also about a third of the vegetables ration. Just enough for the prisoner to survive, it was issued to members of the construction groups and the miners with under 50% productivity. If poor productivity continued for more than three months, the prisoner was transferred to the feather-plucking group (*na peří*). They could not move freely around the camp, their food was brought to them and their lives resembled the lives of an ordinary prisoner. The targets were high and the prisoners were permanently hungry. The penalty for offering them extra food was a minimum of three days' solitary confinement.

The cook was a long-term prisoner who had served ten years by 1958. A friend from his home town, Prachatice, was among the prisoners. They spent a lot of their free time in the cook's room, talking and cleaning vegetables. The room was in the kitchen building, where the guards left them alone. The building was large, with a flat roof and several big boilers, kitchen machinery and a dining room. The camp shop was also situated there.

The prisoners received the coupons that served as the camp currency once a month, and the amount depended on their earnings. At Vojna in the late 1950s, as long as the prisoners had paid all the expenses due to the state, including the costs of imprisonment while they were under investigation and the penalty fixed by the court, they received pocket money at 5% of net pay. The charges for bed and board were high and a sizeable chunk of their pay. Miners working underground were able to earn as much as 15,000 crowns a month, though they paid more for their camp expenses than they would have paid in a luxury hotel.

The security system at Vojna followed the customary pattern established in the Jáchymov camps. The guards usually patrolled the space between the two fences or watched from the towers. Only occasionally, especially when the weather was misty, they were deployed inside the camp. During these times, they moved in close proximity to one another, never leaving each other's visibility. Lights were placed on every second post of the outer fence so that the light fell on the space between the fences; powerful searchlights were placed on the watch towers. In case of sudden power outages, the guards fired sky rockets which lit up the camp until they succeeded in starting the reserve power generator. Rocket signals were also used to declare emergency. All the guards on duty inside the camp wore police uniforms and were unarmed; and they did not use guard dogs.

The inside fence of barbed wire was about two meters high. Between the two fences ran a strip of smooth sand. Guards armed with automatic weapons were posted in the outer zone with watchtowers, about five meters high and standing about fifty meters from each other. Notices warning of the forbidden zone were spaced at ten meters; a prisoner who entered the zone and did not stop and put his hands above his head could be shot after the first warning.

According to the recollection of one of the prisoners,[144] only one failed attempt to escape took place at Vojna between 1955 and 1960. One prisoner was shot dead, another was on public show in the camp, beaten up so badly that it was hard to recognize him. Another prisoner, who served a short sentence, never experienced an escape attempt at Vojna but witnessed one at NPT Libkovice, his other place of detention, which had a milder regime than Vojna. Two prisoners were shot dead while trying to flee the camp. Their decision was hard to understand: they could have escaped more easily from their workplace and one of them had only six more months to serve. When the guards shot at men trying to escape, other prisoners were endangered as well.

A strike took place at Vojna on 4 July 1955, the US Independence Day. The strike came to be known as the "noodle affair." The camp was surrounded by soldiers armed with machine guns. The StB had a free hand in uncovering the initiators of the strike, and commander Vojik was replaced by Košulic.

Though grossly inhuman treatment of the prisoners was less common after 1956, acts of chicanery were still commonplace. One of the guards called "Ginger" liked to needle the prisoners with unpleasant remarks. He told one of the prisoners, a watchmaker, that he looked better with a spade than with watchmaker's tweezers, and that he would have to stick with it. The guard provoked another prisoner, a young man from Ostrava who was a good and disciplined worker. The prisoner slapped him hard in return. After that he was taken to the bureau of the Soviet secret police. It was an unpleasant winter day, with the snow falling and temperature under 0°C. The prisoner was stripped to the waist and chained to the fence, where he stood for four and a half hours, until the end of the shift. The young man returned to the camp after a fortnight, and did not want to talk about his experience. He later confessed that he was poorly treated. He spent time in solitary confinement, where he was kicked by four men. He had water to drink there but was only fed once every three days, and slept on a concrete bed without a cover. No sunlight entered the cell. He was charged 32 crowns daily for bed and board, as if he were in the camp.

Prisoners were not allowed leave, even for the funerals of their close relatives, although prisoners were occasionally transferred for retrials and sick men to hospitals. After much talk about amnesty among the prisoners, it finally came in May 1960. President Novotný declared the amnesty on 9 May, the state holiday marking the fifteenth anniversary of the end of the war in Europe. The amnesty had been preceded by far-reaching administrative measures. Five to six hundred prisoners, who were to be released, were moved to two barracks. The state procurator explained to them what they could expect after their release and answered their questions.

The prisoners then divided into several groups and walked to another house, where they left their things. In the first room they left the blankets and all clothing; then they went naked to another room where they could choose, on credit, socks, underwear, clothes and shoes. The salesmen entered the prices of individual items on an invoice; in another room the prisoners were given small amounts of ready cash. A railway official issued travel warrants to the prisoners in another room and the price of the ticket was added to the prisoner's invoice, and then the whole sum was charged against his credit of accumulated wages. The amnestied prisoners did not bother to check the accounting as their thoughts were elsewhere; their families and homes were within their reach. Before they left, the prisoners were issued with certificates of release, which were to serve them as provisional identity cards.

The amnestied prisoners remained for the time being in their special part of the camp, wearing civilian clothes and moving about freely. The roll calls continued. They were then taken by bus to the Příbram railway station in small groups headed to different destinations, as the camp authorities did not want large numbers of amnestied prisoners traveling at the same time.

One of the former prisoners lived in a high-rise block in Kladno, where his flat faced the flat of Vojíř, the former commander at Vojna. He was then working at the Vinarice camp, where he shot himself dead with his service pistol. Another guard from Vojna, Jaroslav Vančata, a retired major in the SNB, lived in the same block as did Vojíř. He had served at the gate at Vojna, checking prisoners out on their way to work. He was employed at the local self-service store, where he paid off deposits on empty bottles brought in by the customers. He died on 2 October 1998, at the age of 68.

Spy Mania

During the post-war rush for uranium, Jáchymov attracted the attention of intelligence services. The British had taken interest in the production of uranium at Jáchymov during the war and, after the war, they were joined by the French and the Swedes. A variety of semi-professional spies flooded central Europe in an effort to uncover its deep secrets.[145] Some of them worked for the Americans, who did not yet have an established intelligence service. It did not take long before they caught up with the Europeans. In the State of the Union speech on 21 January 1946, Truman complained that foreign countries were making espionage a part of the regular establishment. Truman knew what he was talking about: he was getting reports on Soviet intelligence operations against the Manhattan project.

In comparison with its Western counterpart, Soviet intelligence was much better placed. Immediately after the Red Army entered Prague in May 1945, Soviet security and intelligence agencies began establishing themselves in Czechoslovakia. Their presence was taken for granted by the local authorities and the Soviet security organs were allowed a lot of operational freedom. For instance, the Czechs left citizens of Russian origin, who had sought refuge in Czechoslovakia after the Bolshevik Revolution, to the mercy of NKVD officers.

From the beginning of their stay in Prague, Khozianov and Tikhonov tried to convince Czech politicians that they should invite security advisers from Moscow.[146] They pressured Dr. Štěpán Plaček, the head of the internal security service, to recommend inviting Soviet advisors to Rudolf Slánský, the secretary general of KSČ.[147] Slánský disregarded the request and, during his interrogation six years later, the omission was used as evidence for his trial. On 16 September 1949, Slánský drafted Gottwald's telegram to Malenkov, which the requested that advisers be sent to Prague. These advisors would preferably be experts on the trial against László Rajk, which was being prepared in Budapest.[148]

Likhachev and Makarov, the advisers who arrived in Prague first, failed to find the Czechoslovak branch of an international conspiracy against Stalin and socialism. Instead, they helped the Czechs prepare trials against the functionaries of political parties other than the Communist Party, including the trial of Milada Horáková. A National Socialist leader, she was tried and sentenced to death in June 1950. After Horáková's trial, the search for the perpetrators of the great international conspiracy continued. The investigators and their Soviet advisers focused on Slánský and of other leading communists. A significant part of their investigations followed their activities in the uranium industry.

At the time when the Soviet advisers came to Prague in 1949, Rudolf Slánský was at the height of power. Born into a middle-class Jewish family on 31 July 1901, he joined the Communist Party in 1921, soon after its foundation. He earned his living as a full-time political worker and rose to the leadership of the party in 1929, when he became a member of the Central Committee and the Politbureau. He was disciplined in following the party line at times of violent changes in Communist policy. He was a deputy in the Czechoslovak parliament between 1935 and 1938. After the Munich agreement and the banning of the Communist Party on 20 October 1938, he and his family immigrated to Moscow. Their infant daughter was snatched from her pram in a park and was never found.[149] Slánský acquitted himself well during the uprising in

Slovakia in 1944. Upon his return to Prague in the following year, the party had in Slánský a loyal servant, and Stalin a great admirer. He became Secretary General of the KSČ in 1945, and thus one of the most powerful men in the country.

The party enjoyed great popularity after the war and experienced an unprecedented growth of its membership, even before it managed to take over the state in February 1948. Slánský, as the Secretary General of the party, looked after its organizational structure. He was moving from the bureaucratic shadow into public prominence. He deployed his organizing skills to impose the Stalinist mold on the country, and no more was said about Czechoslovakia's own way to socialism. Slánský helped to position committees of the party so as to control legal system. He became the chairman of the "security five," which consisted of high party officials. Apart from Slánský himself, its membership included Minister of the Interior Václav Nosek, Josef Pavel, Ladislav Kopřiva and Karel Šváb. The group of five took over the court's decisions on all capital cases. By the time Slánský himself was arrested in November 1951, they had passed 139 death sentences.

An efficient organizer, Slánský wanted the party to be a seamless garment, covering the whole country from head to foot. The regional organization in Karlovy Vary, which contained the Jáchymov district, had long caused the KSČ a lot of problems. It was deep in the Sudeten territory, in the region of world-renowned spas—Karlsbad and Marienbad among them— where rich pickings of German property were to be had. Until the departure of the Red Army from Czechoslovakia at the end of 1945 (with the exception of the unit on guard duty at Jáchymov), Karlovy Vary was its regional headquarters, and the first Soviet visitors to Jáchymov mines came via Karlovy Vary. The Czechs, in addition to other problems in the frontier region in the years after the war, had to deal with the Soviet interest in Czechoslovak uranium.

Signs of serious trouble in Jáchymov reached Communist Party headquarters in Prague in September 1947. A letter from the party secretariat in Karlovy Vary, dated 20 September and signed by Svoboda, a party functionary,[150] stated that Svoboda had visited "leading comrades" in Jáchymov who bitterly complained about the running of the mines. An early scrap over control of the Jáchymov mines took place along Czech political party divisions; the communists, dissatisfied with their subordinate position in the management of the NPJ, were convinced that they could manage it better. Svatopluk Rada, member of the permanent commission, was struck by the absence of the Communist Party in the Jáchymov district.[151]

The communist takeover of power in February 1948 did not resolve the problems in the NPJ. The dire situation was further aggravated by the demands of secrecy: spy manias thrived in conditions of tight security. Party officials were frequently changed with crimes. One of them (Acel) shot himself during the investigation of his connections with Western intelligence services. "I should like to say how deep did the penetration of Western intelligence cells reached—as it still does today—into this national enterprise" Svatopluk Rada wrote to the president in December 1949.

In 1948, when another sharp increase in Communist Party members, Rudolf Slánský became prominent in the campaign against the traitors inside the party. "We have to take care of purity in the party," he said on 11 March 1949. "We have to make a cadre assessment of each member, new and old." On 15 September, Slánský added that "We must bear in mind that we have not yet uncovered the whole network of agents in our ranks in the state and economic apparatus."[152]

In the meanwhile, tensions in Jáchymov continued to grow. In January 1949, Gottwald received another complaint from Karlovy Vary, concerning the behavior of party functionaries in two districts of the region. Party members were arrested for no apparent reason: the district secretary had "everybody, who was against him, arrested." Functionaries were called "executioners" and the regional secretariat a "roughhouse" (*syčárna*). The Commission of Party Control (the KSK, *Komise stranické kontroly*) sent a strong delegation to Karlovy Vary, including Jarmila Taussigova and Karel Šváb—both later arrested in the purges—as well as Ladislav Kopřiva, who became the first Minister of National Security in 1950. He remained in the post for two years and was responsible for Slánský's interrogation.

The secretariat in Karlovy Vary was put under a high-pressure police operation. The offices were sealed early in the morning, the telephone lines were cut, and the officials were brought in for interrogation. The secretariat and the houses of party officials, including the flat of Antonín Tannenbaum, the regional secretary, were thoroughly searched. The report to Gottwald, on 28 February 1949, stated "it seems to us that comrades at the regional secretariat resist healthy and constructive criticism, and they do so with the assistance of the security service." A later report, In April, contained a reference to an external enemy: "…we wonder whether a member of the regional secretariat is not working for a foreign intelligence service."[153]

In the highly secretive uranium industry, nervousness and spy manias quickly flared up and spread throughout the Communist Party organization. The communists tried to exclude other Czech political parties from running the uranium industry before coming into conflict with Soviet interests. About the same time as reports on the visit of the control commission's delegation to Karlovy Vary reached Prague, a similar visitation to Jáchymov took place. It resulted in two reports: one of them concerned the situation in the mines, the other the party organization. The reports covered the period when comrade Vrtílko was the district political secretary until the end of 1948. Vrtílko was replaced, in January 1949, by comrade Vaněk.

The Turn of the Screw

> The Jáchymov mines are a problem for the whole district. A secretariat was established in the mining enterprise. Its secretary is comrade Dušek, who comes directly under the control of the regional council. The district however has no right to intervene in anything to do with the mines. Problems which arise for the comrades in the district cannot be solved on the spot and so they remain unsolved. Party secretariat at NPJ counts for very little. The secretary has no jurisdiction, he takes no part in the meetings or negotiations. The party does not play any kind of role here, never mind the leading role. For instance, the party makes a proposal on better division of labor, or it discovers a suspicious employee. The proposal has to be passed on to comrade Čmelínský (sic; in fact Čmelák), who passes it on...The party has no opportunity of controlling how and whether comrade Čmelínský passed on any of the proposals. He is not trusted. Complaints and accusations have been made against him and his activities during the (Nazi) occupation. The security service, when it is approached to investigate, does nothing.[154]

The reports only confirmed Svatopluk Rada's complaints about the state of the party organization in Jáchymov, and added new information on the chaos and mismanagement in the uranium industry. The mines owned 120 hectares of arable land and 560 hectares of pastures. This land remained uncultivated, and a flock of some 200 sheep suffered neglect. Seven tractors, set aside by the Ministry of Agriculture for the farm, were put to other uses. In the nearby village Dolní Žd'ár, the ministry established a model farm, which also passed under the control of NPJ. The stables and the barns became offices and warehouses, and all agricultural production stopped. The communists who tried to do something about the situation had to give up for "secret reasons." The words "secret" and "confidential," the authors of the report complained, were used as a sort of incantation.

Desperate shortage of accommodations for the workers, and expensive adaptations and renovations of property for administrative and technical staff were described in the internal party document, as were the high premiums the officials paid themselves. Responsibility was hard to trace, and maladministration of every kind was rife. People deemed "politically unreliable" were recruited by the management, including those with questionable wartime records. The head of the economic division, a part of the administration about which complaints were often made, drank a lot and lived in a large flat on his own. He allowed his deputy to make orders for completely useless goods. There lay 3,800 carpenters' saws in the stores, as did steel ingots worth some 20 million crowns. Building materials as well as cans of petrol and oil were stacked in the open, unprotected against weather or theft. Working conditions were terrible: "workers labor under medieval conditions," the report stated. The manager of the central workshop was a man "who does not hide his hostility to the present regime."

The American intelligence service, it was further stated in the report, paid vast sums for trite pieces of information. A woman spy confessed that she received a million crowns for a fragment of uranium ore. "She was sentenced to two years in prison, instead of twenty. On the other hand, the policeman, who uncovered her, faced disciplinary procedure, because he apparently had used irregular methods during the investigation," the members of the party commission complained.

The delegation looked into many aspects of management of the company and found fault with most of them. The Russian experts did not interfere with the running of the mines, though they were "concerned with securing production." A Russian engineer apparently suggested to the delegates from Prague that "an able, politically mature secretary, dedicated to the party, who would control the activity not only of our management but of the Russians as well" should be found to run the party organization in the mines. Only he could resolve the crisis: "He should know about everything that goes on in the mines," the report of the commission suggested, despite the fact that it had been told that everything to do with the production of uranium was highly confidential. The reports of party control commissions revealed nervousness within the KSČ, as well as a lack of understanding for the Soviet imperial interests.

It was naive of the Czech communists to assume that the Soviets would welcome additional interference in the running of the uranium industry from the Czechoslovak side. 1949 lay in the shadow of the fiasco of the previous year and, in January 1950, Pohl, one of the two Czech representatives on the joint Czechoslovak-Soviet commission, reported to Gottwald that Volokhov, his Soviet colleague, had returned from Moscow in December, apparently ready to make some concessions to the Czechoslovak side. They mainly consisted of a personnel reshuffle: Simin was to be the new works manager while Razhev—who had taken the blame for past deficiencies—became his deputy; the Soviets also promised to follow Czechoslovak accountancy procedures in future.

Internal party enquiries in the Karlovy Vary region and the Jáchymov district were followed, in October 1949, by the arrests of Bohuslav Hegner, the first manager of the Jáchymov mines, and his deputy and then successor Josef Čmelák. Fourteen other officials of the NPJ were also arrested. The manager of building works, Stanislav Chmela, was sentenced to death on 24 March 1951, but the sentence was commuted to life imprisonment. In the first wave of arrests after the communist takeover of power, eight members of SVS Ostrov, including František Záhrobský and Karel Martínek, also came under investigation.

Chmela and other communists employed by the NPJ were amnestied in 1954. Members of the security police, who had tortured them during the investigations received mild party penalties. Bohuslav Lejsek, the chief interrogator, confessed before a special commission of the ministry of interior in 1963 that he saw "many objects such as sticks, rubber tubing...and truncheons and cat-o-nine-tails which the comrades called 'truth detectors' (*pravdomluvy*)." Lejsek received a "rebuke with a warning" (*důtka s výstrahou*), while the others were merely rebuked by the party.[155]

Towards the end of 1949, security measures were tightened up in regard to any form of information on the mining of uranium. In official documents, everything to do with "U-metal" (*U-kov*), including the monthly production reports for the president, all figures, dates and place names had to be filled in by hand, by one of the members of the permanent commission. On 1 March 1950, Pohl reported to Gottwald that Slánský had asked to be informed on the situation in the Jáchymov mines.

He added that Rada had told him that the "Secretary General, comrade Slánský, expressed the wish to be informed, on a regular basis, about the conditions in the NPJ and about the fulfillment of the plan. I regard it as a matter of course to meet comrade Slánský's wish; as you yourself have however reserved the right to oversee everything to do with the aforementioned enterprise, I should like to ask you for your express consent."[156]

Slánský was arrested on 24 November 1951. Four days after his arrest, Rada wrote to Gottwald "I enclose the last reports which we sent to the former General Secretary of the party, Rudolf Slánský The last report, for which he had asked personally, and which was drafted on the basis of his indications of what he wanted to know, was so important that I thought it necessary to ask you for your personal agreement. When it was given, as comrade Kohlerová [Gottwald's private secretary] confirmed on the telephone, only then did I pass on the document."[157]

After his arrest and during interrogation, Slánský had one of his nervous fits. An interrogator said to him that he would be ready to agree with the bombing of Czechoslovakia with the nuclear weapon in the interest of the imperialists. During his seizure Slánský shouted "give me the atom bomb!"[158] Apart from this single bizarre reference, there is no mention of the Jáchymov connection in any of the studies dealing with the case of the general secretary and with his trial. The public charges against Rudolf Slánský were framed in terms of the ideological war. The trial had strong anti-Semitic undertones, and the charges included an indictment of "Zionism." The Soviet Union was at the time reviewing its previously pro-Israel policy, in which Czechoslovakia had played an important role, as supplier of arms and as training ground for Israeli forces.

The party turned against Slánský, He was sentenced to death, together with ten other comrades, on 27 November 1952. In her memoir, Slánský's widow did not speculate on the reasons for the tragic fall of her husband. She could not believe that Gottwald would be capable of such betrayal while, on the other hand, she suspected him of the crudest mental torture.[159] She wrote: "I lived for years in sincere faith that what the party does is correct, in blind faith in the party. And now: Stalin (that is the Soviet Union) and Gottwald (that is the party) say that Ruda is a traitor, a spy. My inner conviction for the first time came into conflict with my party conscience." She blamed herself that, at the darkest hour in the life of her husband, she was unable resolutely to take his side.[160]

Though Slánský's Jewish origin fitted the Zionist conception of the trial, he was not a suitable candidate for the role of the great traitor and conspirator. His history with the party was impeccable. The meaning of treason for Stalin and his followers was having strong links with the West. Rudolf Slánský had none. There were two shadows of suspicion against him, which could be used, during the investigation, to sow the seeds of doubt in his mind.

The StB, who investigated his case, had evidence in their hands proving that Slánský was planning an escape to the West. It was a trap constructed by Czechoslovak intelligence officers who had fled to the West and who operated from western Germany.[161] More important, from the point of view of the Soviet advisors, was Slánský's interest in the Czechoslovak uranium industry and his eagerness to assert the role of KSČ in the Jáchymov district. He failed to understand that it cut across the Soviet imperial interest.

The Fate of Svatopluk Rada

After Slánský's arrest, the investigation of his activities concerned Jáchymov and his interest in the uranium industry. Since 1949, the StB had plotted two investigations. The first, code-named "Building" (*Budovani*), was to be focused on economic crimes in the uranium industry. Its purpose was to demonstrate that criminal activities in Jáchymov were a part of the conspiracy around Slánský. The other, called Svatopluk, was also concerned with the NPJ.[162] It was deployed against Svatopluk Rada, the highest Czechoslovak official in the industry.

Rada was born in Moravian Zlin on 21 September 1903. He was a mining engineer by profession, who graduated at the High School of Mining at Příbram in 1928. His wife, Bohumila Havelková, came from the Příbram district and was six years younger than Svatopluk. Her father was a miller and his father was the headmaster of a secondary school. At the height of an exceptional career, Svatopluk Rada shot himself in the garden of his house at Smichov (Mozartova 17/1986). His suicide occurred on 18 April 1952, while the trial of Slánský was being prepared. His numerous Czechoslovak and Soviet decorations were displayed at his funeral on 29 April, including the Orders of the White Lion and of the Red Star. Rada was also the first recipient of the Order of the Hero of Labor.

He specialized in the exploration and assessment of mineral deposits, coal in particular. He became an assistant at the School of Mining, and worked in the Silesian mining district of Ostrava, and in the Ukrainian Donbas district, before he turned to the Balkans and Turkey. He worked in Turkey from 1936 to 1937. After this, Rada became an employee of the

Mining and Metallurgy Company of Prague (*Banska a hutni*), and carried out explorations in Bulgaria. He spent a short spell working for the Czechoslovak intelligence service under the cover of the Bata Shoe Company in India from 1941 to 1942. He became a good linguist during his travels, and spoke English, German, Russian, Bulgarian and some Serb.

When his wife joined him in Bulgaria in August 1939, Rada was engaged in organizing an intelligence network, consisting of members of the local Czech community in Sofia. With the help of Rada's company, this network was extended to Yugoslavia. According to the evidence supplied to Czech authorities by his wife long after his death, Rada was in touch with the French secret service before Colonel Ross and Major Elliot became his British contacts. General Zaimov acted as an intermediary with the Soviets.[163]

Rada's secret activities were highly valued by Colonel Bulander, the head of the Czechoslovak military mission in Sofia. Rada, who was able to travel about the Balkans freely, was asked to assess the state of the Bulgarian airfields, the deployment of troops on the border with Greece, and the quality of equipment of the Bulgarian army. He was a "first class resistance worker" in Bulander's view.[164] He left Bulgaria, on the advice of the Czech intelligence service, on 11 February 1941, after the Gestapo established its presence in Sofia. His wife, who did liaison work for her husband and his secret organization, stayed on in Bulgaria. She left Bulgaria in August 1943 with the assistance of the Czechoslovak military mission in Istanbul, embarking on a hazardous journey through Turkey, Palestine and Iran to the Soviet Union.

By that time, Svatopluk Rada had spent more than a year with the Czechoslovak unit in the Soviet Union, stationed at Buzuluk. He had been in touch with Colonel Heliodor Pika, who had negotiated with the Russians over the Czechoslovak military presence in the Soviet Union, and who became the head of the military mission to Moscow in 1941. He was arrested soon after the communist takeover of power in Prague in February 1948, and sentenced to death on false espionage charges in January 1949. At the time of rising confrontation between the East and West after the war, Rada's connections with the Czechoslovak government in London and with the British secret service could be traced in his personal files. Therefore, doubts as to Rada's loyalty could easily arise among the communist leadership.

After his arrival In the Soviet Union early in 1942, Rada did not become a frontline officer in the war. He made contacts with communist politicians and soldiers and, when the new government was being planned in 1945 (as a combination of the London and the Moscow exile wings), Rada's name was mentioned among the leading experts suitable for the Ministry of Industry. After a spell in the arms industry, Rada became, early in 1946, the director general of the Czechoslovak mines. He joined the Communist Party, was made the government plenipotentiary for mining, and a member of the Czechoslovak-Soviet uranium commission. Until the reorganization of the Ministry of Industry, he was its deputy minister between 1949 and 1951.

He had to balance the claims of the strong, traditional coal mining industry with those of the uranium industry, the fastest-growing sector of the Czechoslovak economy. Together with other Czech members of the permanent commission—Kovář, who was replaced by Pohl in 1948—Rada was responsible for the uranium industry at a critical time. The Soviet imperative to produce the largest amount of uranium ore in the shortest time wrought havoc in the mining industry and the economy, as well as in the Communist Party.

As secrecy and paranoia went hand in hand, Svatopluk Rada soon came under scrutiny by the StB. The initial impulse for the security investigation came from Rada's 1949 negotiations, concerning exports of coal from the mines in Komorany and Sokolov, in West Germany. His plans for the mangan-stannite plant (Mangano-kyzový závod at Chvaletice), which was to supply raw materials for the production of sulphur and sulphuric acid, were also reviewed by the StB. The construction and production plans were deemed to have been rushed and incomplete and, in the case of the Karviná-Ostrava district, the government directive concerned with the maximum employment of domestic resources was disregarded. Evidence was also sought that Rada placed men hostile to socialism into top management positions, and that he maintained contacts with them even after they fled abroad.

The most serious charges were connected with the NPJ: Rada's personnel policies were criticized, and he was blamed for the chaotic state of the enterprise. He was suspected of having kept information about uranium at Janské lázně in Krkonoše (*Riesengebirge*) to himself, and that he tried to get hold of the results of the explorations by the Soviet geological service. Rada knew that Soviet and Czechoslovak security agencies had started to concern themselves with the uranium industry. Top managers Hegner, Čmelák and Chmela had all been arrested in October 1949. They were interrogated during the following four months, and sentenced in 1951.[165] The uranium industry investigations involved 41 people, 18 of whom were put under arrest. At that time, Rada drafted a detailed situation report for president Gottwald on the situation in the uranium industry, including an account of the sorry state of the Communist Party organization in Jáchymov and the threat of Western intelligence activities in the industry. Rada placed his faith in being a good communist.

There existed the possibility that complaints against Rada would amount to the charge that he was "hostile to socialist construction," with its necessary adjunct, that he was an agent of Western imperialism. The surveillance of Rada yielded no results, though he was presented, some six months after his death, as a saboteur and an agent of imperialism. The charge was made during the trial with Slánský and others.[166]

The lines of investigation of Rada's work showed the current preoccupations of the security agencies. Rada was a conscientious official who did his best in a fluid and disorderly situation. Labor shortages, the hurried introduction of Soviet practices and Czechoslovak withdrawal from Western markets, were taking their toll. As a member of the Czechoslovak-Soviet uranium commission, Rada bore responsibility for the uranium industry without having sufficient powers. He knew the importance of the uranium industry for the Soviets and, at least occasionally, he tried to defend Czechoslovak interests.

There were meetings with the Russians where Rada felt uncomfortable.[165] He was sent by Gottwald to settle the matter of payments for the deliveries of uranium and to negotiate the production contract for the years 1949–1951. General V. Y. Semichastny of the NKVD was Rada's partner in the negotiations. After returning from Moscow, he reported to Gottwald, on 5 December 1949, that the NPJ fulfilled the plan for the first year of the three-year plan and produced 120 tons of uranium metal. (In the same year, incidentally, Wismut, the company in neighboring East Germany produced 766 tons of uranium.) Rada added that the level of administration was improving, though accountancy for the year 1948 was still causing difficulties.[166] There was a comprehensive report by Rada attached to the brief letter, dated 22 November. It was drafted at a time when Volokhov, the Soviet member of the joint commission, was trying hard to put the NPJ under the sole control of the Soviets. Rada mentioned Volokhov's endeavor, and pointing out that many difficulties of the NPJ were caused by rising costs rather than by bad management.

During the investigation of Rada's case, the main charges against him involved personnel policy. He was suspected of recruiting, for the top management positions, men hostile to the socialist system. The first two managers, Hegner and Čmelák, were among them. Rudolf Slánský had been instrumental in Čmelák's removal. His successor, Simin, was also regarded as unreliable by the StB. Rada was further blamed for the appointment of Bub and Hanzlíček, members of the Czech management team, who were also under suspicion. Other findings of the investigation, such as those concerning access to, and use of the results of geological exploration, show that the Czechoslovak StB were denied the opportunity to carry out the investigation inside NPJ itself. They had to rely on the evidence supplied during the interrogation of NPJ employees arrested in the autumn 1948.

According to one account, the order for Svatopluk Rada's arrest was issued on 2 April 1952. The arrest kept on being postponed for unknown reasons.[169] Another account gives the date of the decision on his arrest as 4 April. The decision was approved by Major Doubek of the StB, and the arrest was to be carried out on 22 April. Rada was put under close security surveillance on 16 April 1952, and shot himself in the garden of his house about 11 o'clock the following night. The surveillance team noticed nothing wrong at the time; his body was discovered the following morning and the cause of death was apparently suicide by firearm.

Rada was forty-eight years old. Until recently, he had been in good standing with Gottwald and the Communist Party. The slightest suspicion against him would have resulted in the removal of his security clearance. He may have been warned of his impending arrest and could no longer bear the pressures put upon his person.

The Russians could have stopped the Czechoslovak security operation against Rada, or at least taken it over themselves. They may have had doubts about his continued usefulness, or he his frequent contacts, in the early phase of the war, with Western and Czechoslovak intelligence may have aroused suspicion. From his long experience, Rada knew more about the uranium industry than any other Czech. He continued to haunt the Czechoslovak StB even after his death. Ludvik Frejka, the head of the economic section of the president's chancery, and Otto Sling, a high party official in Brno, were sentenced to death in the Slánský trial in November 1952. They had been both interrogated in connection with the case against Svatopluk Rada.

Ten years later, in the autumn 1962, the Ministry of the Interior carried out a review of cases of deaths in custody and suspicious suicides. It emerged that Rada's death had been investigated without due care and attention by Kamil Pixa, who was then a police officer. (During the same enquiry it was established that, in the autumn of 1949, Pixa had been responsible for the death of a man held in custody.) According to the evidence of a witness, Rada was visited by two high officials of the

Ministry of the Interior, Antonín Prchala, the deputy minister and Milan Mouăka, the head of investigation, during the evening before his suicide. They had a long animated discussion with Rada. In addition, criminal investigation at the time apparently excluded the possibility of suicide. No member of the surveillance team, which surrounded the garden of Rada's house, reported hearing a shot being fired during the night.[170]

In one of the so-called "Boy Scouts" trials, concluded on 3 April 1953, defendants Jaroslav Švenek and Josef Hettler were sentenced to fifteen-year prison terms. They were fortunate that their trial was held after Stalin's death. A few months earlier, they would have been tried on capital charges.

In the case for the prosecution, there was no mention of either Jáchymov or uranium. Jáchymov was referred to as a "place essential for the defense of the Republic and of world peace." The two defendants had been employed there as geologists, and the charges against them related to 1948 and 1949. Švenek and Hettler were accused of breaching the Official Secrets Act and of supplying information to Josef Anderle, who later fled to the West. Švenek, who pleaded guilty to charges of spying but not of high treason, argued that, as a Boy Scout, he could not tolerate the ways officials in the uranium industry misused their positions for private advantage.

The prosecution condemned the movement to which the two young men belonged:

> The bourgeoisie, which rules in the capitalist states uses, in class struggle against the workers, ideological means in addition to terror and force, so as to keep the workers in subjection to the capitalist exploiters. One of the instruments of the bourgeoisie for spreading bourgeois ideology and blunting class struggle, especially among young people, is the scouting movement. It originated in England and it is a global carrier of the cosmopolitan idea, educating young people of the smaller nations in particular to admire Britain and the USA. In this way it suppressed national self-confidence and educated young people to recognize the superiority of the Anglo-Saxons, who, especially after the First World War, in fact ruled over the whole world. The movement emerged especially in Czechoslovakia as the bearer of cosmopolitanism and of Western ideology. This became clear after the liberation in 1945 and especially after the defeat of the attempt of the reaction to carry out a counter-revolutionary putsch in February 1948. The scouting organizations in Czechoslovakia, under the name of

Junák, adhered exclusively to fascism and Nazism at the time of Munich and of the Protectorate gloom, were nevertheless abolished in 1940, because they were under Anglo-Saxon influences. After the liberation of our country in 1945, Junák was revived by its former leaders so that it could be used against the united and progressive youth movement, which was then being founded."[171]

It was further stated that Josef Hettler became a member of a unit of the Catholic Junák in 1938, and that its leader was Josef Anderle. In the spring of 1948, Anderle came to Jáchymov— the place name was omitted on the charge sheet—where he convinced Hettler that passing on secret information was his duty as a Boy Scout. About three months later, Jaroslav Švenek, also a former member of Junák, came to the second meeting. According to the case for the prosecution, Anderle informed the two men that, because of the threatened ban on the Junák movement, its leadership chose to go underground. He described the situation of Czechoslovak exiles in western Germany, indicating that he had connections abroad, and that the information was to be passed on to foreign powers. They formed an illegal cell and called it Jiskra.

Some time later, in November 1948, Hettler was sought out by one Karel Forst, who asked him for intelligence material. Forst told Hettler that Anderle had sent him. Early in 1949, another meeting took place between the two men, and Forst told Hettler that Anderle had escaped abroad. On this occasion, Hettler received 5,000 crowns from Forst, as well as a pair of nylon stockings for his wife. The defendant pleaded guilty to passing on information, though he maintained that he had committed the offences in consequence of his education as a Boy Scout, and in the belief that the information was not a state secret.

Jaroslav Švenek, who worked in the mineralogy department of the National Museum in Prague, died in 1994. Josef Hettler lived at the time in retirement in a small town in North Moravia. Josef Anderle taught at an American university.

The proximity of uranium-rich regions to the outer limits of the Soviet empire meant that information on the production of uranium reached the western intelligence services and that they were able to form, in the early 1950's, estimates on the quality and quantity of the Jáchymov pitchblende. In the main, however, the Soviets were successful in keeping information on their nuclear weapons program secret. Soviet intelligence was thus successful both in defensive and offensive terms. One of the byproducts of secrecy was the use made of spy manias in the fringe states of the eastern empire to strengthen their ties with Moscow.

Notes

[1] Československý biografický slovník 20. století (CBS), Josef Tomeš (ed.), Prague 1999.
[2] Madeleine Jana Korbel's (later Albright) MA thesis for Wellesley College, Mass, 1959, *Zdeněk Fierlinger's Role in the Communization of Czechoslovakia: The Profile of a Fellow-Traveller*, 3. The thesis contains original testimonies from Fierlinger's contemporaries, as well as the advice of Korbel's father, an academic in the US and former Czechoslovak diplomat.
[3] Korbel, *Zdeněk Fierlinger's Role in the Communization of Czechoslovakia*, 6.
[4] Oral testimony of Arnošt Heidrich, a former senior official in the Ministry of Foreign Affairs, Korbel 8.
[5] Zdeněk Fierlinger, Ve službách ČSR, second edition, Prague 1951, I, 51.
[6] Oral evidence from Arnošt Heidrich; Korbel, 21.
[7] Fierlinger, Ve službách ČSR, sv 1,127.
[8] Fierlinger, Ve službách, 279 et seq.; and Korbel, 35.
[9] The Memoirs of Dr. Eduard Beneš, London 1954, 145.
[10] Dnešni válka jako sociálni krise, London, Nova Svoboda, 1940.
[11] Fierlinger, Ve službách ČSR, I, 341.
[12] Archiv národního musea (ANM) Prague, the papers of Zdeněk Fierlinger, k 9 238; draft of an essay Osvobozenecká legenda, 38.
[13] Korbel, 58; 56 et seq.: "It is interesting to speculate how much Fierlinger's official reports and personal interpretations did influence Beneš in his thinking about future Russo-Czech relations."
[14] ANM, k 36, 1331.
[15] Korbel, 100; also in Recollections and Reconstruction of the Czechoslovak February 1948 Crisis by a Group of Democratic Leaders. A stenographic record made in 1949–1950 of a discussion between Jaroslav Stránský, Václav Majer, Blažej Vilím, Jaroslav Smutný and Lev Sychrava.
[16] Korbel, 101; and Jan Masaryk, memorandum on his conversation with Klement Gottwald, unpublished, Moscow, 21 March 1945, a document kindly provided by Pavel Kosatík.
[17] Václav Černý, *Paměti 1945–1972*, Brno 1992, 185.
[18] Ministerstvo zahraničních věcí, SSSR serie; Box 30: a communication from the Ministerstvo národní obrany, Oddělení pro styk se spoj. armádami, 19 September 1945.
[19] Karel Kaplan and Vladimír Pacl, *Tajný prostor Jáchymov*, 7.
[20] SUA, Archiv ministerstva průmyslu, 683, k 1.48.
[21] Kaplan and Pacl, *Tajný prostor Jáchymov*, 7; the date given here is 19 September.
[22] Státní ústřední archiv (SUA), the Archive of the Central Committee of the Communist Party of Czechoslovakia, AUVKSČ, Fond 100/24, sv 82, a j l031.
[23] Oskar Pluskal, Poválečná historie jáchymovského uranu, Czech Geological Survey Special Papers, Prague 1998, 9.
[24] SUA, ÚVKSČ, fond 100/24, sv 82, a j 1031.

25 SUA, ÚVKSČ, fond 100/24, sv 82, a j 1031.
26 See Annexe I for the text of the treaty and of the supplementary protocol.
27 Vladimír Valenta, Po stopách uranového hornictví na Příbramsku, in Podbrdsko, Sborník státního okresního archivu v Příbrami, IV, 1997, 143.
28 Oskar Pluskal's MSS on the development of the Czechoslovak uranium industry
29 Karel Boček, Ani gram uranu okupantům! Prague 2005, 60.
30 Idem., 61.
31 Století atomové energie? Signed by initials V. S. and F. R., Svět práce, 30 August 1945.
32 Atomová energie a Jáchymov, Obzory no. 30, 1946, 470.
33 SUA, Ministerstvo průmyslu, odbor hornictví, c V-31 tajné, 27 August 1946; and Milan Drápala, Na ztracené vartě západu, Antologie české nesocialistické publicistiky z let 1945–1948, Prague 2000, 274 et seq.
34 Le Coupe de Prague, Une Revolution Prefabriçue, Paris 1949; Czechoslovakia Enslaved, London 1950; and Únorová tragedie, Brno 1995.
35 Pavel Paleček, Ministr Hubert Ripka a jeho osobní archiv, 113.
36 Idem., 120.
37 Otto Friedman, *The Break-up of Czech Democracy*, 133n1.
38 SUA, f 100/24, sv 82, a j 1031; letters of 2 and 13 January 1946.
39 Eduard Táborský, *Prezident Beneš mezi Západem a Východem*, 229.
40 Pavel Kosatík and Michal Kolář, *Jan Masaryk*, 193.
41 Jan Masaryk, Volá Londýn, Prague 1946, 267–268.
42 For instance in Vladimír Vaněk, Jan Masaryk, Prague 1994, and in a brief personal memoir by Jindřich Kolowrat, appended to the biography, 218.
43 US Foreign Relations, Washington, for 1946.
44 Lumír Soukup, Chvíle s Janem Masarykem, Prague [1994], 76–82.
45 Idem., 82.
46 AMZV, Generalní sekretariát (kabinet GS), k 77; the original speech was drafted in English.
47 Idem.
48 SUA, ÚVKSČ, fond 100/24, sv 82, a j 1031.
49 SUA, the papers of Lumír Soukup; op. cit., 77.
50 Idem.
51 Idem., 79.
52 SUA, F1/1-1223/12, the papers of Hubert Ripka; quoted in part in P. Kosatík and M. Kolář, Jan Masaryk, Prague 1998, 240.
53 Idem.
54 Idem.; also, the Soukup papers in SUA.
55 SUA and Soukup, op. cit., 81.
56 AÚTGM, f EB-kor, 76.
57 K. Houdek, Vývoj a současné problémy uranového průmyslu v ČSSR, VSE Prague 1969, 2.
58 30 let uranového průmyslu, SNTL Prague, 1975.

Part 3: The Politics of Czechoslovak Uranium   155

59 Prokop Tomek, *Československý uran 1945–1989*, Prague 1999, Sešity UDV 1, 5.
60 K. Houdek, op. cit., 13; also in F Sorf, Uran, Příbram, 1982, 9.
61 See below, Scarce Labour; Tomáš Dvořák, Těžba uranu versus "očista" pohraničí, Soudobé dějiny, 3-4/2005.
62 Archiv Ministerstva vnitra, (AMV), 304-57-14; I April 1946.
63 Arnošt Schindler, the opening essay in 30 let uranového průmyslu, op. cit.
64 AMV, S-117-11, Jáchymov, 20 December 1946.
65 Interview with Ing. Ladislav Jánský, Trutnov, 8 September 1998. Jánský was an engineer at Eva, one of the pits in Jáchymov, from 1948–1950.
66 AMV, S-117-11, Jáchymov, 20 December 1946.
67 SUA, fond 100/24, 82 a j 1031; L Černík, Ing. Odolen Koblic—využití uranu Jáchymovů.
68 Oskar Pluskal, Poválečná historie jáchymovského uranu, Prague 1998, 13.
69 The memorandum was reprinted in Oskar Pluskal's Poválečná historie jáchymovského uranu, op. cit., 12, who attributes it to "representatives of Czechoslovak industry."
70 Pluskal, Poválečná historie, 13.
71 SUA, Praha V/45, i c 683, k 1448, Záznam pro ministra průmyslu B Laušmana.
72 Archiv Diamo, Příbram. The report of Kovář's proposal is a part of the protocol of the meeting of the Czechoslovak-Soviet commission, on 21 August 1947. Volume 1 of the records of the meetings.
73 SUA V/45, i c 683, k 1448, letter to the deputy prime minister, Z. Fierlinger, 19 February 1947.
74 *Idem.*
75 He subsequently wrote the first article for an academic magazine about Czechoslovakian uranium, "Soviet Russia and Czechoslovak Uranium," *The Russian Review*, 1952, 2, 97–105.
76 For instance 30 let uranového průmyslu, Prague 1975, and A Schindler's introductory essay.
77 Otchet of geologo-rozvedochnykh, geofysicheskikh i gydro-geologicheskikh rabotakh jachymovskovo rudoupravlenia za 1946 god, report in the geology section of the archive of DIAMO Příbram, 77 pp.
78 AMV, slozka 117–11.
79 AMV– slozka 302–161–3.
80 Topografická mineralogie Čech, Prague 1936.
81 Archiv Českeho geologickeho ústavu (CGU), Prague, MS P154, Zpráva o další etapě systematického geologického výzkumu Jáchymovska, 1938.
82 CGU Prague, Zpráva o orientační geologické prohlídce jáchymovských dolů uranových v srpnu 1945, Geofond, MS, P99289.
83 CGU, MS P99289, P2589/1, P2391/1 and 2.
84 Oral evidence from Dr. J. Hettler, one of the Czech assistants employed by the geological service; Zlaté Hory, 1995.
85 J. Kořán, Sláva a pád českého hornictví, Příbram 1984, 224.
86 Rudolf Tomíček, Tězba uranu v Horním Slavkově, Sokolov 2000, 37.

[87] Diamo podnikový archiv, Příbram, k 32, n14.
[88] Ludmila Petrášova, Vězeňské tábory v jáchymovských uranových dolech 1949–1961, Sborník archivních prací, 2, XLIV, Prague 1994.
[89] Oskar Pluskal's manuscript on the Czechoslovak uranium industry, l47.
[90] T. Veselý, Stavba a význam jednotlivých žilných uzlů uranového ložiska Jáchymov, Sborník geologických věd, rada LG c 28, Prague 1986, 11.
[91] Oral evidence from Mr. Ježek, former head teacher in Jáchymov, in 1994.
[92] Vladimír Valenta, Po stopách uranového hornictí na Příbramsku, Podbrdsko, Časopis státního archivu v Příbrami, IV 1997, 159.
[93] Valenta, Po stopách uranového hornictví na Příbramsku, Podbrdsko, vol. IV 1997, 145.
[94] F. Sorf, 30 let naší geologické činnosti na Příbramsku, in Příbramský atom, no. 17, 18/ 1977; in V. Valenta, op. cit., 146.
[95] Valenta, op. cit., 147.
[96] Otchet o geologo-rozvedochnykh, geofisicheskikh i gidrogeologicheskikh rabotakh jakhymovskova rudouprovelnia za god 1947, DIAMO, Příbram, 533–549.
[97] Valenta, op. cit., 149.
[98] Valenta, op. cit., 153.
[99] Valenta, Podbrdsko, vol. IV, 1997, op. cit., 158.
[100] Diamo, Ing. Vladimír Valenta, Závěrečná zpráva, cast IX, Personalní a sociální činnost, Příbram 1995.
[101] Documents on British Foreign Policy (DBFP), London 1985, series I, vol. II, document 193.
[102] Vladislav Zubok and Constantine Pleshakov, *Inside the Kremlin's Cold War*, 44
[103] British Foreign Office FO 371/57094, Bamborough's report to the British embassy in Prague, 4 January 1946.
[104] FO 371/57098, J.No18908, a note by D. E. H. Peirson, 27 March 1946.
[105] FO 371/57098 contains the British reply of 16 May 1946 to the Czechoslovak request.
[106] *Idem.*, Acabrit to War Office, Top Secret Cipher Telegram, l8 March 1946.
[107] Overseas News Agency, Carlsbad Jan. 26, (delayed).
[108] US National Archives USNA, Navy Department, Office of the Chief of Naval Operations, report for the State Department from 3 February 1947.
[109] SUA, ÚVKSČ, 100/24, 82, 1031; 5 December 1949.
[110] *Idem.*, NB date, given as 22.11.
[111] Diamo, Příbram, k 32, n14.
[112] Friedrich Ebert Stiftung Archiv, Bonn, SPD Parteivorstand, Ostbüro, 0072B (Nr 02365), Bericht vom 24.10.1949.
[113] Karel Kaplan and Vladimír Pacl, Tajný prostor Jáchymov, Prague 1993, 32.
[114] Karel Kaplan (ed.), StB o sobe, Prague 2002, 17.
[115] Petrášová, Vězenské tábory v jáchymovských uranových dolech 1949–1961; Sbornik archivnich praci, 2, roc XILV, Prague 1994, 344.

Part 3: The Politics of Czechoslovak Uranium 157

[116] Sborník přednášek a dokumentů, Stálá mezinárodní konference o zločinech komunismu, Praha-Dlabačov, October 1991, 62 et seq.
[117] Petrášová, op. cit., 366.
[118] Stálá mezinarodní konference o zločinech komunismu, op cit, 42–45.
[119] Centrální archiv vězeňské služby (CAVS), sv 6, dokument o organizačním a obsahovém vývoji československého vězeňství 1948–1968; and cf. Petrášová, op. cit., 367.
[120] J Brodský, Řešení gama, place of publication not given, 1970, 69 et seq.; referred to in Vilém Hejl and Karel Kaplan, Zpráva o organizovaném násilí, Toronto 1986, 254.
[121] SObA Plzen, Vězeňský ústav Ostrov, kart 4-29; cf Petrášová, 368.
[122] Petrášová, op cit., 345.
[123] Usnesení vlády České republiky ze dne 16. června 1999, c. 6 09.
[124] For the table giving the number of prisoners as a proportion of NPJ workers see Pluskal, Poválečná historie..., 39. In 1949, the figures given are: 4,630 prisoners out of 13.653 workers, i.e. 33.91%.
[125] Petrášová, 347 and 349.
[126] Kaplan and Pacl, op cit., 42.
[127] Karel Pecka, Motáky nezvěstnému, Brno 1990, 156.
[128] CAVS Praha, Závěrečná zpráva meziresortní komise ministerstva vnitra ČSSR, ČSR a SNV ČSR o výsledcích tzv. nezakonností z padesátých let v bývalých táborech pro odsouzené v oblasti Jáchymov, sv 6; cf. Petrášová, 352.
[129] CAVS Praha, Závěrečná zpráva, sv 26/3, příloha c 7; cf Petrášová, 354.
[130] Petrášová, 360.
[131] Milena Bubeníčková-Kuthanová, Vybledlé fotografie, Střední Evropa, no. 22, 1992.
[132] Petrášová, 401.
[133] Jihočeská Pravda, České Budějovice, 9 April 1968; quoted in Antonín Kratochvíl, Žaluji, Prague 1990, vol. 1, 159.
[134] Karel Pecka, Motáky nezvěstnému, Brno 1990, 222, gives the distance as two kilometers.
[135] Petrášová, 363.
[136] *Idem.*, 364.
[137] *Idem.*, 411–412.
[138] Stálá mezinárodní konference, 72.
[139] Karel Pecka, op cit., 198.
[140] Karel Kukal, Deset křížů, Prague 1993, passim.
[141] Petrášová, 415.
[142] Museum třetího odboje in Příbram, fond I entitled Materiály politických vězňů; the file contains responses from the former inmates to a questionnaire drafted in 1998. The most detailed responses concern conditions at Vojna between 1955 and 1960.
[143] Josef Vácha's evidence for the Museum of the Third Resistance, Příbram, October 1998.

144 Jaroslav Holeček, *Kladno*.
145 cf. for instance W. Stephenson, Intrepid's Last Case, London 1984, 164.
146 Vilém Hejl and Karel Kaplan, Zpráva o organizovaném násilí, Toronto 1986, 62
147 ÚVKSČ, Politické procesy, výslechový protokol S Plačka; K. Kaplan, Der kurze Marsch, Munich 1982, 219 et seq.
148 Vilém Hejl and Karel Kaplan, Zpráva o organizovaném násilí, Toronto 1986, 62; cf. also Miroslav Šesták, Emil Voráček (eds.), Evropa mezi Německem a Ruskem, Historický ústav, Akademie věd, Prague 2000, an essay on the Slánský affair by Igor Lukeš.
149 Josefa Slánská, Zpráva o mem muzi, written in the 1960's and published in 1990 in Prague, p 128 et seq.
150 SUA, ÚVKSČ, Fond 100/1, sv 107, a j 697.
151 *Idem.*, fond 100/24 sv 82 a j 1031, a report for president Gottwald on the situation in NPJ, accompanied by a cover note dated 5 December 1949.
152 Karel Kaplan, Zpráva o zavraždění generálního tajemníka, Prague 1992, 89 et seq.
153 Kaplan, op. cit., 90.
154 [SUA, ÚVKSČ, fond 100/1. sv 107, a j 697].
155 cf. Prokop Tomek, op cit,. 35.
156 SUA, ÚVKSČ, fond 100/24, sv 82 a j 1031.
157 *Idem.* Two of the reports were included in the file, and were dated 24 August and 1 November 1950; the last report, with reference number ZU 12/2/51 was missing from the file.
158 Kaplan, op. cit. 263.
159 Op. cit. 164.
160 *Idem.*, 171.
161 Miroslav Šesták, Emil Voráček (eds.), Evropa mezi Německem a Ruskem, Historický ustav, Akademie věd, Prague 2000, contribution on the Slánský affair by Igor Lukeš.
162 AMV, svazek H-786-5, dokumentace III sektor; cf Tomek, op. cit., 35.
163 Vojenský historický archiv Prague (VHA), fond 255, a note by Bohumila Radová dated 5 December 1968.
164 HVA, fond 255, General Bulander's report of 18 October 1947.
165 AMV, 310–109–2; see above, ...
166 Proces s vedením protistátního spikleneckeho centra v čele s Rudolfem Slánským, Ministry of Justice publication, Prague 1953, 238.
167 J. Kašpárek, a Czechoslovak diplomat who defected in 1948, described Rada's visit to Moscow at the end of 1947; cf Kašpárek's article in the The Russian Review, 1952, no. 2.
168 SUA, ÚVKSČ, fond 100/24, sv 82, a j 1031.
169 Stálá mezinárodní konference o zločinech komunismu, op. cit., an account of deaths in custody and of suicides drafted sometime in 1962, 90.
170 Stálá mezinárodní konference, op. cit., 85 and 90.
171 The document is in the private archive of Dr. Hettler.

Part 4

Wismut AG: A State Within a State

The Consequences of War

The agreement to partition Germany into zones of occupation was completed in November 1944, and confirmed at the conference of the Big Three at Yalta in February 1945. It was not yet clear when the war in Europe would end, nor where the Allied armies would come to rest. While the Wehrmacht put up stiff resistance against the Red Army, the front in the West broke down in the spring 1945 and the US Army made a swift advance to the center of Germany. The whole of Thuringia and a large part of Saxony were occupied by US troops. Churchill started to press Truman to negotiate compensations with Stalin for the withdrawal from "tactical zones," where the Western armies had crossed the previously agreed demarcation line. Truman disregarded Churchill's proposals that the US Army should stay put, and the Americans withdrew to the line agreed in London. Saxony and Thuringia came under Russian occupation; the importance of the US withdrawal for the Soviet nuclear project emerged later. Neither Churchill nor Truman (nor, for that matter, Stalin) were aware of the existence of the vast uranium reserves on the German side of the Erzgebirge, and they played no role in the deliberations.

At the end of the war, shortages of every kind, suffering and hopelessness gripped the Erzgebirge, as well as the rest of Germany. Refugees roamed the country, the economy was in ruins and the Soviets were busy dismantling industrial plants that escaped destruction in their zone of occupation. The mining of colored metals in Saxony seemed to have no future before it.

Late in the summer of 1945, there appeared signs that the occupying power had special plans for the established mines at Johanngeorgenstadt, Schneeberg and Annaberg. Soviet teams went down the old shafts, questioned German scientists, especially those at the mining academy in Freiberg, and perused old archives. They were not interested in rare nonferrous metals, but in one raw material only: uranium ore. Nobody knew at the time how uranium rich those districts were.

The remainder of Sachsenerz-Bergwerks AG miners, about one hundred men, started working several pits on behalf of the NKVD exploration expedition in the autumn of 1945. They had worked non-ferrous metals before and they were ready to mine uranium now. The miners did not mind as long as they could keep their jobs. They had no idea about the future of the new industry. If mining was to be resumed, an additional workforce would have to be recruited. With encouragement by the

Soviets, the mining company began to recruit new workers. The first employment contracts were dated November 1945. A handful of experienced men, mainly from and around Schneeberg, formed the core of the future workforce.

On 13 February 1946, Avramii Pavlovich Zaveniagin, who was responsible for the Soviet atomic project, reported to Beria that "raw material A9," that is uranium, had been found in the vicinity of Johanngeorgenstadt, Schneeberg and other parts of Saxony.[1] Zaveniagin ventured no guess as to the size of the reserves, as the surveys of the territory had not yet been completed and the first reports were treated with caution. For the time being, and in view of the shortage of the metal in the Soviet Union, the short-term exploitation of Saxon uranium seemed worthwhile.

On 4 April 1946, the Council of Ministers of the Soviet Union decided to place the extraction of uranium in Saxony under the control of the NKVD. The Saxon "search and survey expedition" was to be transformed, as of 1 April, into the "extraction and survey expedition" and put under the first directorate of the NKVD.[2] The expedition, similar to the geological service the Soviets introduced in Czechoslovakia, was headed by geologist N. M. Khaustov. It was to explore the Johanngeorgenstadt district and carry out a survey of other promising sites in Saxony. Johanngeorgenstadt mines were expected to yield at least one ton of uranium by the end of 1947.[3]

The decision of the Council of Ministers referred to Johanngeorgenstadt as the only production site. It was close to Jáchymov, and it was assumed that there existed a continuity of deposits running underneath the state border. A further edict of the council, no. 9372, issued on 29 July, announced that the expedition would be called Sächsische Bergbauverwaltung. Regular mining operations for uranium started in the Soviet zone of occupation in the summer of 1946, about seven months after the foundation of the NPJ in Czechoslovakia.

So as to keep the activities of the small NKVD enterprise in Saxony secret, it received its own field-post number: no. 27304, and the mining of uranium was publicly presented as the excavation of bismuth and cobalt. Pioneer brigades of the Red Army set to work alongside the German firms. About 130 specialists, who had worked for the Transport Ministry or had built the Moscow metro, arrived from the Soviet Union before the end of 1946. Beria appointed NKVD Major General Mikhail M. Maltsev, an experienced manager of forced labor, to head the enterprise. Maltsev's direct superior was Colonel General Ivan Serov, Beria's deputy in the Soviet zone of occupation.

In January 1946, the "Vereinigt Feld im Fastenberg," a mining business in the neighborhood of Johanngeorgenstadt and parts of the Sachsenerz Bergwerks AG, was occupied by units of the Red Army.[4] The mines, which had been neglected during the war, were quickly restored; they had been equipped with electricity, compressed air ducts and rails as early as the 1920s.[5] At Johanngeorgenstadt, the Frisch Glück, Hoffnung and Schaarschacht pits were among the first to be inspected by the Soviets.

Six miners from the Schneeberg district were sent to Johanngeorgenstadt, and the first pitchblende was brought from the Frisch Glück pit in the spring 1946.[6] The ore was lifted from the pit in rucksacks and Martin Vogel, the foreman of the miners, reported daily to the Soviet geologists. The foundations of Object 1 were laid down.*

The initial mining operations were carried out in the old town, with the exception of the Günther pit, located about 4.5 km from Johanngeorgenstadt. The revival of mining was a mixed blessing for the inhabitants. No other town in the Erzbebirge grew as quickly: home to 7,000 people in 1939, its population rose to 33,000 inhabitants in 1950, a figure that excludes those who lived in the hostels. The uranium industry created new jobs and brought money to the district, but also damaged the countryside and the old town itself. By the mid-1950s, much of the old town had to be torn down.

It took only three years to put 26 mines on stream. Object 1 reached its maximum size in 1949.[7] It became the first Wismut AG object to produce uranium for the occupying power, and had more employees than any other part of the company. The mines at Breitenbrunn and Zschorlau were also taken over by the Russians in autumn 1946.[8] The Breitenbrunn district was added to Object 1, while the Zschorlau wolfram mines were put into the care of the Soviet metallurgy company (SAG).[9]

Objects 2 and 3 grew out of the operations in the vicinity of Schneeberg, Schlema and Alberoda. There were no signs of activity there until the summer of 1946. The turning point came early in August, when the Oberschlema spa was closed down and Red Army units occupied the main spa buildings. Geological probes were made in the immediate neighborhood Oberschlema, and in the famous Semmler-Stollen, directly

---

* The term "object" was used for relatively independent units in the uranium industry. They reported directly to the board of Wismut directors. In common speech, terms such as "enterprise," "factory" or "works" were used instead of "object." As a rule, several mines formed an object.

underneath the spa house, the Russians struck a solid seam of uranium. It was extracted by "medieval instruments," and each shift advanced 25–35 centimeters along the seam. Martin Vogel wrote that "after about two months we had compressed air at our disposal, which went down the old pump shaft, where the water used to be pumped for the spa guests. The compressor was located directly under the reading room of the hotel. From then on it all went faster, but treatment in the famous radium spa was over."[10] The Soviet management of Object 2 moved into a wing of the hotel and ordered a survey to be made of the Gallus tunnels and the slag heaps in the vicinity of the old silver mines. The Gallus complex was approached through the cellars of the town houses, and soon the miners struck another rich seam. The disused pits in the vicinity of Oberschlema were gradually all opened up.[11]

In the early days of Object 2, some 3 kilometers of galleries were accessible. At the beginning of 1948, their length amounted to 30 kilometers. "Objekt Nummer 2" had 17 pits in Ober- and Niederschlema in the early 1950s. By then, the characteristic features of the Saxon mining districts, wooden hoist towers, scattered among town houses and business quarters had disappeared. Wismut AG showed as much concern for old buildings as for property rights. Property necessary for the uranium industry was confiscated and its inhabitants were expelled. Mining was carried on everywhere, even in the backyards of houses. The chronicle of Wismut refers to the years 1946 to 1953 as to a period of "total mining" in Schlema, when the region became a single, vast complex of mines.[12] Surface mining for uranium was also started, and became especially damaging to the countryside. The inhabitants of the Schlema region had to endure soil subsidence, or breaches in gas and water pipes. Enormous slag heaps grew up and slime ponds were created. When uranium mining was finally suspended in 1989, damage to the landscape seemed to be virtually irreversible.

In May 1946, the Saxon mining administration (Sächsische Bergbauverwaltung) requested small companies in Schneeberg and Neustädtel to search for uranium especially in the Weißer Hirsch, Beust and Ritter pits.[13] In September 1946, the mines belonging to the "Gewerkschaft Schneeberger Bergbau" were occupied by the Red Army and added to Object 2. In April 1947, they formed the new Object 3. It contained a large number of pits (13 old and 11 new mines), with modest production.

In March 1947, Object 4 came into being in the neighborhood of Annaberg.[14] Until the end of the war, the state of Saxony ran a mining operation there under the name of Sankt Anna vereinigt Feld, and the pits were used by the new mining enterprise. Mining took place close to the center of the town in Annaberg as well. By 1949, Object 4 consisted of 19 pits.

The mines in the Marienberg district came under Soviet control in spring 1947. Old mines in the immediate neighborhood of the town, which had collapsed or had become waterlogged, were opened up. The Marienberg company was registered in the summer of 1947 as "Vitriol-Werke AG," and was later transferred to Wismut as Object 5.[15] Wismut AG had surveys carried out on about 450 slag heaps in the district, and low quality uranium from the heaps was utilized. Object 5 became one of the poorer sources operated by Wismut AG.

No formal acknowledgement of the change in ownership of the mines was ever made. Companies taken over by the occupying power were referred to, by the government of Saxony in 1946–1947, as "C-enterprises," or as "the Soviet mining enterprises." Nobody worried about property rights. The nuclear project was so important that the Soviets acted first and allowed the new situation to be formally resolved later.

On 8 May 1947, the government of Saxony decreed the nationalization of all mining companies, and it passed the appropriate law,[16] providing for mining enterprises still in German hands to be transferred to the state. The mines confiscated by the occupying power were, of course, excluded from the new ruling. On 10 May 1947, the Council of Ministers in Moscow passed order no. 1467-393, on the establishment of a new joint stock company. It was to be a subsidiary of the Soviet colored metals industry, called Wismut. The order was signed by Stalin himself.[17] A further order, issued on 16 May, stated that the main office of Wismut was to be in Moscow, with a branch office at Chemnitz. The purpose of the company was given, in the first paragraph, as the "extraction of colored metal." Like the Czechoslovak-Soviet secret uranium treaty, the document made no mention of uranium. In subsequent documentation, including internal company communications, references to metal "A9" or to "the ore" were made.

German law on limited companies was to apply to Wismut. The value of the investment capital, as well as the main offices of the company (the main company board, the board of control and the general meeting), were to correspond to German norms. The Soviets used similar practices in the case of joint-stock companies in the German zone of occupation,[18] just as they used the institution of joint-stock companies in other countries of Eastern Europe.

On 26 May 1947, the commander-in-chief of the Soviet forces of occupation in Germany (SMAD) issued order no. 128, to the effect that six mining operations and a processing plant would be transferred into Soviet ownership. It was confirmed by order no. 131 of the head of the military administration in Saxony.[19] The date of 26 May 1947 may therefore be regarded as the foundation day of Wismut AG. The mining districts of Johanngeorgenstadt, Schneeberg, Oberschlema, Annaberg, Lauter and Marienberg, as well as the enrichment plant at Pechtelsgrün, passed into Soviet ownership. On behalf of the Saxon government, Freiberg mining engineers Dr. Oskar Oelsner and Karl Hahner formally transferred the property to a Soviet commission under the leadership of Major General Maltsev. The property valued at approximately 9.4 million marks, and the act of transferring it was confirmed in February 1948.[20]

The mining districts in Soviet ownership formed a region named "Mittelfeld," registered at the administrative court at Aue.[21] "Mittelfeld" was marked out in the land registry maps as the area where Wismut AG carried out its operations. It had 1.6 million inhabitants at the end of 1948,[22] and became a special region—a sort of uranium province—within the Soviet zone of occupation.

It came under the absolute rule of Wismut AG, an unusual joint-stock company with only two shareholders. Both located in Moscow, they were the Main Directorate for Soviet Property Abroad of the Council of Ministers and the State Company of the Colored Metal Industry (Copper). The first formal meeting of Wismut AG was held in Moscow on 5 June 1947,[23] and the only point on the agenda was the nomination of the directors. Mikhail M. Maltsev became the director general.

Wismut and War Reparations

At the Yalta and Potsdam conferences in 1945, the Allies did not reach an agreement on a common reparations policy, or on the total sum of reparations due from Germany. A compromise was reached at Potsdam, which envisaged different reparation regimes for the several zones of occupation.

Soviet authorities were to decide the amounts of the Russian and Polish claims against Germany. The original demand for $10 billion was to be paid exclusively by the Soviet zone, and was to be settled within twenty years. Soon after the end of the conference at Potsdam, the Russians began dismantling industrial plants on a large scale. The middle of 1946 marked a change in their approach, and they began to claim a share in the current production. The German SAG companies carried the main burden. The Soviet zone of occupation and (after its establishment) the German Democratic Republic received reparation credits (Gutschriften) from the various transactions.

Early in its existence, Wismut AG could afford to ignore its financial results. Paper money was in plentiful supply as the Nazi war economy had been financed by creeping inflation. Prices of goods and services were set, as were the wages. The situation continued after the war, and the Allied Control Commission (ACC) did not want to aggravate the chaotic economic situation by freeing wages and prices. A lot of paper money was chasing too few goods, and confidence in the mark lay at a low point. For instance, cigarettes became an alternative currency as did other goods. Payments were also made in hard currencies, especially the dollar. In large towns a black market flourished and barter trade was carried on everywhere. Rewards in kind, or food in the works canteen, were more important for workers than wages.

A currency reform was therefore unavoidable. Just as the Big Three at Potsdam had not been able to agree on the amount of reparations, the occupying powers found it impossible to design a common currency reform. In summer 1948, separate reforms took place in the western and the eastern zones.

The Soviet occupation authority financed work in the uranium province from several sources: confiscated banknotes, newly printed occupation currency, and deductions from the state or zonal budgets. Investment in the uranium industry was kept strictly secret so that the size of the industry would not be revealed. The finance minister of Saxony himself, Gerhard Rohner, knew nothing of the financial situation of the company.[24] The general director's annual reports were also reticent about the amount and source of money passing through Wismut. It may be assumed that Wismut was almost exclusively financed by German taxpayers.

In view of the total expenditure and the vast mobilization of labor for the sake of the uranium industry between 1946 and 1953, Wismut AG became the biggest source of reparations in the 20th century. The Party of Socialist Unity (SED) therefore pressed the occupation authority to take into account uranium deliveries as reparation payments. Since the Soviet zone of occupation was obliged to make sacrifices and put enormous resources into the development of the uranium industry, it should at least help to reduce the reparations obligations: so ran the argument.

It is hard to establish when the Soviet authorities acknowledged the German argument. The so-called "Pieck protocols" contain hints that uranium deliveries were accounted for as reparations. The first president of the GDR, Wilhelm Pieck, made his own notes on the negotiations with the Soviets. One of them concerns a meeting in which Pieck, Prime Minister Otto Grotewohl and Walter Ulbricht, the deputy chairman of SED, took part. The Soviet side was represented by Semionov, the political adviser to the Soviet control commission in Germany, NKVD General Vsevolod Merkulov and the head administrator of Soviet property in Germany, Bogdan Kobulov. The meeting took place on 26 November 1949 in Berlin.[25]

The timing itself of the meeting was significant. Stalin's bomb had been successfully tested in September and, in October, the GDR came into being. At the meeting on 26 November, a secret protocol to regulate further activities of Wismut AG was signed. The protocol guaranteed the Soviets continued and exclusive access to uranium production of the newly formed state. Uranium was not to appear on the export statistics, and the two sides agreed on providing Wismut AG with labor and food on a preferential basis. Deliveries of uranium were to be priced in US dollars ($1 being the equivalent of 11 marks) and charged to the reparations account. For the Germans, the protocol was a considerable achievement, as it was clear at last that the export of uranium would help to pay the reparations.

In Moscow, several committees of the foreign ministry and of the supreme planning office (Gosplan) tried to calculate reparation costs between 1948 and 1950.[26] In internal analyses, which have never been published, the value of the uranium deliveries was assessed at $7.6 million in 1946–1947, $37.8 million in 1948 and $77.8 million in 1949, for a total value of $123.2 million. In comparison with the estimated sums for uranium deliveries, the reparations deliveries until the end of 1949 amounted to $1.2 billion, i. e. almost tenfold the amount. According to this calculation, the share of uranium in the reparations accountancy would have been rather modest. The notional credit items on the reparations account afford little insight into the real value of uranium exported to the Soviet Union, nor do they indicate the real economic consequences of the uranium industry for the GDR.

More than six months after the foundation of the GDR in October 1949, the Soviet Union declared its intention to waive a part of its reparations demands. In fact, they started falling back in 1951. Reparations accountancy was to be carried out on the basis of world market prices for the year 1938, in US dollars. The Soviet control commission passed the annual reparations bill, amounting to $211.4 million, on to the German government. The sum included supplies of industrial goods as well as of uranium from Wismut AG and, in contrast to the previous years, it set a ceiling on the deliveries. The deliveries of uranium acquired a special position in the accountancy of reparations, providing more than half the acknowledged reparations deliveries.

In 1950, total value of the GDR product amounted to 30.66 billion marks. Out of that amount, nearly 5 billion marks had to be set aside for the "special use" (Sonderverbrauch) account. The account contained all the budget items connected with services rendered to the occupying power. However, these expenses were not openly included in the budget. Thus, in the year 1950, the account amounted to 4.9 billion marks, with the costs of occupation at 2 billion, reparations 1 billion, Wismut AG 1 billion, SAG investments 248 million and miscellaneous at 547 million marks.[27]

The uranium industry in 1950 used up about 3.5% of the national income of the GDR, and Wismut AG sent 1,224 tons of uranium to the Soviet Union. A notional amount of $36 million was put on the credit side of the reparations account, or nearly $30 for a kilogram of uranium.

In regard to the sum mentioned, The Pieck protocols are on the whole a trustworthy historical source. However, Pieck understood little about the conduct of the economy, and it is possible that he confused the figures. If they are reliable, then the Soviet Control Commission in Germany put a much lower value on uranium deliveries than Gosplan did in Moscow. While Gosplan worked with the sum of $78 million for 1949, the Control Commission valued the higher deliveries for 1950 at only $36 million. Either Pieck's figures were wrong, or the Soviet Control Commission calculated deliveries of uranium at political prices.

For the following year, 1951, the Germans were credited with $89 million for 1,675 tons of uranium—that is, at about $53 per kilogram. The world market price for uranium was unavailable at the time. The Americans had spent about $35 a kilogram of uranium from the Belgian Congo, and in the early 1950s, the US government paid up to $92 per kilogram of American uranium. After the rich deposits at Mi Vida were opened up, the price of uranium sharply fell to as low as $16.[28]

About 9,500 tons of uranium were delivered to the Soviet Union between 1945 and 1953, and about $350 million were written off. The figures yielded a notional average price between $37 and $45 per kilogram of uranium. The $350 million credit had to be balanced against expenditure of more than 7.3 billion GDR marks on the uranium industry. This meant that the GDR spent 20 marks on uranium production for each "reparations dollar" it earned. If indirect costs were included, pro-

Table 4. Export of Uranium to the Soviet Union by Wismut AG, 1954–1990

Year	Export	Year	Export	Year	Export
1954	3967	1966	7070	1978	6130
1955	4607	1967	7110	1979	5261
1956	5248	1968	6948	1980	5242
1957	5278	1969	6412	1981	4870
1958	5302	1970	6389	1982	4622
1959	5345	1971	9485	1983	4486
1960	5356	1972	6627	1984	4444
1961	5991	1973	6721	1985	4470
1962	6371	1974	6777	1986	4090
1963	6730	1975	6884	1987	4059
1964	6983	1976	6695	1988	3924
1965	7091	1977	6358	1990	3800

Source: Final Report by Wismut AG, Chemnitz, 1991.

duction of uranium would have been even more expensive for the GDR. According to Soviet accountancy, the GDR had not yet met half the reparations costs by the end of 1953. In spite of that, the Soviet Union required no further payments from the Germans as of the year 1954. There were good reasons for Moscow's generosity.

The policies of the ruling SED ran into severe difficulties, culminating in the June 1953 uprising. For the first time, Soviet tanks were used to save an unpopular regime in the Eastern bloc, and Moscow made financial sacrifices so as to save a satellite regime. On 22 August 1953, the Soviet government announced that reparations payments would be terminated as of 1 January 1954.[29]

As a part of the concession, the Soviets restituted 33 large enterprises to the GDR. The "Protocol about the Cancellation of German Reparations Payments" of 22 August 1953 was never published, and contained no reference to either Wismut AG or to uranium. It was later announced that, on the basis of an agreement between the two governments in 1954, a Soviet–German joint stock company, Wismut, would be created. Its product was referred to as "bismuth ore."

It was stated that the activities of the company would be carried out on the basis of equal partnership. The German side acquired the right to participate in making decisions, and the conduct of the enterprise was subordinated to the provisions of German law. The company was no longer subject to the law on limited companies as Wismut AG had been.[30]

The Soviet side handed over to Wismut SDAG assets valued at 2 billion marks, which became the share capital of the company. GDR undertook to pay its share, 1 billion marks, within five years. The provision in paragraph 5 of the statute was hardly advantageous for the Germans. They were to pay for Wismut twice over: they had paid for its development out of their taxes, and they were now to pay again. The Minister for Heavy Industry, Fritz Selbmann, failed in his intention to negotiate raw materials barter with the Soviet Union.[31]

When the new company was established, its shares were issued in the names of physical persons. It was owned by two German and two Russian officials. The board of directors (Vorstand) of SDAG consisted of four people, and Fritz Selbmann became the chairman of the company. The Vorstand fulfilled control rather than executive functions, while the general director was solely responsible for the routine conduct of the business. Valentin N. Bogatov had replaced Maltsev in 1952 as the general director. According to the agreement, he was to be followed by a German general director after five years. The company had a Soviet general director until 1986.

The treaty of August 1953 assigned monopoly rights to explore and extract uranium on the GDR's territory to Wismut. The costs arising from the ownership of the mining rights were to be borne by the German side. In an exchange of letters following the conclusion of the treaty, SDAG Wismut was determined to be exempt from taxation, with the exception of taxes on wages. The company activities were to be shown neither in the economic plans nor in the statistics; no customs controls were to apply either to the export of uranium or to the import of equipment and other material for Wismut.[32]

One of the benefits the GDR derived from the treaty was that the costs of the mining of uranium were to be carried jointly by the two governments. Tensions arose in the latter half of the 1950s, and then again in the 1970's, when the price of uranium was calculated. Ulbricht and Honecker took part in the negotiations, and tried to convince the Russians to take into consideration the rising production costs. Finally, paragraph 12 of the treaty provided for the transfer of assets no longer used by Wismut to the GDR. Land and other property was to be restored and cleaned according to the standards required by GDR laws. During the 1950s, little was done about the extensive ecological damage that resulted from the mining operations.

Maltsev and His Team

The life of the first director general of Wismut AG, Mikhail M. Maltsev, spanned an unquiet era between the time of the revolution in Russia and the emergence of the Soviet Union as a contender, with America, for world primacy.

Born in 1904 at a railway station of Nikitorka in Artemoska district, he attended a school for railwaymen until 1918. The railway was then a symbol of progress, and in a critical sense it defined Maltsev's place in life. Revolution and the civil war put an end to Maltsev's school years. He joined the Red Army when he was fifteen, fought in its ranks, and commanded a platoon until the end of the war in 1922.

Maltsev became an electrician at Scherkassi, joined the Communist Party when he was twenty-one, and was sent to the party school at Kiev. He served as a functionary in the youth organization, Komsomol, for three years, before returning to Scherkassi refinery as party secretary. He remained there until 1930, when Stalin's first five-year plan had just begun. Industrialization needed people like Maltsev, and the party gave him another chance to study, this time at the Dnepropetrovsk energy institute. He stayed there for five years, until 1935. It was a time when the industrial infrastructure of the Soviet Union was being laid down, some of it under the supervision of the Ministry of the Interior (NKVD). The NKVD mobilized the specialists, provided the materials, and built its own industrial empire with the use of forced labor. The empire needed specialists from both industry and the universities. Maltsev joined the NKVD after the conclusion of his studies.

The "Great Terror" began the following year and did not spare the NKVD. Nikolai Yezhov, its head between 1936 and 1938, was himself arrested and shot. "Yezhovchina" meant that nobody was safe. Some 1.6 million people were arrested in those years, and of these more than 680,000 were executed. The Terror was aimed at the educated elite inside and outside the party. In the first half of 1936 alone, 14,000 managers in the industry were arrested for alleged sabotage. Stalin wanted to create a bureaucracy dedicated solely to him.

Maltsev became a beneficiary of the mayhem. He belonged neither to the "former people" (*byvshi cheloveki*), the men and women connected with the old regime, nor to the newly established Bolshevik leadership. He survived the Great Terror unharmed. In the meanwhile, the forced labor camps grew fast. The numbers of convicts doubled between 1935 and 1941 to almost two million. Vast new construction sites came into being, and Maltsev served his apprenticeship as the master of forced labor in several of them. He moved from Volgograd to Kaluga and then to Briansk. Maltsev was appointed to his first commander's post at Kaluga, becaming one of the functionaries who ran the murderous gulag system. In March 1943, Maltsev took over the command of one of the largest NKVD camps in the Vorkuta coal district. He also served as deputy to the Supreme Soviet; his wife was the state prosecutor at Vorkuta. The Maltsevs thus enjoyed absolute power over the inhabitants of a province inside the polar circle.[33]

Maltsev's technical education and experience running large forced labor camps proved his ability to run field post no. 27304 in Saxony. In addition, he was the recipient of the highest state decorations, and in 1945 was raised to the rank of major general. Beria needed such men, well versed in the spirit of Stalinism, for the most important post-war project: the construction of the atomic bomb. It is probable that Beria himself chose Maltsev for the complicated task in Germany. At the beginning of January 1947, Maltsev came to Chemnitz to take over command of field post unit no. 27304. After the foundation of Wismut AG, he became its first director general and remained in the job until May 1951. Ivan Serov, Beria's proconsul in the Soviet zone of occupation, was Maltsev's only superior. The military authority in the Soviet zone had no powers to interfere in the uranium business. Maltsev had with him no detailed instructions from Moscow. The amount of uranium produced was the only criterion of the success of the industry.

In summer 1947, the uranium program in Germany was run by a board of Soviet directors, and its title referred to a branch of the Soviet colored metals company in Germany (Hauptverwaltung der Zweigniederlassung der staatlichen AG der Buntmetallindustrie in Deutschland). The company's seat was located in Aue until 1949, and then at Siegmar-Schonau near Chemnitz, a town that became Karl Marx Stadt. The personal composition of the board changed often, and consisted exclusively of Soviet officers.

After the foundation of Wismut AG, Lt. Col. Baraniuk, who ran the Personnel Department, and Lt. Col. Einbinder of the Information Department were best known to the public. "Department S" was the intelligence arm of the special committee for atomic questions and the NKVD. The department was run by Beria's associate, Bogdan S. Kobulov. Its task was "to detect and frustrate people suspect of espionage, terrorism and diversion, as well as defense against enemy activities, aimed at hampering the production program..."[34] The Soviet military employees of Wismut were supervised by a specially established intelligence department under the Minister of State Security, Viktor Abakumov. He was on bad terms with Serov, and tried to acquire more influence on the nuclear project. As a result of animosity between the two men, Abakumov's intelligence unit came into conflict with Department S.

In the first months of Maltsev's regime, the harsh conditions in the uranium business showed that he wanted to apply Vorkuta methods in Germany. He wanted to fulfill the target and the industry was run accordingly. But Maltsev soon found his bearings in the new environment. In remote Vorkuta, he ruled unrestrained by regard for the outside world, whereas in a populous region of Erzgebirge in the center of Europe, it was impossible to apply all the Vorkuta methods. Even so, his officers did not hesitate at the beginning to resort to practices such as withdrawing food rations from miners who did not fulfill their quotas, or the use of military tribunals in cases of alleged sabotage.

Until 1954, Wismut top management consisted only of Soviet civilians or military, whereas from 1950, the Germans could be found running the mines. The first Germans, who moved into general management in 1955, were Fritz Kroll, who became the First Deputy of the General Director, Josef Wenig, who ran the Labor Department, and Schröder, the head of the Personnel Department. Wismut offered German managers good living conditions, including cheap accommodation, family houses, as well as special shops from the mid-1950s. They also received better food rations. Privileged employees had university educations, and were technical-engineering staff, in a few cases company physicians, or they were state-decorated workers.

There is one name which repeatedly appears in contemporary eyewitness accounts: Johannes Schmidt, mining supervisor (Obersteiger) at Schneeberg.[35] Recruiting skilled workers presented the industry with immense problems. In the spring of 1946, Beria's secret service went talent spotting amongst the miners working for Sachsenerzberwerk GmbH. The local people kept on mentioning the name of Johannes Schmidt, who was nowhere to be found. Soviet counter-intelligence had arrested him and confined him to a special camp on the suspicion that he was an active Nazi. Maltsev personally appealed to Ivan A. Serov so as to secure the release of Schmidt and other skilled miners.[36] Schmidt was released and became the supervisor at Objects 2 and 3 and, later, at Object 9 in Oberschlema. His income was high, he was allocated a new flat and a car, and was given officers' rations. He knew that a suspicion of sabotage could mean a death sentence, and worked hard to became the "soul of the enterprise."[37] Schmidt came from a miner's family and attended the Zwickau mining academy. He had practical experience of a number of jobs in the mines of Harz and Thuringia. He moved around a lot during the economic crisis in the early 1930s, until settling down, at the end of 1933, in the Schneeberger Bergbau. He eventually became pit supervisor (Obersteiger), and joined the Nazi Party in 1937. He was in charge of the whole Schneeberg district in 1944, and after the war was accused of having treated the forced workers poorly. The company employed about 150 Frenchmen, Ostarbeiter (workers from east Europe), and Russian POWs. They had their rations reduced or were beaten when they failed to fulfill the target. As supervisor, Schmidt was responsible for their treatment.

His skills were in demand, so Wismut gave Schmidt a second chance. He did his job as best he could and kept out of politics. SED membership was, at the time, becoming essential for Germans keen to advance in their jobs. Schmidt and other "bourgeois specialists" who were not party members were sidelined. The management of Wismut AG continued to value them, but their careers were overtaken by SED members.

The German Party of Socialist Unity (SED) began to establish itself in the uranium industry in the summer of 1947. The new party came into being after the Communist and Social Democrat parties had merged a year earlier, when the principle of parity between them was agreed.[38] At the Party's II Congress in September 1947 it was clear that it would go down the Stalinist road. Open debate at the congress was replaced by homages to Stalin and to the Soviet Union. Immediately after the congress, Otto Grotewohl declared the Soviet Union to be the model for the SED.[39] As tensions between the Soviet Union and the Western Allies rose, the vision of a special "German road to socialism" faded out.

The party's recruitment drive resulted, by the end of 1947, in membership of some 1.8 million people. Most of them had not belonged to either the old Communist or the Social Democrat parties. The SED became an authoritarian party, run on the lines set down for the Bolsheviks by Lenin. Its organization was centralized and strictly hieratic, designed to implement the intentions of its leaders with a "will of iron." The party's members had no influence on its decisions, and anyone who stepped out of line was at once put down. The SED came to think of itself as the "avant garde of the working class".

Until the summer 1947, the SED was conspicuous by its absence in the uranium industry. Party leaders in Saxony had addressed themselves from time to time to the developments in Wismut, without being able in any way to influence them. Early in 1947, the mining districts became a closed military zone, a measure that also hindered access by the party to the workforce. In the mining districts, the SED was organized locally. The district committees, partly because of the demanding working routine of the miners and their daily struggle to find enough to eat, found it difficult to reach their members and convince them to take part in political activities.

The situation began to change in summer 1947, when SED groups began forming inside the company. In the autumn, special representatives of the Saxon party committee at Wismut AG were appointed. They were Kurt Böhme, Robert Kessler, Herbert Pomp and Ernst Wabra, and they formed an organization that became known as District Party Organization Aue II. As all four representatives were old communists, the Russians trusted them.

There were communists in the top ranks of the SED who had been active in the resistance against the Nazis, whereas the background of party members and the middle ranks of its hierarchy was more heterogeneous. In the regions, there existed small groups of communists, and a somewhat stronger representation of the former Social Democrat Party. The majority of new SED members had not belonged to any socialist organization before 1945, with the obvious exception of the National Socialist Labor Party, or NSDAP. Its former members were welcome in the ranks of the SED. Fritz Selbmann complained to the party representatives at Wismut that up to 90% of the party functionaries of the "first hour" were former Nazis.[40] Though District Party Organization Aue II questioned that figure, which appeared to have been plucked from the air, it was apparent that Selbmann had touched a sore spot.

In December 1948, the local organization at Silberstein bitterly complained to Lt. Col. Einbinder that several Nazis occupied high positions in the warehouses and in the SED itself. "Personnel policy pursued by Wismut AG in…Objects 2 and 3…is unacceptable to all the employees. We are especially…disappointed by the attitude of several officers of the Soviet occupying power at Wismut AG, who protect the fascist personnel, despite our arguments."[41] In September 1947, about 2,400 SED members worked at Wismut AG, of whom only a few took part in its political life.[42] By November 1949, SED membership in Wismut had risen to almost 21,000, or about 13% of all employees, a figure still considerably below the average for the whole country.

The occupying power allowed access to Wismut to one party only; the two middle class parties—the Christlich-Demokratische Union (CDU) and the Liberal-Demokratische Partei Deutschlands (LDPD)—were unable to establish representation inside the company. Political control over Wismut objects lay in the hands of the officers of the NKVD, who dealt with all important issues between Wismut AG and its German employees. NKVD officers took part in the important meetings of the SED leadership and controlled personnel changes within the party.

Soldiers of the Party

After the proscription of social democrats in the SED party organization at Wismut, it came to be run by functionaries whose loyalty to the party line was beyond doubt. At Wismut, party centralization and party discipline came to the fore more than anywhere else. The secretaries of Aue II were all communists who had resisted the Nazi regime. The party at Wismut regarded itself as an elite group in a party, dedicated to the Soviet Union and working for it in a vital industry.

In December 1947, Kurt Böhme became the first chairman of Aue II. Böhme was born in Deuben, Dresden in 1913, and grew up in a working class family. He joined the Communist Youth League in 1928, and the party in 1930. In the same year, he visited Moscow for the first time on the invitation of the Communist Youth International. After Hitler came to power and the KPD [Kommunistische Partei Deutschlands] was outlawed, Böhme worked for the underground organization. He was sentenced to four years and seven months' imprisonment in 1936. After the war he became a political secretary of the KPD in Freital, and early in 1947 he was appointed instructor in the cadre department of the Communist Party District organization.[43] There can be no doubt that Soviet political officers had a decisive say in Böhme's appointment, and the same can be said of other leading functionaries' appointments to the Wismut party organization. Böhme, who had first visited Moscow when he was seventeen years old, was regarded as being absolutely reliable by the Soviets in charge of Wismut.

The special status of Wismut as in fact a military enterprise under NKVD control made Böhme's position virtually unassailable. The connections between the Wismut party organization (Aue II) and the regular district committee of the SED at Aue (Aue I) were slender. Aue II was unpopular not only at Aue, but at Annaberg, Auerbach and Marienberg as well. Eventually, Böhme had to defend himself against charges that he was the chairman of a "party within a party."[44] He was indeed closer to the Soviet management of Wismut than to his own party.

The party functionaries resented the secretiveness of their comrades at Wismut as much as they resented their privileged access to scarce goods and services. They complained of "Bonapartist attitudes" and "hollow bathos" of the party functionaries at Wismut AG.[45] Their complaints made no impression upon the Soviet management. Party leaders in the end accepted the special position of the SED organization at Wismut, which was put under the direct control of the Politbureau.

The drive to make the SED a Stalinist party sped up in the summer of 1948, and the former Social Democrat Party members came under inquisitorial pressure. The so-called "Schumacher people" suffered most; anybody suspect of "social-democratism" was liable to be arrested. The mayor of Johanngeorgenstadt, Alfred Franz, fled to the West in the summer of 1949. The deputy chairman of the SED district committee and a former social democrat, Konrad Vesely, was expelled from the party at the beginning of 1950.

At each meeting of the Aue II committee, many cases of enemies of the party were dealt with. The suspects were not only the social democrats, but also the communists who opposed the Stalinization of the party.[46] Böhme himself came under pressure at the beginning of 1951. He left the uranium industry after he was relieved of his post in March. He went to the party school in Moscow and subsequently became an officer in the police (Kassernierte Volkspolizei, or KVP). His Soviet friends prepared for him a soft landing; he later became deputy head of the political department of the Ministry of National Defense, and then the military attaché in Moscow. He dropped out of the party hierarchy in 1954, when he lost his function as a candidate member of the central committee.

Horst Dohlus, hairdresser by trade, became for a short time the head of the party organization at Wismut. The central committee then chose Günther Röder, who was a coppersmith by trade. In contrast to his predecessors, Röder, born in 1923, had not taken part in the resistance during the war. He joined the Communist Party after he left an American POW camp. He became a youth movement functionary in his hometown of Zeitz. He studied at the party high school from 1949–1950, and became the head of the engineering department in the central committee of the SED in Berlin. Ulbricht gave Röder his support.

Röder appointed his own men to offices in the party organization, a former member of the Waffen SS among them. He survived criticism on this count but was unpopular among the miners for trying to raise production norms.[47] Serious accusations of corruption against Röder reached Ulbricht, who had himself run into severe trouble in consequence of the uprising in the GDR in summer 1953. Ulbricht let his protege go early in 1954. Röder and five of his colleagues from the SED organization at Wismut appeared before a party tribunal. Röder was dismissed for "attitudes harmful to the party," and became a worker at a machine tractor station. Alois Bräutigam, a master bricklayer and miner, followed Röder as the head of the party organization at Wismut, where he stayed until 1958.

In June 1950, the uranium industry was taken out of the IGB (Industriegewerkschaft Bergbau), the miners' trade union, and IG Wismut was founded. It was the result of the joint effort by trade union officials and by Soviet political officers and against the opposition by the IGB. The process separating IG Wismut had started at the end of 1949, when the SED formally created its own organization in the uranium industry.[48] IG Wismut thus referred directly to the national executive committee of the FDGB, the trade unions congress, and Richard Leppi became its first chairman. The son of an Upper Silesian miner born in 1904, Leppi was a miner as well. He joined the Communist Party and in 1931 became the Secretary for Agitation and Propaganda in the Upper Silesian district. Under Nazi rule, Leppi spent several years in prison and concentration camps, and in 1941 he was sent to work as a miner at Freiberg.[49] After the war, he belonged among the party activists at Wismut, who had helped to develop the party organization. In April 1948, he took over the leadership of the district committee of IGB, replacing Richard Gründel.

The FDGB had enjoyed, until 1950, some independence from the SED, whereas the Wismut trade union came, under the influence of the SED from the very beginning. The chairmen of the Wismut trade union organization were appointed and dismissed by the SED. The usage was introduced in 1948, at a time when the trade unions still enjoyed the status of a more or less independent organization.

Leppi soon discovered the limits of his ability to influence events. He tried to pursue independent policies, and neither the SED leadership, nor the management of Wismut, would tolerate that. He did not oppose the party line or the Soviet management, yet he was not prepared, without reservations, to accept its every whim. Soon after his appointment, Leppi opposed the introduction of Sunday shifts, and advocated voluntary

Sunday work. The trade union was defeated and, in the spring 1950, Sunday shifts were introduced. The decision was supported by the newly established IG Wismut, which thus lost some of the miners' confidence.

Leppi's enemies tried to have him sacked and, in the middle of 1949, the Party Executive Committee for Saxony decided to dismiss him. The decision was simply ignored by Aue II. About a year later, another opportunity to get rid of him emerged, after he slapped a woman at the party school. His enemies, including Kurt Böhme, used the incident to give Leppi the final push. He was dismissed in December 1950, and a year later he had to leave the party because of "shameful conduct." For Leppi and his wife, who worked for the Saxon party committee, the incident was fatal. She lost her job and the party ordered her to leave her husband and children.[50] Richard Leppi achieved a postponement only of the complete loss of independence by the trade unions.

Among SED leaders, Fritz Selbmann stood out as one of the strongest personalities. Born in 1899 at Lauterbach in Hessen, Selbmann was the son of a coppersmith. He trained as a miner and joined the Communist Party of Germany in 1922. He commanded communist paramilitary units in the Ruhr district in 1925, and became one of the most influential party functionaries in the coal region. The party sent him to Moscow to study at the Lenin School in 1928, and after his return Selbmann represented the communists in the Rheinland provincial diet before becoming the head of the party organization in Saxony and Upper Silesia. He was a member of the Prussian diet between 1930 and 1932, and was arrested in April 1933, after Hitler took over power. Selbmann was sentenced to seven years' imprisonment in 1935 for "plotting high treason." During the war, he passed through several concentration camps. He partially lost sight during nine months' solitary confinement in darkness and fled the Dachau camp towards the end of the war.

Selbmann worked in Leipzig after the war, on the revival of the KPD. He was briefly the president of the labor office before he became the Minister of Industry in Saxony, which was economically the strongest land in the Soviet occupation zone. Selbmann pressed for the introduction of a planned economy and the nationalization of industry and trade. He sometimes went too far in his endeavor to improve society, even for the Russian masters of the zone. He fought hard, on the other hand, the Soviet demands for reparations. He tried to reduce them because he was convinced that they held back economic revival. The Soviets appreciated Selbmann's self-confidence, and offered him the post of the minister of industry for the whole zone. He turned down the offer as he wanted to complete his job in Saxony. When the GDR came into existence in 1949, Selbmann became the minister for heavy industry. He was responsible for prestigious construction projects, including the Eisenhüttenkombinat Ost, the iron foundries conglomerate.

In 1952, when the Hungarians and the Czechoslovaks were seeking, with the help of the Soviet advisers, the "enemy within," it seemed that the SED as well would organize Stalinist purges. A former member of the Politbureau, Paul Merker, was described as a "subject of US financial oligarchy" and remained in custody until 1956. In May 1953, Franz Dahlem, one of the last great rivals of Ulbricht, was excluded from the Central Committee.[51] It is likely that Ulbricht considered Selbmann as a suitable defendant in a show trial. The minister had apparently made mistakes during the construction of the iron foundries and his contacts with "bourgeois experts" bordered on sabotage. Before any trials could be staged, Stalin died in March, followed by the uprising of 17 June 1953. Those events apparently saved Selbmann.

He was the only high state functionary who tried to discuss grievances with the workers during the uprising on 17 June. The men had heard too many unfulfilled promises. In the end, economic compromise combined with Soviet assistance helped the SED to stabilize the situation. The cancellation of the reparations debt as of 1 January 1954 and the transformation of Wismut into a German-Soviet joint stock company helped to improve the economic situation in the GDR. Selbmann, who had negotiated the transformation of Wismut, became the chairman of the board of directors of the new company. He tried to barter uranium for higher deliveries of raw materials from the Soviet Union, or to get a better price for the metal. Selbmann achieved partial success.[52]

At the beginning of 1955, Ulbricht decided to divide the Ministry of Heavy Industry into three parts. For the ambitious Selbmann, it was a severe setback. As the Minister for Mining and Metallurgy, Selbmann played the decisive role at the beginning of the atomic industry in the GDR.[53] In 1955, he became the head of the Office for Nuclear Research and Technology, and in 1956 he conducted negotiations with Moscow on the construction of an atomic power plant in the GDR. Ulbricht wanted the Office for Nuclear Research to become a ministry; Selbmann opposed the idea, as it resembled too closely a similar office in Bonn, with Franz Josef Strauss as the minister.

Selbmann's political career faltered towards the end of 1957, and Wismut provided the setting for Selbmann's fall from grace. SED leaders liked visiting Wismut headquarters at Chemnitz. They felt free there from the restraints of Berlin, and speculated on political developments. Gossip and drink flowed freely. At a time of internal party strife, it was an unsafe pastime. At a party at Chemnitz on 9 December 1957, discontent with Ulbricht's policies came out into the open.[54] Fritz Selbmann as well as Gerhard Ziller, the Central Committee Secretary for the Economy, were explicit in the matter. Their choice of the time and the place was ill-considered, as there were, in the management of Wismut, several dedicated supporters of Ulbricht. He summoned members of the Politbureau to meet him on 12 December, and Selbmann and Ziller were ruthlessly hunted down. Ziller took his own life a few days later.

Other real and imaginary opponents of Ulbricht soon had to leave their offices.[55] Karl Schirdewan, one of the last two former concentration camp inmates (Selbmann was the other) among the leaders of the party, Fred Oelssner, the party theorist and Ernst Wollweber, the Minister of State Security, were excluded from the Politbureau on the grounds of "factionalism." They suffered mild punishment: they were offered jobs whose political insignificance was compensated for by high remuneration.

The purge at the top was followed by a cleansing of the entire party organization. SDAG Wismut was also affected. Alois Bräutigam, the first secretary of the SED at Wismut, left to become district party secretary at Erfürt. Deputy Director General Schröder, was invited several times to Berlin to explain himself.[56] He was dismissed from his post, and the Politbureau charged him, on 11 March 1957, with the misuse of power and chicanery.[57]

Fritz Selbmann in the meanwhile soldiered on. He made a self-critical speech at the fifth SED congress to save himself from taking a fall. He retained high positions in the state economic apparatus until 1964, but had but lost much of his former influence and retired at 65. He then tried his hand at writing. In his papers, Wismut, the most important enterprise Selbmann was responsible for, was mentioned only in passing. The uranium industry may have been outside his main interests, or Selbmann was unable to break the secrecy that surrounded it.

The power struggles inside the SED and the Wismut management remained hidden from the public and from the miners. They in any case had other concerns, including, in the years after the war, the struggle to survive.

Migration into the Erzgebirge

The possibility of using the German workforce as a form of future reparations was discussed by the Allies during the war.[58] The economic value of labor reparations was thought to be less important than their political effect. Despite some objections, the Americans agreed to labor reparations at Yalta in February 1945. At Potsdam in July and August, the subject remained shelved, which did not mean that the Soviets had forgotten about it. They insisted on the formation of German "labor battalions," and the Allied Control Commission released a declaration on the subject on 20 September 1945. It stated that Germany was bound to supply "workforce, personnel and specialists and other services to be employed in Germany or elsewhere, as Allied representatives may order."[59]

In October, the Soviet occupation authority issued order no. 43 on the provision of "labor force for industrial enterprises," including the threat that those unwilling to work would be refused food ration cards.[60] From then on, German labor exchanges in the Soviet zone controlled the movement of labor.[61] A supplement to the order followed on 29 November, as order no. 153, which allowed compulsory direction, by the occupying power, of unemployed persons without regard to their qualifications.[62] There was no appeal against the decisions of the occupying power.

The practice of compulsory labor recruitment was confirmed by Allied Control Commission (ACC) order no. 3, issued on 17 January 1946. The uranium industry in the Soviet zone recruited most of its workers on that basis. In the following months, British members of the control commission pressed for the introduction of compulsory service in the coal industry.[63] The British and the French military authorities used the order so as to provide labor for the black coal industry, which was essential for industrial reconstruction. In the Ruhr mining district, the method of compulsory labor failed and three out of four workmen left their jobs.[64] Where compulsion failed, improvements in food supplies succeeded and, in the autumn 1946, the British authorities concentrated on improving the situation of the miners.

The Soviets, who knew of the disappointing results of compulsory labor in the coal industry in the western zones, continued to use it, on a large scale, between 1945 and 1949. The enterprises contributing to reparations payments—the wharfs and the machinery and chemical industries—all employed forced labor. The uranium industry recruited more workers by compulsory orders than any other business.

By 1948, all the key positions in the Administration of Labor were held by SED members.[65] On the initiative of the Saxon SED, the state government established, in August 1947, a branch office of the Ministry of Labor at Aue. In 1949, it became known in 1949 as the "Special Department for Ore Mining" (Sonderabteilung Erzbergbau). It was headed by Max Weber, a miner.

The German authorities had little notion of how much labor would be needed. The personnel department of Wismut gave the Russian occupation authorities some rough estimates of the requirements, which depended on the progress of the search for uranium. Enquiries from the Saxon state government, and from the Deutsche Wirtschaftskommission (DWK), as to the possible duration and extent of the uranium industry remained unanswered.[66]

The military administration in Saxony began issuing compulsory orders in April 1946, when the labor exchange at Aue was required to provide 800 workers for Johanngeorgenstadt.[67] Despite the transfer of workers from neighboring enterprises, and the assistance of the Red Army, the demand was not met. After this initial failure, Wismut started its own recruitment drives. The refugee camps in particular became the favorite hunting ground for NKVD officers, and for German middlemen who received a premium for each worker. A Soviet officer sent 800 refugees from a camp at Hoyerswerda to Aue by rail. "The people were not (medically) examined, nor were they miners by profession...the transport consisted of amputees, sick people and old men and women."[68] Maltsev put a stop to independent initiatives in the autumn 1947.[69] The head of the migrants department at the military administration, Lt. Col. Volodin, issued an order on 23 October 1947, which addressed itself to the recruitment practices of Wismut. Interference by NKVD officers in the work of the German labor exchanges was to stop forthwith.[70]

Apart from spontaneous recruitment initiatives, the recruitment of forced labor for the uranium industry by the labor exchanges increased. In the last quarter of 1946, Saxony was required to provide 4,400 workers for the industry. Despite protests by the head of the state labor exchange, Dr. Heinze, the military authorities persisted in the recruitment drive. Miners were withdrawn from the coal districts of Oelsnitz and Zwickau. They formed, together with miners from Sachsenerz-Bergwerk AG and the refugee miners from Silesia, the core of Wismut's skilled workforce.

Wismut AG informed the military of its requirements, which in turn passed on the order to the German labor authorities. Many more workers were usually asked for than were actually required. Personnel officials at Wismut hoarded labor because compulsory contracts ran for 6 to 24 months, and replacements had to be continually sought out.

Labor exchanges were obliged to fulfill Soviet demands at any cost, and were assessed according to their ability to meet the set requirement. The tense situation was reflected in a letter of 4 September 1947 from the Minister of Labor of Brandenburg to a labor exchange: "The recruitment of voluntary labor is to be preferred, but if the numbers were insufficient, they have to be supplemented by compulsory labor. Labor exchanges have to inform by telephone the state labor office straightaway, on a daily basis, how many workers were provided for the ore mining industry."[71]

In their anxiety to meet the targets, labor exchanges press-ganged many thousands of unsuitable workers into the uranium industry. At the same time, the regulations issued with the order no. 153 provided a way of avoiding the draft. All the prospective miner had to do was to get a medical certificate. Doctors in the mining districts had to make a decision between obliging the labor exchange and protecting their patients. They examined the new arrivals and placed them into one of three categories: workers suitable for mining underground, workers suited for jobs on the ground only, and people unfit for mining. Because of food shortages, up to about a quarter of the men examined were placed in the third category.[72] They sometimes tried to browbeat the doctors to declare them unfit for work in the mines. If the doctors let too many men go, they ran the danger of being arrested by the Russians, as was a physician in Annaberg in June 1947.

Complaints about compulsory employment varied. Recruits who were transferred to Wismut from another job often referred to order no 3, which applied to unemployed men only. The DZVAS (Deutsche Zentralverwaltung für Arbeit und Sozialfürsorge) declared that, in urgent cases, the order could be used in the case of employees as well. The president of the Thuringian high court wrote to the prime minister that "The labor office…has been compelled by the occupying power to provide workforce for the mining industry which could not be recruited in full numbers on the basis of the existing laws. A direct order from the occupying power to use employees on a compulsory basis in case of need is however...not available. Yet the labor office has no other way out than to extend compulsion to suitable employees."[73]

In the meanwhile, demand for labor in the uranium industry grew by leaps and bounds. On 1 August 1947, the Soviet military administration released a secret order for the recruitment of 20,000 men for the industry. The Saxon Ministry of Labor called an emergency meeting. The districts of Aue, Annaberg and Marienberg had to be alerted and be ready to receive the new miners. However, accommodation presented the local authorities with an insoluble problem. On 5 August 1947, the Saxon government issued a ban on any other migrants into the three districts. Only miners were allowed entry. "Whether their families would follow later depends on how long the ore mining at Aue will continue."[74] The town Aue was commonly known as the "gateway of tears." The local labor exchange was the last stop for the recruits before they were sent to their places of employment. Fresh recruits arrived at Aue in large groups and under guard. They were sent from the railway station to Auerhammer, an inn converted into a transit camp with appalling sanitary and hygienic conditions.[75] Registration at the labor exchange took a few days, and then the workers were sent to the Wismut objects. They reported to the personnel department and to the accounts and insurance offices. A medical examination was carried out, and the recruits received permission to deregister at their home food office. After doing so, they received the improved Wismut rations. If they had not found private accommodation, they were sent to the transit camp at Eibenstock.

After the breakdown of the volunteer recruitment campaign, the occupying authorities turned to the local organizations of the SED and, eventually, to the party leadership. In October 1946, the first consultations took place between the military authorities in Saxony and the SED organization.[76] In addition to the SED, trade unions and youth organizations were used in the drive to recruit volunteers.

Labor and Its Shortage

Compulsory orders resulted in a wave of flights from the uranium industry. Tens of thousands of people ordered to work for Wismut fled to the western zones of occupation in Germany. Bad working conditions as well as the "fear of the Russians" upset the new miners, and there was a widespread rumor that the uranium industry was merely a waystation to Siberia.[77] The living conditions in the uranium districts were far from attractive, and the accommodation crisis reached its peak in 1947–1948.

The German authorities tried to repair the bad reputation of Wismut AG, and questioned the truth of Western press reports concerning the "Aue uranium hell."[78] Despite press campaigns and the effort to provide sufficient food supplies, the flights continued. As propaganda failed to stem the flood of refugees, the authorities and Wismut tightened up the controls. The miners had to hand over all their personal documents to the Wismut administration, receiving works passes in exchange. It helped Soviet and German policemen keep an eye on the workers, and there were frequent checks of personal documents in towns and railway stations.

In October 1948, the president of the DWK central office for labor and social welfare, Gustav Brack, instructed the state governments of Mecklenburg and Brandenburg that the transports of workers to the uranium province should bypass Berlin.[79] The city, divided into four sectors, was a popular escape route, and the former Wismut miners living there were ready to help. DWK reported the incidents to the military government, and a new route for the transports was worked out.[80] Despite the new provisions, the government of Mecklenburg registered the flight of some 2,000 recruits in January 1949.[81]

It is impossible to ascertain the exact number of men who fled before they reached their place of employment. According to several estimates —and they may well be overestimates—of 3,000 workers directed to Annaberg during 1947, about 2,700 escaped.[82] According to the figures of the State Ministry for Labor, the military authority procured 52,000 workers for the uranium industry, most of them from Saxony, before the end of 1947.[83] It is not certain how many of them actually joined Wismut. The data of the personnel department of Wismut indicate that it had scarcely 19,000 employees. This means that at least every other man either fled or was categorized as being unable to work in the mines.

Table 5. Fluctuation in the Numbers of Directed Labor at Wismut in 1947

Required numbers of directed workers	52,500	
Actual numbers	43,600	100%
Unsuitable workers	6,500	15%
Sick workers	3,500	8%
Escapees	14,600	33%
Actual number of employees	19,000	44%

Because of the failure to fulfill the production plan, the general director's office sought to justify itself in the annual report for 1949. Much of the blame fell on the inexperienced workforce.[84] In 1949 alone, 109,000 new workers were hired, and 48,000 left Wismut within a year. Termination of contract was the most common reason: 25,000 men, or 18% of the workforce were involved, and 6,300 men, or 5%, fell sick. There were in addition 17,000 men, or 12%, who did not complete their contracts. In other words, they fled their employment, either because they were not up to it or because they could not stand the conditions in the industry.

In the first three years of the uranium industry's existence, about 50,000 men fled and about 15,000 workers left employment for reasons of health. The development of the uranium industry was demanding for the post-war society. In addition, mass flights of workers to the West further damaged the poor reputation of the occupying power and discredited the SED. The uranium industry was condemned in hundreds of articles and wireless reports in the western zones of occupation. Despite the wild exaggerations contained in reports about the conditions in the uranium industry—or perhaps because of them—the reports had a direct effect in the Soviet zone. They strengthened the peoples' distrust of the uranium industry, as well as their determination to leave the zone.

The Soviet administration was concerned about mass flights from the uranium province as early as 1947. Col. Tulpanov, the head of the propaganda department in the Soviet zone of occupation, complained in a letter to Suslov, a member of the Politburo,[85] that even workers with left-wing sympathies and SED members preferred to flee to the West than be forced to work in the mines. In the West, compulsory recruitment in Ruhr coal mines had ceased in 1947, and the Soviet authorities opted for a more flexible approach to the procurement of labor for the uranium industry.

It was not easy for the NKVD to pay attention to local conditions and give up their ingrained habits. It had controlled the gulag system since its inception, and the use of convict labor for the most strenuous tasks in the key projects of the five-year plans became an established practice. It was therefore only a matter of time before someone in the uranium industry remembered the pool of prison labor.

In autumn 1946, the Soviet Ministry of the Interior wanted to establish a camp with some 3,000 internees in the vicinity of the mines in Saxony.[86] The Soviet internment policy was meant to remove former Nazis from public life and stamp out opposition to the new regime. The internees were to travel to the Soviet Union as "reparation labor." Yet the hopes of the Russians remained unfulfilled. Although at least 150,000 men passed through the camps, after a few months' internment only a small part of them were still able to work. Poor food and appalling living conditions took their toll.[87]

In the second half of 1948, the internees were to be released from the Soviet labor camps and employed at Wismut. A minimum two-year contract with Wismut was to secure an early release. Some 24,500 internees and 2,500 POWs were to be handed over to Wismut AG. The plan, developed by Sergei Kruglov, the Soviet Minister of Interior, was never realized. Only the POWs, who were about to leave Jáchymov, were sent to Wismut AG. The reasons for the failure of the plan can only be speculated on. Before the spring in 1950, of the 158,000 internees in the special camps in the Soviet zone of occupation, some 30,000 persons had been deported to the Soviet Union. Some 43,000 internees died in the camps and 46,000 were released.[88] It meant that nearly all the remaining internees, according to Kruglov's plan, should have gone to work at Wismut.

The Soviet policy of internment came under review in the summer of 1948. In Moscow, the Politbureau agreed on 30 June to release most of the internees and abolish many of the special camps. The decision was made in connection with a new approach to the de-nazification policy, but as Cold War tensions increased, de-nazification became sidelined. From summer 1948, Wismut AG had to look elsewhere for extra labor.

The Germans who fled or who were expelled from the countries of Eastern Europe offered a promising source of labor. The term "Heimkehrer" was a euphemism for the Germans who, in a large wave of reverse migration occurring immediately before and after the war, were forced to leave their homes in Poland, Czechoslovakia and territories further east.. The recruitment of the "returnees" played an important role in the considerations of the personnel department of Wismut. There were skilled miners in particular among the refugees from Upper Silesia. but the departures of the Germans from Eastern Europe had been so chaotic that it was at first impossible to identify the specialists among them. They found Wismut themselves: some came as volunteers to start a new life in occupied Germany. Beginning in autumn 1946, larger groups came with the "miners' transports" from Upper Silesia or Bohemia.[89]

Representatives of the DZVAS visited the camps, and posters appeared in prominent places: one promised that "Germany is waiting for you!" The posters stressed the opportunity to earn good wages, claiming that the industry had "...the best food supply conditions, deliveries of textile goods and payments in kinds of cigarettes and alcohol."[90] The campaign was deliberately aimed at those refugees who had lost their families. Most of the returnees from the Soviet POW camps were in a poor physical state. The Soviets released the sick men first, as they could not be used in the Soviet Union. For instance, it was estimated that, at the returnee camp at Fürstenwalde, only 10% of the inmates were capable of any kind of work, and still fewer could be employed in the mining industry.[91]

In trying to meet the targets, the recruiting officers disregarded the physical condition of the returnees, and sent them to Wismut without medical examination. "On 4 December 1947 50 homeless returnees—who had been found suitable for work in the mines at the camp in Pirna/Sonnenstein—were examined at the Oberschlema surgery for suitability for work in the ore mining industry, and 36 of them were found to be unsuitable. They suffered from severe undernourishment, weakness of the heart muscles, etc. They insisted that they had been only superficially looked over..."[92] There remained an unanswered question: who wanted to work for the "Russians" on a voluntary basis? Many recruits turned their backs on the uranium mines at the earliest opportunity and fled.

The results of the first recruitment campaign among the refugees and the returnees were disappointing. Between July 1946 and July 1947, the transit camp at Frankfürt an der Oder registered about 218,000 returnees, among whom there were about 38,000 homeless people. Only 58 of them agreed to work for Wismut.[93]

In February 1948, the Saxon Ministry of Labor carried out a survey of workers directed to the uranium industry. Between October 1946 and December 1947, 43,590 skilled and ancillary workers from the whole of the Soviet occupation zone were passed on to Wismut.[94] There were between 1,900 and 2,236 "homeless returnees" (heimatlose Heimkehrer) among them, or about 5% of the workforce.

Conditions in the uranium industry also acted as a deterrent. Accommodation in the dormitories was compared to Soviet POW camps.[95] Weak and sick people were dismissed and forced to find another job. The communities adjacent to the uranium districts did not want to accept "such human wrecks," as the district council at Glauchau complained.[96] The government of Saxony condemned driving "such unwanted intruders" from one place to another.[97]

Skilled miners had the best chance to succeed at Wismut. Many came from Upper Silesia in 1948 and 1949, and provided the majority of pit supervisors (Obersteiger). One of them, Max A., described his first impressions: "Our transport consisted of about 1,000 people, among whom there were about 350 miners, whom we met at the railway station on 5 July (1948). I was directed to Johanngeorgenstadt mines...I became a driller's (Hauer) mate; he was about 17 years old and presented himself in a superior voice as a Hitlerjugend member. He could not drill, I had to teach him..."[98] Max A. later became a supervisor (Steiger). His working party prospered: production rose and the miners received special rations, the so-called "Stalin parcels." He was sent to a mining school at Freiberg and became a drilling instructor. He also took part in the public life of a new community of Rittersgrün, and in October 1950 was elected to the local council, where he served until 1974. Since 1953, he was under medical supervision, suffering from silicosis.

Observers in Germany and abroad were amazed when they discovered that so many women were employed in the uranium industry. An American press release in August 1950 spoke of "...the many women and girls who carry out the heaviest work. They are commonly known as 'ore angels' (Erzengel) and 'miner women' (Bergfrauen). Many of them perform every kind of men's jobs underground, with the exception of drilling."[99]

The reports on the employment of women, all of them drafted by men, contain hints of the creeping immorality that affected the mining districts. The women apparently came to Wismut not because they wanted to do a day's honest work, but because they wanted to pull money out of the mens' pockets. In a report September 1948 from the DWK's Labor Department, it was stressed that "Among the women who come to Aue voluntarily, the greater part are morally and ethically corrupt."[100] The women were also blamed for the sharp rise of venereal diseases in the mining districts. "Doctors in Aue say that about 60% of the incoming women are infected with venereal diseases. At the time of the workers' arrival a thorough examination is impossible, as there is never enough time...As an urgent measure, the state labor exchange of Saxony will send a telex to all other state labor exchanges in the zone to the effect that, as of 25 September 1948, no worker should be sent to Aue without a certificate of an examination for venereal diseases.'[101]

None of these reports however mentioned the exceptional situation, namely that there existed a mining enterprise in Central Europe that employed up to 30% women. For the Soviet occupation authorities, on the other hand, employment of women underground was nothing new.

In the course of industrialization of the Soviet Union, women were employed for the first time in jobs traditionally performed by men.

Mining meant hard work underground, and had always been a male preserve. Women in the mining industries usually worked as typists or administrative personnel. In Wismut, women worked underground and assisted in the mining operations as radiometrists, engine drivers, or in other jobs. Many labored above ground on the slag heaps, washing the ore, or in the lamp rooms and cloakrooms. They never reached the higher rungs of the company hierarchy. The taboo on employing women underground, which had survived until the end of the war, was broken by Wismut AG.

Many protests were made against the employment of women by the local SED organizations and the government of Saxony. The protests concerned women who performed heavy work, such as moving boxes containing pitchblende. The German authorities were also concerned with the accommodation offered to women in the uranium industry. Conditions in the transit camp of Auerhammer were appalling: "This camp is so overcrowded that the girls take turns in the same beds."[102] The civil administration of Saxony proposed construction of accommodation for single women and youths. The plan could not be realized, nor did the authorities succeed in putting off the recruitment of women until suitable accommodation was available.

In relation to the males, the proportion of women volunteers was initially higher, and better rations probably attracted single women. In the fourth quarter of 1948, the proportion of women in the workforce was put at 30%. Reliable figures exist for Object 2 at Wismut, where 1,327 women worked at the end of December 1949. The total number of employees at the Object was 22,407, of which 7.6% were women.[103] The actual number of women may have been higher at that time, though it would not have exceeded 10%, or the same proportion of women employed by the brown coal industry in 1950.

For the sake of comparison: in the black coal industry and in salt mines at the time, the proportion of female employees was 4%. The high proportion of women at Wismut and the brown coal industry can be accounted for demographically. Many millions of men had been killed, or were still prisoners of war. After the abolition of Nazi labor legislation, there were no legal barriers against the employment of women in mining. The labor exchanges could have never met the high demands for labor in the uranium industry had they not turned to women who were seeking employment.

In October 1947, order no. 239 banned the employment of women underground, but it took about a year before it was actually applied in all the pits. Women still worked underground at the beginning of 1949, when they either left the industry or remained above ground.

The Soviet Secret Police and Its Assistants

The NKVD management of the Saxon mines initially opposed, for security reasons, the employment of former Nazi functionaries in the uranium mines. They feared espionage and the rise of a Nazi underground movement. Former SS, SD, SA and Gestapo members were excluded from employment in the uranium industry, as well as persons "who participated in the crimes against the USSR."[104] The policy was upheld for few months only.

This created some bizarre situations. While the Soviets did their best to promote service in the uranium mines as honorable employment, the Germans considered it to be a punishment. The declared policy of Saxon labor exchanges was to send former members of the NSDAP to the mines, a policy that was opposed by Soviet managers.[105] The management at Johanngeorgenstadt and Schneeberg refused to employ former Nazis,[106] and the labor exchanges were criticized for regarding compulsory service in the industry as a penalty. A refugee bitterly complained to the labor office, explaining his resignation from the SED by his experiences in the uranium mining industry: "It would be right to employ Nazi swines in the mines, so that they would realize what they have caused."[107]

With the hectic growth of Wismut, the criteria for employment slackened. The pit bosses could not afford to observe the rules: production targets had to be met, and thousands of extra workers were needed. In contrast to other Soviet-owned enterprises, Wismut had no history behind it, and the workforce was recruited from all walks of life and every part of the Soviet zone of occupation. It offered its employees the chance of avoiding de-nazification procedures.

It is hard to establish how many former NSDAP members worked for Wismut, and information on the political adherence of the Wismut employees is scarce. A picture of the situation at the two Wismut pits in November 1947 exists, but it is a blurred snapshot, as the information comes from low-quality questionnaires.[108] 13% of the miners put down the SED as their political party, and 12% admitted their membership in the former NSDAP; it may be assumed that there were some Nazis in the SED organization. About a half of the workers in the two pits were trade union members.

At Object 1 at Johanngeorgenstadt, the majority of the miners were thought to have been Nazis in the first post-war years.[109] The officially tolerated organizations, CDU (Christlich-Demokratische Union Deutschlands) and LDPD (Liberal-Demokratische Partei Deutschlands), were banned from Wismut from the very beginning. The idea that there prevailed two strong groups among the uranium miners, members of the SED and former Nazis, is misleading. The NSDAP no longer existed, and its members wanted to forget the past. There were only a few resolutely recidivist Nazis among the new miners. Most of the former NSDAP members tried to survive as best as they could and come to terms with the new situation.

Protests against the employment of the Nazis were the order of the day in all the Allied zones of occupation. Former Nazis who succeeded in slipping into Wismut's administration, managed the miners' camps or made good in business became the chief targets of discontent. In hard times, trade in scarce goods in particular offered unusual opportunities for exercising influence. The camp in Zwickau district, Silberstrasse, was run by Kuschminder, a former Nazi who enjoyed the trust of the Soviets. As late as December 1948, the SED organization at the camp asked Lt. Col. Einbinder, the head of the personnel department of Wismut, to remove him.[110] As far as the SED was concerned, even the cleaning women had to have a spotless past. SED enterprise groups (Betriebsgruppen) resolved that "Personnel policy of Wismut AG in the two camps in Objects 2 and 3 is opposed by all the functionaries...We are especially...disappointed by the attitude of some of the officers of the occupying power at Wismut, who protect the fascist workers, despite our protests."[111] No record of Einbinder's reply is extant.

At the end of 1948, the mayor of Annaberg took it for granted that former Nazis were obliged to work underground and that they formed the majority of the workforce.[112] As long as they worked well, nobody minded them. As a rule, conflicts started when it came to the allocation of premiums and of other privileges. For a long time after the war, local officials blamed former Nazis for social tensions, corruption affairs and rumors. In a December 1951 report by the headquarters of the German People's Police (DVP), it was stated that in "the mining districts there live a comparatively high number of former active Nazis and officers of the fascist Wehrmacht...In addition to their general inimical stance there exists serious danger...that they will try to use existing difficulties for the sake of provocations."[113]

The former NSDAP members were generally regarded as security risks and kept under close supervision. In one of the districts, police looked after 27 "reactionary and enemy elements" at the end of 1951.[114] They were largely former NSDAP and SS members, as well as former prisoners in the Soviet special camps, who had found employment at Wismut. They were accused, among other things, of having organized meetings of veterans, and were thought to be "fanatic enemies of GDR." The constabulary at Marienberg worked on 363 cases, of whom 199 were classed as "enemies of the democratic order."[115] Six years after the end of the war, the police still used categories such as "fanatical enemies," or "adherents of the old order."

From the beginning, the Soviets introduced a strict security regime at Wismut. Guard duties at the pits and other objects, as well as the protection of the uranium province, were taken over by special units of the Soviet Ministry of the Interior (NKVD or MVD), accounting for up to 15,000 men.[116] They were under the command of the MVD chief at Wismut, Capt. Raikin. There were in addition heavily armed military units stationed in the uranium districts. They were treated as a closed military zone. All approach roads had military checkpoints, where special rights of passage applied. The mines were surrounded by wooden fences and watch towers, and access to them was possible only through a guarded gate.

Since Wismut was a part of the Soviet atomic complex, it was not surprising that several NKVD departments were in competition in the uranium province. Department S made the general director's office its headquarters, and its members were exclusively secret servicemen. Until the beginning of the 1950's, its head was Col. Sukhanov, who controlled maintenance of the security regimes. There were NKVD posts at district or town levels at the fourteen Wismut objects. A special NKVD group commanded by Major Malygin was of particular importance, as it worked in all the pits and plants of Wismut. Malygin reported directly to Serov, and the group investigated cases of espionage and diversion.

The NKVD initially feared that the Nazis would follow the example of the Soviet partisans. The mining districts seemed to be suitable for underground activities. The security service built up a network of informers and, by June 1947, 189 Germans had been recruited.[117] In the first ten months of 1947, the NKVD investigated more than 2,000 cases of "fascist and anti-Soviet elements," and claimed to have broken up several underground organizations. Col. General Bogdan Kobulov, who was the deputy head of the office responsible for Soviet companies in Germany, wrote to Beria on 7 August 1947 about the elimination of "three fascist groups." One was called "Texas," consisting of eleven persons, and planned murders of Russian officers. Another group wanted to kill the Soviet manager at Oberschlema. The third, "Edelweiss," was led by a former mines accountant, allegedly planned to blow up some of the Wismut objects. On insufficient evidence, 47 people were arrested in 1947 and charged with "anti-Soviet underground activity." Eleven of them were handed over to the court martial and received sentences ranging from 8 and 25 years. 36 other miners were interned in a special NKVD camp.

The fear of an organized Nazi underground movement proved to be false, and the Soviet secret police concentrated on maintaining discipline among Wismut employees. A November 1947 NKVD analysis stated that, among 18,000 employees, there were 3,000 "harmful elements." Its own statistics show the direction of NKVD's work among the miners took: between autumn 1946 and November 1947, 1,592 people were arrested, of whom 1,571 were accused of offences against work discipline. Only five miners were arrested for "anti-Soviet agitation," and the NKVD always gave hearing to rumors of strike plans.

While the mining objects were under construction, the occupying power just about managed to look after their security. The day-to-day running of the pits, as well as the rising criminality, presented a difficult problem for the Soviets. A large part of the miners made to join the Wismut workforce wanted to leave as soon as possible. A few tried to obstruct the building of the uranium industry, or at least pass on information to the West. The SPD Ostbüro was the most frequent recipient of such information, followed by the American or British secret services.

To reduce employee flight and frustrate espionage activities, the occupying authorities turned to the local Germans for help; the language barrier alone insuperable for the Soviet security forces. On 15 February 1947, Gotthard Schudy, a former adjutant of Field Marshal Rommel, was asked by MVD Lt. Col. Sukhanov to establish a special police force for the mines (Bergpolizei).[118] The twenty-seven year-old Schudy was a capable organizer. He was given the title of commissioner and put in charge of three large mining districts: Johanngeorgenstadt, Annaberg and Schneeberg. Schudy's mines police were formally attached to the local mining authorities,[119] and its taskmaster was the NKVD and its paymaster Wismut AG. The mines policemen were better paid than the ordinary police force, and wore civilian clothes until the end of 1949. Several units of the mines police had, however, started to wear uniforms and carry arms before the whole force became uniformed.

Its headquarters were established at Aue in February 1947, and gradually every Wismut object became a police district. In 1949, the headquarters were moved to Siegmar-Schönau. The police kept a close watch on the Arbeitsbummelanten, or work shirkers, and dealt with criminal offences while trying to keep peace and order. The large number of flights by the miners kept worrying the Wismut management, and every day the investigation service (Ermittlungsdienst) received a list of the miners who had failed to turn up for work from the pits. The policemen tried to trace them and, if possible, bring them back to work.

In the first years of the existence of Wismut, the rapidly growing number of employees, transport problems and generally chaotic conditions created opportunities for the miners to avoid police attention. As the police became more closely knit, the ability of the Bergpolizei to control the mining community improved. If a miner failed to turn up without a good reason, his name was entered into a register. If he committed a criminal offence, a report on him was passed on to the headquarters at Siegmar-Schonau. If the suspect could be found, he was taken into custody at Chemnitz and tried by a special law court. If the miner had visited one of the western zones or West Berlin, his case was investigated by either the MVD or the K5, the predecessor of the East German security service.

The mines police guarded the Wismut objects and the borders of the mining region, and carried out passenger checks at the railway stations. Together with the MVD, the police were responsible for the control of visitors, who had to have permission from the Soviet command either at Chemnitz or at Zwickau. The policemen also helped to keep watch on the personal lives of the miners. As the policemen worked closely together with the MVD, they did not enjoy much popularity among the miners. There were frequent reports of maltreatment of the miners by the policemen.[120] In September 1949, the police force was reorganized on the basis of an agreement between Lt. Col. Sukhanov and the state police authorities in Dresden. The force was to be called "Works Police A" (Betriebsschutz A), and become a part of the state police authority. Later, after the abolition of the historic states in the GDR, it came under the authority of the Central Directorate of the People's Police (Hauptdirektion der Volkspolizei, HDVP).[121] The works police was equipped with standard uniforms and arms. Due to a scarcity of policemen, it was again reorganized a year later, and the 35 precincts were transformed into 12 larger districts. At the end of 1950, the commander of the works police, Schudy, fled to the West and was succeeded by Chief Commissioner (Oberkommissar) Lössel.[122] After the foundation of SDAG Wismut, the mines police acquired the status of a district police force, and reported directly to the Ministry of the Interior. From mid-1950s, it patrolled the pits and accompanied uranium transports to the Soviet Union.

Finally, at the end of 1957, the Soviet occupation authorities arranged for all guard duties to be carried out by the Peoples' Police.[123] At that time, the police force consisted of 2,300 men, and the "sixteenth district," or the Wismut uranium region, disposed of twice as many men as did the force policing the whole Dresden district. At the end of 1958, there were almost 13,000 works policemen in the whole of the GDR, meaning that one out of five of them was on duty at SDAG Wismut.[124] In November 1948, new personal identity cards were introduced for the entire Soviet zone of occupation.[125] The measure was aimed at reducing the variety of personal documents, and possibly strengthening the position of the DWK in regard to the occupation authorities. The management of Wismut AG was determined to keep its well-established system in place. On joining the company, every worker handed over such documents as were in his possession to the management. He was then issued with a special identity card. It became the only document the miner had during his employment at Wismut. When he left, he was given back his old papers.

The practice of issuing special documents was common in the Soviet armaments industry and in the state administration. It was intended to make control of access to sensitive objects easier, and to show who employed the holder of the special pass. For Wismut management, it was important to try and reduce the number of miners who deserted their jobs. If the company could not stop them leaving, it could at least make it more difficult for them to live in the Soviet zone of occupation.

The pit managers used the system of special passes to put pressure on the miners to stay on in their jobs beyond the agreed period. When the term of service ran out, the miners were denied their personal documents. The miners who refused to be intimidated ran the risk of being unable to register with the police at home, and as a result not being issued food ration cards.

The DWK failed in their attempt to introduce the new identity cards for the whole Soviet zone of occupation. The Ministry of Labor complained about the attitude of the Wismut management in a letter to the state governments in November 1949: "This untenable situation has made it necessary for us to turn urgently for help to the Ministry of the Interior...Wismut AG believes that it cannot accept these measures for serious reasons."[126] The management of Wismut AG at least promised that it would hand over the documents of miners no longer employed by the company to the police at Chemnitz.

General Maltsev insisted that new employees should exchange their identity cards for company passes. The chief of the police in Saxony accepted Maltsev's position and made a special agreement with Wismut. The Saxon Ministry of Labor and Social Welfare contested the right of the police to conclude agreements of that kind.[127] As late as the end of 1949, a police chief told trade union representatives that no such special agreement existed,[128] nor did the Saxon Minister of the Interior want the practice to become publicly known. Protests against had taken place at several Wismut objects, and the miners insisted that the practice was unconstitutional.[129]

On 11 February 1950, a meeting took place between the representatives of the Ministry of the Interior and the head of the Wismut personnel department, Lt. Col. Einbinder. He angrily condemned the resolution of the miners at Object 176. Einbinder insisted that the documents of the miners who had fled should remain at Wismut, and that every miner should "personally collect his own papers."[130] There was nothing the Ministry of the Interior could do.

Another exchange of personal documents, started in March 1952, also didn't affect the practice at Wismut. The exchange was made necessary by the mass exodus of people—with their old identity papers—to West Germany. The old identity cards were withdrawn and, according to the regulation of 15 March 1952, identity cards held by the police and by Wismut workers became equivalent to the new documents.[131] The practice, first introduced in 1947, was retained. It highlighted the special position of Wismut as a part of the military complex.

From Compulsory to Voluntary Employment

Works councils had existed before Wismut AG was officially established, and they tried to improve the working and living conditions of the miners. They turned to the president of the state labor office, describing to her the conditions in Johanngeorgenstadt.[132] Appeals for help from the works councils caused the FDGB committee in Saxony to draft a letter to the "Russian trade unions" on 15 April 1947. The Russian trade unions were addressed for tactical reasons; the Soviet management of Wismut AG and its superior agency, the NKVD, were alone in being in the position to improve the conditions of the miners. The FDGB committee hoped that the letter would, in the end, reach the right people; that is, Beria and General Maltsev.

The letter stated that the political parties in the Soviet occupation zone and the FDGB had tried to recruit volunteers for the industry, and that they had not been able to establish contact with field post no. 27304. "Unfortunately, FDGB, despite its efforts, has not until now succeeded in establishing communication with the relevant Soviet authority. The miners are embittered because FDGB and SED, as organizations of the working people, could not make their influence be felt in correcting the abuses."

Reasons for discontent were named: wage scales set on 1 September 1946 were disregarded by the management, and there was no minimum wage. Directives on wages were issued, which were not a part of the wage scales. The unavailability of pneumatic hammers, poor ventilation, irregular payment of wages, tough penalties for insignificant misdeeds, insufficient supplies of work clothes and no washing and changing rooms were mentioned.

The FDGB made six demands: the application of wage scales and the setting of a minimum wage and appropriate leave; the guarantee that those workers, resident elsewhere, would be able to visit their families; the use of only financial penalties to punish offences by the miners; a supply of sufficient work clothes and boots; support for the works safety commissions by the management, so that working conditions would be improved and accidents avoided; full cooperation with the works councils according to the existing German laws.[133]

The chairman of the Soviet trade unions, V. V. Kuznetsov, passed the letter on to Molotov. Molotov made a marginal note on the letter and sent copies to Zaveniagin, Merkulov and Sokolovski: "Please acquaint yourself with this and agree on the necessary measures with comrade Beria."[134] Beria reacted swiftly. At the end of June, Molotov remarked that "Measures have been taken: comrade Beria reported to comrade Molotov." Maltsev's actions point to what the measures probably were.[135]

On 23 or 26 April 1947, the first meeting between the representatives of the central committee of IG Bergbau and General Maltsev took place.[136] The trade unionists presented Maltsev with the demands, formulated in the letter to the Soviet trade unions, and handed over several copies of wage scales for the uranium industry agreed upon by the occupation authorities. Maltsev secured several improvements in working conditions for the miners: from May 1947 onwards, warm meals were to be served in all the Wismut objects. Five rest homes for the miners were put at the disposal of the trade unions. The trade union was to appoint its own secretaries and labor safety inspectors.[137]

In the middle of May, the deputy chairman of the Saxon committee of IG Bergbau, Fritz Schreiber, visited the Marienberg and Annaberg districts. He and other trade unionists were received by Maltsev on 30 May 1947. He assured them that a uniform pay scale would be introduced for the uranium industry. The talks between Maltsev and the representatives of IG Bergbau would continue at the beginning of June. Maltsev opened the discussion at one of the meetings by expressing his disappointment with the behavior of the trade unions, who had not addressed their complaints to him first and instead turned to Soviet trade unions. He added that "We should let bygones be bygones, we, the management of the mining enterprise shall now cooperate closely with the trade unions."[138]

Maltsev deftly linked concessions to the trade unions with a recruitment campaign for the FDGB and SED, as he wanted their organizations to help run Wismut. In regard to the question of leave, Maltsev referred to order no. 115, according to which the same provisions were to be made for the uranium industry, as for the rest of the mining industries. The trade unions achieved partial successes in the matter of wage scales. The absence of uniform rules had resulted in large fluctuations of wages. Whereas some miners earned up to 1,200 marks a month, many worked for about a tenth of the amount. A minimum wage was now guaranteed, and a special commission, together with the pit managers, established uniform wage norms. The question of the continued existence of the works councils remained open, as Matsev was not ready to compromise on the matter.

The SED leadership was aware of the existence of severe difficulties in the Saxon uranium industry beginning in the spring 1947 at the latest. The party executive formed the "Büro Lehmann," which was to deal with social problems in the industry. Helmut Lehmann was a leading SED expert on social policy. A former social democrat, Lehmann concerned himself with the resettlement of refugees, the organization of health care, the search service of the German Red Cross, the organization of victims of fascism and POWs.

Lehmann had received disturbing reports from the Saxon mining districts, and had a comprehensive collection of materials prepared for him by the Central Office of Health Care (Zentralverwaltung für Gesundsheitwesen, DZVG) and by the Office of Hygiene at Chemnitz.[139] The chairman of the SED, Wilhelm Pieck, had been informed of the arbitrary measures taken by the occupying power in connection with the recruitment of the workforce for the uranium industry.[140] Lehmann and Pieck dealt with the problems of the uranium industry on the Politburo; the responsibility later passed on to Otto Grotewohl, the Prime Minister, Walter Ulbricht as the head of the SED and Fritz Selbmann, the minister of industry. They carried out the decisive negotiations concerning uranium mining with the Russian military and civil authorities.

Reports on conditions in the uranium industry, most of them drafted by Max Weber, also reached SED headquarters in Berlin. Their tone was sharper than that of the reports from the labor office in Saxony. Lehmann succeeded in getting the problems of the uranium industry on the SED agenda in August 1947. He wrote that "According to available medical reports, the working conditions of the employees are most unfavorable. Many physically and medically unsuitable persons were directed to this work...provisions for them are totally inadequate. Promises about special rations have not been kept. Safety at work is insufficient and the effect is worse as most workers are unskilled. The doctors...are required, against their better judgment, to declare men as fit for work who are not."[141]

Lehmann was concerned with the conditions of work and their adverse effect on the standing of the SED. The submission was briefly discussed by the SED central secretariat on 15 August.[142] The party shelved the problem in the hope that local authorities would keep and eye on the problem. A special party commission, with Paul Merker as its head and Lehmann as one of its members, remained inactive for a long time. It was denied access to the uranium province and the party was unable to do anything about it.[143]

Though Merker's commission remained ineffective, the problems of the uranium industry returned to the Politbureau in October and November 1947. Max Weber, the delegate from the Saxon government for Wismut, was invited to present his case. He spoke of the social problems of the miners, especially the difficulties concerning their accommodation. "Many people did not take up employment in the mining districts, but left for the West...Many of them who were hostile when they arrived, escaped at the right time. The conditions nowadays are at any rate different, as the mining...is well organized and the management of mining by the occupying authority is so arranged that orderly conditions have been achieved. The reports of the workers who left for the West are naturally strongly exaggerated."[144] Weber referred to the financial advantages enjoyed by the miners and played down the dangers of their work.

The party leaders were calmed down by Weber's report, and remained passive with regard to the Soviet authorities. Neither Ulbricht nor Grotewohl were the decisive factor for the improvement of conditions at Wismut. Instead, a group of trade union leaders in Saxony helped the miners. They were lucky that their letter ended up on the desks of Molotov and of Beria.

Maltsev was under strong pressure to succeed, and made use of the willingness of the trade unionists to help him. In contrast to the practice in the Soviet Union, he assured the Germans that minimum welfare standards would be applied. These standards were derived from the traditional regulations in the German coal industry. In return, Maltsev expected increases in productivity. The tendency to introduce the gulag system in the Saxon uranium industry was stopped, and Wismut AG did not develop into "Vorkuta on German soil." Only later did it become apparent how important the change of attitude to the uranium industry, which began in the summer of 1947, really was.

When the policy of compulsory employment, especially as it existed in the uranium industry, ran into criticism, order no. 234 of the Soviet military administration offered the possibility of a change of course. In place of indiscriminate rationing, a production-linked system of awards was to be used.[145] It was important for the mining industry that the Trade Union Congress (FDGB) and the German Economic Commission (DWK) pressed for the reduction in compulsory employment on the basis of the order no. 234. On 2 June 1948, the German Economic Commission (DWK) released new regulations for the protection of the employees' rights. Compulsory orders could be used only in exceptional cases and required an agreement by the FDGB. The Wismut management showed little understanding for the new policy. The SED leadership and the miners' union (IG Bergbau) also tried to circumvent the new rules. They feared that the limit of six months on the employment of labor would lead to further destabilization of the fluctuating workforce.

Many pit managers tried to resolve the problem in their own way and did not let the good workers go after their contracts expired. They threatened criminal prosecution, or refused to return the miners' personal papers or pay their premiums at the end of the contract period.[146] The pit managers and their assistants found the new rules hard to accept, and labor exchanges and other offices also did not observe them. Many illegal methods of meeting high labor requirements existed: people at work were put into the compulsory system, pressure was put on state employees to work for Wismut and nationalized industries and communal offices received notices on how many employees they should release for the uranium industry. Heads of personnel departments sent people who were politically out of favor to Wismut.[147] Instead of direct, indirect pressures started being used.

On the other hand, in 1948, the number of voluntary workers began to grow. Labor exchanges put the proportion of volunteers as high as 60% for 1948, and 90% for 1949. In 1950, all the new Wismut workers apparently came on a voluntary basis.[148] The actual share of the volunteers was probably lower, though SPD Ostbüro informers reported an increase in immigration to the uranium districts by voluntary labor in 1949.[149] Wismut wages were so attractive that the railway authorities in Aue paid premiums to railway men for simply staying in their jobs. Beginning the middle of 1948, broad based recruitment campaigns were organized throughout the Soviet zone of occupation. Labor exchanges ran "persuasion commissions" (Überzeugungskommissionen) consisting of a representative of the SED, the FDGB and other mass organizations. After the foundation of the GDR in October 1949, the Soviet Control Commission started passing labor requirements on to SED as well as the Ministry of Labor, and labor for Wismut appeared on the SED Politbureau agenda.

After the results of the recruitment campaign in spring 1950 proved disappointing, and Marshal Chuikov complained to Pieck,[150] the SED decided to give the recruitment drive top priority. In April it established the "Aue Commission," which was to examine the ways and means of fulfilling Soviet labor requirements. The commission mobilized the party apparatus, and its functionaries were sent to nationalized factories to publicize the attractions of the uranium industry among the workers.[151]

Every newspaper contained articles on the theme "the ore-mining industry needs you!" They extolled Wismut's excellent working conditions and the possibility of high wage. Some of the descriptions of the uranium industry were surprisingly open. On 18 September 1951, the magazine *Film-Funken* informed its readers that "...very valuable uranium ore is being mined here. The ore is essential for our economy, as it is a barter article and contributes to the reparations deliveries. The enterprises of Wismut AG are largely under the control of the German people's police, which protects it against saboteurs and enemies of the people. Only a few years ago the mining industry employed people driven by the lust for adventure or those who believed that they could lead fickle lives. It was clear to everyone that such people could not achieve positive results..."

Since a long perspective emerged for the uranium industry, and the compulsory employment of labor reached its limits, both the Soviet authorities and the SED discovered the advantages of a stable workforce. It was better for productivity, which was the main concern of the Soviet management. For the SED and the trade unions, other motives came into play. Repeated recruitment campaigns had done little to improve their reputation, and it was hard to find party and union members among the ever-changing workforce. In addition, investment in education and acquisition of appropriate qualifications by the miners could be planned only on a long-term basis.

The FDGB had been founded in 1945 as a politically independent association of trade unions, which was to defend the interests of workers and employees.[152] It soon came under the influence of the KPD/SED, and made their political and economic interests its own. The supporters of a united trade unions organization, which would act as a "transmission belt" for SED policies, succeeded during 1947. In the uranium industry, the change from an organization that meant to defend the workers' interests to an organization dominated by the SED took place faster than elsewhere in the East German economy. It did not mean that it always took the side of the Wismut management; on single issues, different interests came into play.

Early in January 1950, the trade unions' executive committee passed on to Wismut AG "Recomendations on the Overcoming of Fluctuations in Wismut AG",[153] which proposed a list of measures aimed at stabilizing the fluctuating workforce. Accommodation for the miners was assigned top priority. The building program was to be helped by the state, and plans for better transport took second place. Faster construction of the railway line Johanngeorgenstadt—Schwarzenberg to Aue—was recommended, as well as a new line between Johanngeorgenstadt and Eibenstock. The Wismut vehicles park was also to be improved. On paydays, which were to be regular, the sale of alcohol was to be forbidden.

Once again, the trade unions tried to reach an agreement on wage scales. The only ruling had been announced in the autumn 1947, in Maltsev's order no. 115. The managers nevertheless continued to make decisions on production norms, which determined wage levels. In the middle of January 1950, trade unions asked the ministry for planning to support the recommendations.[154] The Wismut board of directors turned down the proposal for agreed wage scales once again. A new labor contract was to be drafted instead, and the Soviet management of Wismut AG remained averse to concluding binding agreements of any kind with the trade unions. They were content with the established system of production incentives and the so-called "combat plans" (Kampfpläne). The latter, they believed, were the ideal means for increasing production, which was their abiding concern. This should be achieved through improved organization of labor, further mechanization and voluntary increases in work norms. The management held out hopes for further advances in the material and cultural life in the province, but was loath to make any firm undertakings. A minority of miners refused to give in to the massive pressure from the SED, and the trade unions and took no part in the "voluntary increases in work norms."

So as to attract a stable workforce, the department for mining of the trade unions congress (FDGB) pressed for the repeal of the special legal position of Wismut AG.[155] The head of the Department of Labor of the Soviet Control Commission, Pavel V. Morenov, had in fact declared many times that Wismut should come under the GDR legal system.

However, the situation in the uranium industry was singular. Wismut management ran all aspects of the camp and issued all the rules and regulations. These were interpreted by the bosses of the objects in different and sometimes contradictory ways. Trade union influence had no place in the system, and trade union leaders continued materially to depend on the good will of the pit managers. The situation, still in existence in the early 1950s, was quite unlike that of any other enterprise in the Soviet zone of occupation.

As demand for employment at Wismut AG exceeded the number of jobs on offer, a more rigorous system of personnel selection was introduced. The change started taking place when the prospects before the industry were very good. It was completed in 1949–1950, when discoveries were made of important reserves of uranium. The assumption that the industry had only a short life before it was abandoned once and for all, and short-term improvisations became unnecessary. In the meanwhile, labor hoarding had led to overstaffing at several objects, and it was estimated that up to 30% of the miners were underemployed.[156] As the mechanization of production improved, some labor became surplus to requirement. The proportion of miners employed on compulsory orders and voluntary workers changed in favor of the latter, which gave the party and the trade unions a chance to improve their hold on the industry. Beginning in June 1948, compulsory labor orders could be issued for periods of up to six months only. In about two years, Wismut gave up the practice of compulsory employment altogether.

While good workers were offered attractive inducements, weaker workers started being dismissed for minor infractions of the regulations or for low productivity. Beginning in May 1949, Maltsev's order enabled "work shirkers" (Bummelanten) to be immediately discharged. Those miners who wanted to leave Wismut made good use of the order: they failed to turn up for several shifts and waited to be dismissed. Still in 1949, for every miner who left, two more were taken on. This trend continued into the first half of 1950.[157] In 1950, the proportion stood at 64,000 hired workers to some 28,000 who left employment.[158]

The first temporary stop to hiring labor, and the first great wave of dismissals, took place in autumn 1950. About 20,000 workers were made redundant as some of the Wismut objects were closed down. Another wave of dismissals followed in the spring 1951. By the end of June, more than 50,000 workers had left Wismut employment, almost two-thirds of which were made redundant.[159] The new stricter personnel policy included a ban on employing persons who had returned as POWs from the West, or persons in general who came from the western zones of occupation.

From the beginning of the 1950s, it became easy for Wismut AG to attract voluntary labor. For the second quarter of 1951, more new workers were hired than had been planned. In the spring 1952, more workers were, for the first time, dismissed than hired.[160] The time of workforce growth was definitely over.

An Overview of the Wismut Workforce, 1946–1953

Tracing the movements of the workforce at Wismut between 1946 and 1953 is an extraordinarily complicated matter. The company grew fast, and its individual objects and pits ran themselves on an autonomous basis. The statistics of the labor exchanges include only those workers, whom they directed to the uranium industry—some of them were unsuitable for work in the mines and many others did not turn up. They were described by the authorities as the "missing recruits" (Fehlvermittelte), and their numbers were especially worrying in the first three years of the existence of Wismut AG.[161] In addition, the company used the services of private headhunters, whose activities were not recorded by the labor exchanges.

Requests for the provision of labor by the occupying power provide better guidance.[162] Until the beginning of 1952, Soviet authorities requested more than 447,400 workers for the uranium industry, and probably over 400,000 men were actually provided. Such enormous numbers may be explained by the fast growth of the industry rather than by the fluctuations in the workforce. About 60% of the workers came from Saxony in 1948.[163] In later years, the share of Saxon workers increased. Between autumn 1946 and the end of March 1951, according to the figures of the state government, Saxony provided Wismut with more than 216,000 workers.[164] At the beginning of the 1950s, the number of Wismut employees reached the highest level, numbering nearly 200,000. (These figures are from Soviet annual reports, some of which were entered by hand: this signified the highest level of confidentiality in German and Czechoslovak uranium industry documentation.)

The figures given were valid for the end of each year: 2,257 in 1946, 18,775 for 1947, 63,383 for 1948, 139,587 for 1949, 195,906 for 1950, 153,112 for 1951 and 100,998 for 1953. The number of Soviet employees at Wismut was given for the last three years only: 6,897 in 1950, 10,925 in 1951 and 8,752 in 1953. There exist detailed figures for 1953.[165] According to them of about 101,000 employees in December 1953, 93,307 worked in production, of this number 2,692 were Soviet civilian employees and 6,060 were military. 7,691 persons worked for the administration.

Despite many open questions, it is beyond doubt that the uranium industry grew in those years at an extraordinary rate, reaching its highest number of employees in autumn 1950. With more than 200,000 workers, including the Russian contingent, Wismut AG became by far the largest enterprise in the Soviet zone of occupation. It had easily overtaken other mining sectors and its size could be compared only to the black coal mining industry of the Ruhrgebiet.

So as to develop the uranium industry in the shortest possible time, human and material resources from the Soviet zone of occupation were pumped into a comparatively small area in the Erzgebirge. The initial direction of compulsory labor was supplemented later by the use of indirect pressure and, finally, by a system of production rewards. The occupying power could use Wismut workers at its discretion, without regard to the law, or it could dismiss them, when it became necessary, at short notice. For most workers, their Wismut service was a short, though decisive episode that they generally regarded negatively. The other miners stayed on at Wismut on a voluntary basis and, after some time, formed the core of a reliable and well-rewarded workforce.

Becoming a Model Enterprise

Wismut AG was by no means an ordinary company. The largest reparations enterprise of the 20th century, it was built in a short time out of nothing. In the first years of uranium industry, men worked in primitive conditions, resembling the industrial practices of early modern Europe. Huge numbers of workers made up in part for the missing technology, and increases in production followed in the second phase.

The miners worked under virtually military discipline for several reasons. Wismut operated in a defeated country, and was a part of the Soviet nuclear industry under the supervision of military units. As late as the mid-1950s, Soviet management ran Wismut on the basis of orders. In any case, over the centuries mining developed its own hierarchies resembling those in the military, with their fixed ranks of officers, non-commissioned officers and men. The miners had supervisors and chief supervisors, and a newcomer started as a "navvy" (Fördermann), irrespective of his previous employment and training. The cutter (Hauer) set the rhythm of work, and the navvy, who set his sights on becoming a miner, was his mate. The men and non-commissioned officers worked underground. The officers consisted of the company's management, the pit bosses and some technical staff, and were to be found mostly on the ground.

Training at Wismut was initially conducted by a few supervisors (Steiger) who had originally come from the coal mining industry. Miners who wanted to fulfill the norm had little time to look after the new arrivals. Their skills were decisive for the productivity, and thus the earnings, of the team. Those of them who wanted to gain extra qualifications joined the Wismut training center, which was established in a disused Freiberg barracks in 1948. It schooled middle-ranking workers: supervisors, geologists, administrative workers etc. Between 1948 and 1952, about 9,400 employees attended courses at the center;[166] opportunity for advancement helped to keep many people working for Wismut AG.

For many workers, the day began with a long journey on foot or by bicycle. The development of the public transport system was slow and did not keep up with the growth of the workforce. Until the beginning of the 1950s, the railways were in a chaotic state, trains and buses were overcrowded and serious accidents were commonplace.

At the beginning and end of each shift, the miners' passes were examined at the Soviet control posts. Every miner had to go through a gate with a Geiger counter; when it sounded an alarm, the miner had to undress. Sometimes their workclothes were contaminated, or the soles of their shoes contained trace amounts of radioactive matter. The use of bags or briefcases was forbidden for a time, and only string bags were allowed.

After they passed the gate, the miners collected the tools from the store. They carried lamps, drills, picks and spades. The store was eventually moved underground and simple repairs were carried out there. In the early years the miners used ladders to get down the pit. The supervisors allocated the work, which usually began with clearing the rubble in places where blasting had taken place. The trolleys were loaded and taken away. It was hard work, the toughest part of the shift. After that, holes were drilled, filled with explosives and the fuse was lit. When all went well, the shift ended after the blasting. There were often difficulties—there was no electricity, compressed air, or boring rods.

The supervisors, who enjoyed a lot authority, also carried a lot of responsibility. The Soviet pit managers, the so-called "shift Russians" (Schichtrussen), relied on them to keep an eye on the fulfillment of work norms. When the miners struck a uranium seam, special regulations came into force. The seam could not be blasted, but taken out with pick and axe. The ore was packed into boxes and placed in a strictly guarded store. For the miners, a rich seam was a feast-day. High premiums were paid for each box. The desire for premiums led to carelessness and dishonesty. "We always kept a box of good quality ore," a miner remembered, "and put it into boxes with lesser ore. Like that one earned higher premiums...It was not unusual for the premiums for the ore to be considerably higher than monthly wages. This financial incentive also made the miners invent every kind of trick, so as to make extra money. We for instance often worked disused or closed off galleries in dangerous circumstances and we kept quiet about the place of origin of the ore..."[167] A special system of premiums was introduced for the cutters (Erzhauer). They often worked alone on old sites, mining the residual ore. They were exposed to high radiation, and received monthly premiums of 1,000 marks.

In the years after the war, food supplies were more valued than wages. Food was difficult to find and all the zones of occupation ran a rationing system. They functioned badly and were administered by a huge bureaucracy. Germany was a "rationed society," with the ration card being the average person's most important document. The coupons could be exchanged for food at fixed prices. Six categories of rationing were established, based on the Soviet example. Heavy work was rewarded by the first category, whereas categories 5 and 6, also known as "cemetery rations," were set aside for children and pensioners. No family could survive on rations alone. Hunger drove people into the countryside to search for food, while the black marketers moved into towns. A miner who earned 60 RM a week had a hen that laid five eggs a week. He kept one and exchanged the rest for twenty cigarettes, which cost 160 RM on the black market. The hen earned three times the wages of its owner.[168]

In June 1947, a miner received the following basic rations: 13.5 kilograms of bread, 1.2 kilograms of meat, 1.5 kilograms of noodles or pulses, 15 kilograms of potatoes, 9.6 kilograms of vegetables, 2.4 kilograms of cheese, 6 liters of milk, 750 grams of sugar, 600 grams of fat, 900 grams of jam, 2,000 grams of soap and 125 grams of coffee.[169] Canteen meals, production or risk related coupons nearly doubled basic rations. As a much-quoted saying by Bertold Brecht ran: "The grub comes first, the morale later." After initial difficulties, the far above average food supplies brought a stream of volunteers to Wismut, including men and women who had sought work in the western zones. Daily rations in the uranium industry amounted to about twice the rations offered to industrial workers in 1949, and uranium miners stood at the top of the caloric hierarchy in the Soviet zone of occupation. The system of production incentives was later improved with coupons for consumer goods such as clothes and shoes, in addition to extra food coupons.

In the accounts of the post-war years, Wismut workers remember two things: "miners' ruin" (Kumpeltod) and "Stalin parcel" (Stalinpaket). Every Wismut employee who worked underground received two to five liters of duty free schnapps. The miners enjoyed the strong drink and used it for barter. Retired miners fondly remember the contents of the Stalin parcels. They were the reward for fulfilling the norm by 150%. They contained meat, sausage, flour, cheese and tinned food. A miner remembered that "They were productivity parcels. They could be had every month. When a miner worked well, he would receive a Stalin parcel. Half a Stalin parcel comprised six large jars of fruit preserves, either apple pure or cherries. A whole packet represented twelve jars. And coupons could be used for purchases at the butcher's. That was what was so good about Wismut."[170]

The parcels were good propaganda, and the equivalent of US Care parcels in the western zones. The reconstruction of the Ruhr region, after discussions about socialization, was carried out under the slogan "bacon instead of socialism." From autumn 1947, the British military government started to issue the Care parcels, which originated from US

Army stores. Each parcel contained at least 40,000 calories, and its award depended on the miner's productivity.[171] After the currency and rationing reforms in summer 1948, the living conditions of the miners in the Ruhrgebiet began to improve continuously, and the socialization of the Ruhr mining industries was becoming less probable.

In the Soviet zone of occupation, on the other hand, all the mining industries were nationalized, or in the case of Wismut AG, the company became the property of the Soviet state. Scarce resources in the Soviet zone were channelled to the reparations industries, and to Wismut especially. The occupying power used the formula of "bacon and socialism," and the magic power of calories exercised its pull. While tens of thousands of forced workers "voted with their feet" against Wismut in 1947–1948, and fled to the western zones of occupation, there soon were more applicants for jobs than Wismut could use.

The work hours, wages and awards at Wismut were similar to the rest of the German coal industry, but with some Soviet adjustments were added. Wismut AG had an eight hour working day, from Monday till Saturday, with a Sunday free. As a rule, three shifts were worked round the clock. The early shift began at 6 AM, the day shift at 2 PM and the night shift at 10 PM. The new element was the introduction of work on Sundays and feast days. The miners referred to Sunday work as "subotnik," a term used in Lenin's time for voluntary and unpaid labor. Although Soviet pit managers paid well for Sunday work, the trade unions initially disapproved of it. Under pressure from the management and from the SED, the trade unions started promoting "voluntary" Sunday work in August 1948. The miners called the trade unionists "Russian serfs," and continued their opposition against the innovation. In 1950, the management used an order to enforce it, and the numbers of miners reporting sick at the weekend increased.

In October 1945, the Allied Control Commission (ACC) froze wages so as to avoid inflation. It meant that low wages in key industries, especially in coal mining, could not attract enough workers. The ACC then issued directive no. 41 of 17 October 1946, which allowed wage increases in the coal mining industry of up to 20%. New tariffs were introduced in the mining industry of the Soviet zone of occupation as well, promising increases of 25% and improved payments for piecework.[172]

From the beginning, the uranium industry used a productivity-related tariff. The minimum wage was paid only in the case of the fulfillment of the pertinent norm. Shortages of tools and materials often made the norm difficult to fulfill, and there were pit managers who made the miners stay on until they met the target. Initially, while the works councils existed, they resisted the pit manager's pressure and demanded agreement on wage tariffs. The Soviet management turned the demand down, because it wanted—in contrast to the rest of the mining industry—to introduce a broader range of wages.[173]

Since there was no agreement on tariffs at Wismut, wages were regulated by orders. Order no. 115 of 3 June 1947 was followed by order no. 163 of 7 July 1947, concerning severance pay. Orders no. 239 of 20 October 1947, and no. 250 of 1 November 1947 concerned productivity and other awards. The last order introduced the "Wismut extra charge" (Wismut-Zuschlag), which provided for every worker underground to receive 40% extra wages, and workers on the surface between 10% and 30%. Most of these orders usually followed initiatives by the trade unions and the SED.

This complicated system was supplemented by detailed regulations made by individual objects. Model labor contracts and other orders also existed, which impinged on the welfare of the workers. Finally, the order of 5 August 1948 provided for "loyalty premiums" which helped to create a stable workforce. A premium of 5% of gross yearly wage was payable after the completion of six months' service, rising to 20% after two years. On 1 August 1949, extra rations for Wismut employees were promised, together with a supplementary card for members of their families.[174]

The productivity-related social policies of Wismut AG began working. Wages rose to as much as 4,000 marks per month, and Wismut employees became the best paid workers in the Soviet occupation zone. Rumors of fabulous earnings spread throughout the country and awoke wonder and envy. Comparisons with the gold rushes in Alaska and California were made, and the Erzgebirge Klondike gained in reputation as a place of easy pickings. There were no "uranium millionaires" on the "Texas Express," as the crowded trains of Saxony became known. They could be found in Colorado or in Canada, but in the Soviet zone of occupation, there only existed a state company which paid the miners unusually high wages.

In return for these high wages, a great deal was expected from the workers. In *The Man of Marble* and the *Man of Iron*, film director Andrzej Wajda commemorated the enthusiasm of the years of construction in Poland and the bitter disappointment that culminated in the riots in 1956. The central character of the films was a worker who surpassed the work norm, and became a folk hero with the help of party functionaries. In the

GDR, the popular novel *The Wonder Worker* (*Der Wundertäter*) by Erwin Strittmatter tells the story of a miner who worked a record shift, thus earning the respect of his colleagues. Such activists were honored everywhere in the Soviet bloc in the late 1940s, and the Soviet miner Stakhanov was their model.

In the Soviet zone of occupation, Adolf Hennecke, a miner in the black coal district of Oelznitz, was chosen for the distinction. On 13 October 1948, after thorough preparations, he achieved 387% of the day's norm. He entered the hall of Soviet-type heroes, but not all his colleagues were as pleased with him as the party media were. Surpassing the norm was not an undisguised blessing for them.

The reaction of SED leaders at Wismut to the news of Hennecke's achievement was initially surly.[175] Some months before Hennecke, Wismut miners had achieved better results. Wilhelm Luck, Horst Radecker and Josef (Sepp) Wenig were among them. The management of Wismut failed publicly to promote them. A secret Soviet enterprise, where some of the best workers were former Nazis, did not make the ideal subject for a propaganda campaign; and there was the unsuitable name for a "Wenig" publicity.*

At Wismut, high productivity by teams of miners was preferred. The "shock brigades" began forming at the end of 1947, also based on the Soviet model. The increasing mechanization required teamwork, and the best brigades garnered the highest rewards. The brigades created a new social context—the team subordinated its members, and their income depended on the overall result. The teams created a sense of solidarity among the miners and provided a counterweight to the bosses. The best teams met at conferences and exchanged their experiences. The team spirit extended to the miners' free time as well, and lasting friendships were made.

The reverse side of high achievement soon became apparent. The management used it to legitimize the setting of higher norms: for 1949 alone, the increase amounted to 10%.

---

* The word "wenig" means "few" or "less" in German. That is why the authorities didn't want to start a campaign with Mr. Wenig as their hero. It would have produced endless jokes.

Uranium Towns

The great uranium rush after the end of the Second World War transformed landscapes and towns, as well as their inhabitants. The uranium province experienced dramatic population growth. The towns in the territory of the first big Wismut objects were the most affected. Many were old silver mining towns which had known periods of prosperity: Johanngeorgenstadt became Wismut Object 1, Schneeberg became Object 3 and Annaberg Object 4. The uranium industry helped them to recover some of their former prosperity, albeit at a considerable cost.

The small community at Oberschlema had 2,600 inhabitants at the end of the war. At the end of 1948, there were some 15,000 miners working there, of whom almost 3,000 lived in Oberschlema. Aue as well as Zwickau belonged among the most industrialized towns in Saxony, with machine, textile and paper industries. In Zwickau, the textile industry was prospering, with strong black coal mining and automotive industries (Horch and Audi). In addition, it became an important center of ancillary industries for Wismut (such as Grubenlampe Zwickau, where miners' lamps were made), and a temporary dormitory town for the miners.

Aue had a dense industrial structure with Wismut Object 2 in the vicinity. The Aue labor exchange district included Schwarzenberg, Schneeberg and Johanngeorgenstadt, as well as the communities of Breitenbrunn, Niederschlema and Oberschlema. They all had an important stake in the uranium industry. In three years, the number of inhabitants of the Aue district increased by more than 80,000 people, with the official data including permanent residents only. As accommodation was scarce and labor contracts short, it may be assumed that official figures underestimated the number of people working in the district.

Many Wismut employees lived in temporary hostels or private rooms, giving other addresses as their main residence. Whereas the districts of Annaberg, Freiberg and Marienberg had only a small share of workers with a different permanent residence, about 295,000 inhabitants out of which 76,000 had a different permanent address in Aue at the end of 1951.[176] Zwickau town and district had an even higher number of outsiders: according to the August 1950 census, it had about 391,000 inhabitants, while there were more than 493,000 persons registered at the district office of the peoples police in November 1951.[177] Finally, there were people who avoided registration altogether. Several hundred men lived illegally at Wismut hostels—in autumn 1951, the Aue police had lists of more than 300 people who were not officially registered.[178] Their real numbers were much higher. The number of inhabitants at Aue and Zwickau, the two important mining districts, were until 1951 far above the figures given in official statistics.

The industry had no long-term perspective, and no provisions were made for any permanent accommodation of Wismut workforce. Everything, including accommodation, was organized on a makeshift basis. Beginning in the summer of 1947, the district councils and Max Weber, the head of the special office for metallurgy at IG Bergbau, tried to deal with the accommodation shortage by banning newcomers or by resettling old residents.

When German authorities realized that they could expect the arrival of a large workforce for the uranium industry, they stopped receiving refugees and expellees. Only refugee families who had a Wismut AG connection were allowed to remain. For instance, the district council of Annaberg proposed to eject newcomers who "were not connected with the district by their employment, that means in the main former farmers and agricultural laborers, who [would be]…moved to agrarian regions."[179] A precedent was created in Marienberg at the end of 1947, where an unstated number of persons "without employment" in the district were evacuated. [180] These people received a letter from the council warning them that "In the next few days you will be informed when the transfer will take place to Olbernhau; it is certain that you will like it there, as the town was not affected by the mining industry…"[181] The district council even considered the possibility of evacuating all the inhabitants of Marienberg, so as to make space for the incoming miners. However, the plan was never realized.

This did not mean that plans for transferring people were shelved. In autumn 1948, Seydewitz, the prime minister of Saxony, stopped the evacuations. He suggested that people should move only if they wanted to, and that they should receive state support.[182] The Ministry of Labor was put in charge of the assisted moves and, by the end of 1948, some 200 families, consisting of some 1,600 people, had decided to leave the mining districts. The Department for Resettlement objected that the departure of the families should be held back until "accommodation for the last transports of the expellees from eastern territories is secured."[183]

Various estimates give the figure of about 1,000 persons who had left the mining districts on the basis of the assisted scheme before the end of 1949. This was far fewer than the authorities had hoped for. People were not interested in making the move, and had nowhere to go. As mass evacuation of refugee families proved to be impossible, industries and farms standing in the way of Wismut operations had to move elsewhere. The idea of planned evacuation was revived: "The evacuations have been voluntary up to now and they will have to be helped along. People hide behind the bourgeois parties. Bourgeois parties oppose evacuation."[184]

The growth of the uranium industries was a singular burden for local government. Military authorities and Wismut management demanded that "temporary miners" should be accommodated in close proximity to the mines. The miners who had been the first to arrive often found private accommodation. The offices for housing acquired extraordinary powers, and could demand accommodation in other people's houses and flats, or even their complete evacuation. They used the Allied Control Commission order no. 18—which allowed occupying powers to vacate accommodation for an urgently required workforce—to house the miners. In Johanngeorgenstadt alone, several hundred apartments had to be vacated.

While perspectives before the uranium industry were unclear, and the possibilities of the Saxon administration were limited in the extreme, state-supported building schemes in the mining districts could not be considered. Wismut AG and the local government offices looked for provisional accommodation: schools, inns and other larger buildings were either leased or confiscated. Dozens of provisional dormitories were constructed. At the end of 1949, Wismut employees lived in more than 120 towns and communities. At that time, 137 large hostels and camps existed, some of which had been built for the Nazi labor service (Reichsarbeitsdienst) during the war. Others had been used for foreign workers or POWs. Most of the mass accommodation consisted of regulation barracks with about 40 rooms, each room housing four people.

It was hard to establish how many miners lived in emergency quarters and how long they stayed there. They came and went, presenting the local authorities with considerable difficulties. They had different personal documents from the rest of the population, which did not make the work of the police any easier.

Social tensions were highest in towns with the largest emergency accommodations, and until 1951 the growing cries for help brought little response. Complaints to the Ministry of the Interior, which were passed on to the Premier Otto Grotewohl, were explicit enough. In an "Assessment of the Political, Social and Cultural Situation in Zwickau" in February 1950, it was stated that "the political and cultural signs of decay in the town and district of Zwickau are closely connected with the existence and the behavior of Wismut AG."[185] The Ministry of the Interior suggested that, instead of emergency buildings, priority be given to the construction of houses for miners, and for the necessary infrastructure.

The first building program was carried out by Wismut AG as early as the end of 1947. Some 1,000 apartments were built in Johanngeorgenstadt, Schneeberg and Niederschlema. IG Bergbau supervised the program, coining the slogan "For the Best Workers, the Best Apartments." From 1948, two-floor miners' houses (Wiener Häuser) were built, as well as wooden, single-family houses (Berliner Häuser).[186] The building program was no more than a drop in the ocean. In 1950, some 1,000 company apartments were available for a workforce of over 200,000 people.

Despite repeated attempts to include housing in the economic planning of Wismut AG, very little changed. Only in summer 1951, during the Saalfeld riots when the miners protested against the conditions in the industry, the SED, under pressure from the occupying authorities, reacted. It combined offers of improvements in living conditions with threats of repression. The program for the building of new accommodation was announced at the end of 1951. It provided for the construction of 5,000 apartments and of community buildings.[187] Emergency accommodation remained in use. With the beginning of the decline in the number of employees at Wismut, and with more ambitious building plans, the lowest point of the accommodation crisis was left behind.

On 1 October 1952, Wismut AG and the Ministry of Building concluded an agreement on the construction of apartments and houses of culture. Wismut AG was to build 552 houses in Ronneburg, 355 in Schneeberg, 158 in Johanngeorgenstadt and 183 in Dresden. In addition, ten new hostels were to be built, as well as many kindergartens, swimming baths, shops etc. About 34.5 million marks were set aside for the building program.[188] After the revolt of 17 June 1953, the building program was again upgraded. At the end of 1953, Wismut allowed building cooperatives to be formed and, within three years, 24 such cooperatives were formed, with 2,350 members.

Justice in the Uranium Province

The harshest times for Wismut miners, in spring and summer 1947, corresponded to a sharp rise in criminality. Thefts concerned the necessities of life, such as food, coal and wood, rather than property.

The industry—like all mining industries in history—attracted a select company of criminals and opportunists. The legal section of the military administration of Saxony pointed out that "Hence the increasing number of embezzlements, misappropriation of pieces of clothing, thefts etc. High wages have led to alcohol abuse in the case of a part of the workforce and to a large number of everyday crimes."[189] Conditions in the uranium province, rather than mass worker migration, lead to higher criminality.

The situation remained critical in the following years and the decline in criminality, first noted in other parts of Saxony in 1948, did not apply to the mining districts. In comparison with the second half of 1948, the number of thefts in Zwickau increased by 34%, and in Chemnitz by 88%, in the first half of 1949.[190] The growing number of attacks on state organs worried the Soviet military authorities, who recommended German jurisdiction to improve "mass enlightenment," to punish acts of violence more rigorously and to organize show trials.

The occupying authority created a new legal framework for the zone of occupation, and it took care of the most serious cases itself. The term "criminality" included acts as well as attitudes of political opposition. Conflicts concerning labor laws could also be criminalized. The range of available penalties and repressions was broad, from reduction of wages to long-term confinement in a labor camp in the case of political opposition, or to the death penalty for alleged espionage.

Flights of workers in the uranium industry were classed as criminal acts against the occupation regime. Miners who left employment unlawfully could be tried on the basis of the order by the Allied Control Commission No. 3, paragraphs 19 and 20. In the Soviet zone of occupation, the DVI (Deutsche Verwaltung des Innern, DVI) added, on 11 June 1948, further detailed regulations to the order.[191] According to the regulations, it was the duty of the mines police to establish whether an offence against order no. 3 was committed, and whether or not the miner left Wismut in breach of his labor contract. The police were also obliged to investigate other criminal offences committed by the miners, such as removing their work clothes from the place of employment.

In cases of miners who fled and stole company property, police authorities at Zwickau were informed. From there a report was passed on to the central criminal bureau of the state where the miner had last lived. In Zwickau, files were kept on the refugee miners and the station reported them to the appropriate state criminal bureau.[192] In case the miner was not found within three days, he was entered into the register of missing persons. The volunteers that left employment before their time were not to be investigated by the police, but the company was free to file a complaint with the local law court. The register of missing persons contains several thousands names, though makes no mention of the miners who returned.

It is impossible to tell how many fugitives were stopped in their flight. The German courts were reluctant to deal with miners who had broken their contracts. The court at Salzwedel, for instance, sentenced seven Wismut miners, who had finished their six month term of duty, at the beginning of 1948.[193] The pit manager wanted them to stay on, and threatened them with confinement in a Soviet prison. The miners used the first opportunity to return home, unaware of any offence they committed. They were sentenced on the basis of order no. 3 of the Allied Control Commission, and the sentence came under criticism by DWK.

Despite harsh sentencing policies, the management of Wismut insisted that German law courts were too lenient,[194] and pressed for special jurisdiction. The DWK opposed the views of the management in vain. In many cases, the miners were not passed on to German law courts. They were either punished internally by the company, or tried by a Soviet court martial.

Since its foundation, Wismut AG had its own security organization and justice apparatus. Beginning in 1947, a special military tribunal (SMT) for Wismut AG was established at Siegmar-Schönau. Between 1945 and 1955, SMTs existed also in other parts of the occupation zone, established on the basis of article 8 of the Soviet military legal code of 1926. It provided for the use of court martial "in territories where, due to extraordinary circumstances, no ordinary courts exist."[195] The tribunals usually consisted of a military judge, who was in the chair, and two military members.

The Soviet security authorities could also invoke the Allied agreement that provided for the prosecution of Nazi criminals and the protection of the occupation regime. In the Soviet zone of occupation, the range of charges was broader than elsewhere. In addition to the prosecution of "Nazi and war criminals," SMTs dealt with cases of charges of "counter-revolutionary crimes." In the first months of their existence, in autumn 1947, the charges became broad and political, and people who opposed the new regime could expect draconian punishment.

Interrogation of suspects took place, usually at night, in the investigation cells of the NKVD. Various forms of mental and physical torture were used and confessions were extorted from the prisoners. Many of them signed documents written in Russian, without being aware of their contents. The basis for the proceedings by the SMTs was usually provided by paragraph 58 of the Soviet criminal code, which related to "counter-revolutionary crimes."[196] The tribunals worked fast and passed, in secret sessions, unusually high sentences. Confessions became the prime proof of guilt, witnesses for the defense were not admitted and the presence of a defense advocate was rarely allowed. The defendants had little chance of coming out of the proceedings as free men. The number of Germans sentenced by all the SMTs in the Soviet zone of occupation has been estimated at 40,000 to 50,000 persons.[197] Beginning in 1948, people sentenced to more than fifteen years imprisonment were confined in the special camp at Bautzen. It is impossible to tell how many people sentenced by the special military tribunals were deported to a camp close to Vorkuta in the Soviet Union.

In the uranium province, the military tribunals seem to have been stricter than in other parts of the occupation zone, and passed about 1,500 sentences on Wismut employees. They did not concern Nazi offences, but everyday criminality or "counter-revolutionary crimes."

As far as it can be established, the military tribunal at Wismut dealt with cases that, from the point of view of the occupying power, were the most serious, including the theft of uranium ore. In more than a half of all the cases, the prosecution rested on the accusation of theft. This was unusual in itself, and differed from the common practice of tribunals in the Soviet zone of occupation. This may be explained by the high numbers of breaches of contract and flights from the uranium province. For instance, when a man on a compulsory labor contract fled in his workclothes, he was regarded as a thief. In the first years of the existence of Wismut, Soviet management tried to stop the breaches of contract by harsh sentencing, which would stop the miners from considering flight.

Political offences, including espionage, sabotage, theft of uranium, "subversion" and attempts to strike formed the second largest group of cases before the Wismut SMT. Sentences were often passed on absurd charges or after admissions of guilt. Miner Gerhard Fieker, for instance, declared in the court that he himself had invented the evidence concerning his connections with foreign secret services.[198] He had in fact criticized safety measures at work, and passed on the information to the American broadcasting station in Berlin, RIAS. The SMT sentenced him and Axel Weidenberg, a geologist, to death in spring 1951. The charges included the collection of information on Soviet troop movements. The sentences were carried out on 1 November 1951. The third defendant, Hans Gerd Kirsche, was sentenced to 25 years' imprisonment in the Soviet Union. He was killed in an uprising at Vorkuta on 1 August 1953.[199] It can be assumed that capital sentences were passed and carried out in cases of other employees of Wismut. It is equally unknown how many long-term sentences were passed by the military tribunal.

The case of Ludwig Hecker, a miner, shows the effort by the secret police and SMT to suppress workers' opposition. Hecker talked about the possibility of a strike with his mates on the night shift. The miners decided to lay down their tools and complain about poor organization and low safety standards at work. They were promised that their complaints would be examined, and the miners made up for the shift they had missed. That was not the end of the matter. A member of the mines police came to wake Hecker up at his house after a night shift and told him "I have to take you to Barenstein, to custody…" Then the rest followed—solitary confinement and nighttime interrogations. "'Who asked you to strike?' 'Nobody.' He said 'you asshole, Gecker, you asshole! When they had strikes in Germany you were still shitting in your pants. How do you know how to make a strike?'…Then he said to me 'I am going on leave. We shall never see each other again. When I return you will not be around here. You are now going to be sentenced.' I said 'Major, sir, could you please tell me what I shall get? According to German legislation I should not be here at all.' Then he said to me 'What do you think? You'll be punished.' So I said 'three months, four months?' 'Oh Gecker' he replied, 'you are an asshole. We don't reckon in months, we don't reckon in years, we reckon in decades. We decide the fates of men!'"

On 31 May 1950, Hecker was sentenced at Kassberg. At the desk in a small room there was an officer who acted as the judge, with an assistant on each side, and an interpreter opposite him. "Behind me by the wall there stood a man with a Kalashnikov. One of the assistants read a book, the other snoozed. Sentencing was a matter of ten minutes. I had no lawyer. It was clearly and curtly explained to me that I was guilty of an act of sabotage and that I would be sentenced for sabotage according to paragraph so and so, article so and so. It was all rattled off, there was no consultation or anything. The sentence had been agreed long before, they brought it with them. And then he said 'Ten years' labor camp.' The punishment was thought to have been light, those ten years."[200]

The victims of trials by SMTs also lost their property. The possessions of three miners sentenced in August 1948 were handed over to representatives of Wismut for their personal use.[201] The confiscated items were usually modest pieces of furniture.

In coal mining, the authoritarian style of management had been common until 1945 and the term "pit militarism" was applied to it. Rigid hierarchies and harsh penalties, including punishment for low productivity, were a part of the everyday lives of the miners. The system started breaking down after the war, as a result of the process of de-nazification.

Wismut AG had different problems than the rest of the mining industry. German employees had to accept the authority of Soviet managers who, in turn, found it difficult to impose strict discipline on inexperienced and often reluctant workers. The managers relied on a combination of premiums and penalties. When a worker did not fulfill the norm, he did not get the regulation warm meal. The more severe penalties included shorter leave, reduction of wages and of premiums, or detention in labor camps or special internment pits (Strafschachten). The miners called them "concentration camp pits" or "death pits"[202] There were however no pits where the miners were deliberately mishandled or exposed to mortal danger. The punishment pits were usually parts of an object with especially difficult working conditions, including insufficient technical equipment, bad ventilation and high radiation risk. The Siebenstein pit at Filzteich near Schneeberg was among the most perilous workplaces. It produced rich pitchblende and the galleries became easily flooded. Miners were allowed to work a maximum twelve shifts in succession at Siebenstein, a unique regulation in the industry during 1947–1948.[203]

Confinement in a labor camp sometimes followed infringements of labor discipline or a criminal offence.[204] It meant that the offender escaped a more severe punishment. It would be misleading to regard the labor camps or special pits as an attempt to import Soviet methods into the occupation zone. The brutal ways of maintaining discipline were used for a comparatively short time.

The labor camps belonged under Saxon jurisdiction.[205] In view of the high crime rate, full prisons and personnel problems in the legal system, both the Ministry of Justice and the Ministry of the Interior pressed for new ways of dealing with petty crime in 1949. "Labor camps should be created as soon as possible. Only recidivists, the truly criminal and antisocial elements and serious offenders should be confined in institutions."[206] The Ministry of Justice used social and educational arguments for the establishment of labor camps. The camp at Aue had several branches [the documentation for the camp is missing] and 270 inmates in the autumn of 1949; their number apparently rose, in a short time, to 600. It is impossible to establish how long the labor camps existed and how many miners they contained.

The uranium industry seemed to be destined to employ convicts from German penal institutions. On 1 September 1947, the German Administration of Justice (Justizverwaltung) and the DZVAS issued regulations on the employment of convict labor.[207] Little was said about what the prisoners would actually do, though it was assumed that they would perform some form of physical labor. Labor exchanges were to decide their places of work and the People's Police (Volkspolizei) was to guard them. The convicts (Bewährungsarbeiter, B-Arbeiter) were to receive the same wages as the civilians, and had a chance of having their sentences reduced.

The personnel department of Wismut opposed the employment of prisoners, though they were employed by other large enterprises such as the railways or on the construction of the dam at Sosa. Wismut remained an exception. Individual objects of Wismut AG, the labor exchanges as well as the Ministry of Justice in Saxony, were for different reasons reluctant to employ the B-workers in the uranium industry. For some, the industry did not fulfill the criteria for the employment of prison labor, while others hesitated over whether working for Wismut should be a privilege or a punishment. The main reason for reserve on the part of the German labor exchanges was their fear that they would discredit the ongoing recruitment campaign for the uranium industry.[208] As the freedom of movement for the employees of Wismut was reduced, there existed the danger that the dividing line between ordinary workers and the prisoners would disappear.

The law courts were unaware of such considerations, and managers of individual Wismut objects held different views. A conflict flared up a between the Ministry of Labor and Wismut. At the end of June 1950, the ministry demanded that no more prisoners should remain in the employment of Wismut. While several of the object managers retained them, the ministry could do nothing about it. The Ministry of Justice also received complaints that some of the prisoners used the uranium industry as a device for avoiding serving the full sentence.[209] If the prisoners produced good results, and if they were in good standing with their Soviet superiors, they could be certain that they would be protected against German justice.

The increase in the thefts of pitchblende meant trouble for the management of Wismut AG. Such cases were always handled as either sabotage or espionage. In March 1950, Federal Director Maltsev issued a secret order concerning the "insufficient security standards in the mines and the strengthening of measures against the theft of ore, explosives and building materials."[210] All the pits were provided with police guard round the clock and the workers were searched upon leaving the pits;

the use of briefcases was forbidden on the grounds of Wismut objects. In addition, all Germans who had joined the company before August 1950, and without the agreement of the Soviet Ministry for State Security, had to leave Wismut.

There was discussion among the lawyers about the nature of uranium theft: was it an economic crime only? Or was it to be classed as an act of espionage or sabotage, as the state prosecutor at Chemnitz wanted?[211] Some of the sentences were too high for the president of the GDR high court, who argued that "...the necessary protection of Wismut mines in the GDR is not always exercised within a strictly legal framework."[212] He cautioned, on the other hand, before proffering charges of economic crime only when it came to the theft of uranium ore.

In the summer of 1954, a miner was sentenced by the Wismut court at Chemnitz to twelve years' imprisonment. He was apparently expected to bring a new instrument for the detection of uranium ore to Berlin, where he was promised the payment of 6,000 DM from the British secret service.[213] By that time, the tightening up of the security system and high penalties helped to reduce the thefts of uranium. Trade in pitchblende across the Iron Curtain started to fade out in the mid-1950s, when uranium was no longer regarded as a rare ore and the interest of the secret services was satisfied.

Apart from uranium thieves, the mines police and the NKVD concentrated on hunting down "spies" and "saboteurs." Between 1947 and 1949, seven or eight people a month were interrogated on the suspicion of working for Western intelligence services.[214] It seems that Schudy's mines police were more successful in tracking down spies than the NKVD. An informer of the Ostbüro gave the figure of 53 arrests by the mines police for the first half of 1948, 92 arrests for the second half and as many as 167 arrests for the period between January and August 1949.[215] After his flight to the West, the head of the mines police force, Schudy, reported that the arrests were carried out on the grounds of mere suspicion. Confessions were extorted from the suspects, and many innocent people were sentenced to long terms of imprisonment.

In addition to offences against the security regulations of Wismut AG and espionage, there were many cases of so-called "sabotage." They usually arose out of lapses in the miners' work, lapses that were not politically motivated. Human errors were criminalized, and the managers responsible could explain the failure to meet production targets by the "wrecking activities of the class enemy." Thus for instance in July

1955 seven Wismut employees received long sentences, one of them life imprisonment, on charges of "economic sabotage."[216]

It was difficult for German law enforcement agencies to operate in the uranium districts, where Soviet authorities were all-powerful. Sensitive cases, from the point of view of the occupying power, were dealt with by the special military tribunals. With increasing frequency however, petty offences were passed on to the Germans. The courts however were understaffed, and the court in Chemnitz, known as "Abteilung C," made frequent complaints to the ministry of justice.

So as to contain criminality in the mining districts, the court at Chemnitz passed comparatively short sentences, which were to be enforced immediately. Within a few months, prisons became so overcrowded that further sentencing had to be held over. "...the consequences cannot be overlooked. Even nowadays it is laughable when the police, often on foot, take the detained persons from one place to another, without being able to place them anywhere...Every month the number of sentences to be carried out grows by the hundreds and the whole police and justice apparatus is found to be illusory, as the penalties cannot be carried out."[217]

Early in 1948, the Saxon Ministry of Justice issued an order that the less serious breaches of law were to be punished with monetary penalties, in agreement with Wismut AG.[218] The law officers further recommended that the offenders should be put on "probation work" in the mining industry.[219] It meant that the courts virtually surrendered responsibility for people working for Wismut, where the management decided how the offenders should be punished.

All these measures could at best solve some acute law enforcement problems; over the long term, they would hardly be effective. That was one of the reasons why Maltsev, on 1 February 1949, wrote to the head of the military administration of Saxony, General Dubrovski, about the creation of special jurisdiction for the mining districts. It was desirable "on the grounds of the strong concentration of German population in the ore mining localities, because of numerous thefts of working clothes, timber, coal and potatoes, as well as desertions of the persons liable to work in the mines..."[220]

The writer of the letter, Maltsev, chose his words carefully. He described mass flights from the pits as "desertion." If only a part of the escaped workers were captured, it would have meant many hundreds of prosecutions. Publicly conducted cases, in the view of the Soviet management, would have been harmful to the reputation of Wismut and would have revealed the working of an organization classed as strictly secret. There was the need for a separate German jurisdiction, protected from the view of the outside world.

The military administration of Saxony recommended to the Ministry of Justice that a special department for the uranium industry be created in the city court at Chemnitz. The appropriate regulation was issued on 26 February 1949,[221] and the department of the court for mining (Bergbau) was to deal with offences in the judicial districts of Annaberg, Aue, Brand-Erbisdorf, Chemnitz, Marienberg, Schneeberg and Schwarzenberg. In March 1950, the districts of Augustusburg, Limbach, Stollberg and Zwickau were added.

Labor law cases were also transferred to a special tribunal. Though the Russian directors of Wismut often insisted that German laws applied in the mining districts, the managers of the individual Wismut objects usually paid the promise little regard. Until the foundation of the GDR, the legal department of Wismut did not feel itself obliged to bring infractions of labor laws before German tribunals.[222] In the case of disputes between individual Wismut objects and German authorities or enterprises, Wismut did not hesitate to put the judges under pressure.[223] The managers of objects appealed to the decisions of the management of Wismut and refused to follow the instructions of the German labor tribunals. In spring 1949, the Saxon Ministry of Justice shelved all cases concerning labor laws until the appearance of the new regulations. Arbitrariness in the interpretation of labor contracts thus received official sanction.

The DWK tried to clear up the matter with the occupying authorities. It took up the proposal of the Ministry of Labor that all labor law cases within Wismut should come under a special tribunal of the court in Chemnitz and that all members of the tribunal would be nominated by Wismut; the proposal was supported by the labor department of the military administration.[224] Only after the establishment of the GDR did Wismut begin to concern itself with the proposal. On 3 December 1949, Maltsev met the representatives of the Ministry of labor and Social Welfare.[225] The subject of discussion was the incorporation of the labor inspectorate of Wismut into the state inspectorate, and the establishment of the new chamber of the labor tribunal at Chemnitz to deal with labor disputes with Wismut.

Wismut representatives turned down the proposal of a public tribunal out of hand. "Before all else there exists the danger that the documentation would come into the hands of people who would misuse it to the disadvantage of the Soviet Union."[226] For this reason, Noserov, the Wismut legal representative, advocated the establishment of a special tribunal in Chemnitz. An agreement was reached on 14 July 1950, according to which all labor law litigation would come before "Department W," Wismut's own tribunal in Chemnitz. It worked "in complete isolation from all the other departments" of the court. The Ministry of the Interior and the Soviet Control Commission (SKKD) appointed the judges and members of the tribunal, who were to be nominated by Wismut and the FDGB. Of these people Wismut nominated two thirds. With the exception of one among the thirty-four co-opted members nominated in 1950, all were SED members. Documentation before the tribunal was declared to be secret, and the legal department of Wismut was to oversee that the regulations were adhered to.[227]

After the establishment of Department W, the number of cases went up sharply. They concerned payments for work on public holidays, questions of leave, serious accidents, the terms for termination of employment, or withholding the workers' personal papers. In 1950, most cases ended in a settlement, and very few verdicts were passed—clear breaches of contract by Wismut were put out of court. The tribunal was unsatisfactory because it was dependent on the party and the trade unions inside Wismut. It nevertheless meant a small advance, because the miners could for the first time complain against the arbitrary decisions of the management, though with little chance of success.

The Cold War of Words and Spies

Towards the end of the 1940s, uranium mining at Wismut AG became the focus of the confrontation between the East and the West. The more people that came into the gravitation field of the uranium industry, the more debate it invited. In the first years of its existence, Wismut acquired a bad reputation because of the system of forced labor, poor working conditions and the many social problems created by the chaotic growth of the company and its style of management.

The Soviet management appeared at first to be little concerned with the reputation of Wismut and the reports reaching the West. General Maltsev simply issued a ban on reporting on Wismut. In a conversation with the representatives of the IGB in the middle of 1947, he said that there should be no "reporting on the conditions in the ore mining industry either in the press or at meetings."[228]

The prohibition only held for a few months. Wismut AG carried on its business close to the East-West fault line, and it could not be sealed off in the same manner as other parts of the Soviet nuclear program. Beginning early in the autumn of 1947, criticism of the ways and means of the uranium industry in Saxony became widespread in the media of the western zones and in West Berlin. The Soviets as well as the SED leadership were put on the defensive. Reports on the "ore-mining industry" started to appear in the local press, in particular in the *Freie Presse* in Chemnitz, the *Sächsische Zeitung* in Dresden, the *Neues Deutschland* and in the official organ of the occupying power, *Tägliche Rundschau*. There were also occasional broadcast reports.

In the western zones of occupation and in the West in general, the first press reports appeared soon after the start of the East German uranium industry. They became more frequent in the summer 1947, soon after the first wave of compulsory labor recruitment for Wismut. The reports were largely based on evidence supplied by refugees who had worked in the Saxon mines. Social Democrat newspapers in West Berlin were especially well-informed, as they benefited from their proximity to the uranium districts and from the porous nature of the Berlin borders. The incoming information was collected by the SPD Ostbüro, and was available to anyone who cared to use it. After the establishment of the two German states, reports on "slave labor in the Zone" somewhat declined after the mid-1950s. After the transformation of Wismut AG into SDAG Wismut and the first summit meeting in Geneva, the theme disappeared from the headlines altogether.

The first propaganda confrontation about the uranium industry took place in the autumn 1947. Against the background of the Cold War, media on both sides did not pull their punches. Criticism of Eastern practices culminated when the chairman of the SPD, Kurt Schumacher, spoke of "slave labor in the uranium industry" to American trade unionists in September 1947. The material for his speech came from information supplied by the refugees and from the reports of the Ostbüro.[227] Schumacher's tough anti-Stalinist stand, and especially the words "uranium slaves" struck home in East Berlin.

On 12 November 1947, Max Weber addressed the SED leaders, and Kurt Schumacher's speech was on the agenda. The SED feared that the accusation of the party providing cover for the employment of slave labor would result in a loss of face, both at home and abroad. The party, they believed, had done its best for the miners. It had nominated its own representative for the Saxon uranium industry only a few months ago. General Maltsev had agreed to the introduction of wage scales, and

Büro Lehmann concerned itself with the shortcomings and malpractices in the industry.

Max Weber was acquainted with Western criticism and he knew what the situation in Wismut itself was. He dismissed the reports as "part of a vicious campaign against our Zone."[230] When Pieck interrupted him and asked what the situation was with regard to the so-called "slave labor," Weber replied: "We can say with confidence that no miner works as a slave laborer." When Pieck asked him about the number of volunteers at Wismut, Weber replied that "no workers have been ordered" to work at Wismut. He pointed to some 4,000 refugees, of whom about 40% became voluntary employees at Wismut. On the other hand, reports from labor exchanges for 1947 indicate that, with the exception of a few skilled miners, no volunteers were employed by Wismut.

It can be assumed that Weber tried to construct a defensive position. In any case, he succeeded in putting the minds of the SED leaders at rest. Pieck believed him and issued a brief press statement on 13 November 1947, which denied that forced labor was used in the uranium industry. It was easy for the social democrat press in West Germany to discredit that statement. Western reports concerning forced labor or the living conditions in the mining districts were easy to verify and were, for the most part, correct.

On the other hand, journalists relied largely on rumors when it came to accident rates at Wismut. They did so at the cost of their own credibility, and the counter-propaganda by the SED could represent Western newspapers as dealing out a pack of lies. In Berlin on 1 November, *Neue Zeitung* reported an accident at Wismut on 22 October when eighty miners lost their lives. The report was an invention from beginning to end.[231] Max Weber addressed himself to the problems of Wismut in public. After rumors about severe accidents in the Saxon uranium industry were picked up by Western media, he became responsible for counter-propaganda from the eastern zone. In an interview given at the end of 1947, Weber denied conjectures about soil subsidence at Oberschlema and reports about a mining disaster: "It is a fact that no pit disaster has happened in the ore mines of western Saxony since 1945. The reports that soil subsidence in Oberschlema claimed many victims are also completely plucked from the air."[232]

In defensive moves against "Schumacher propaganda," Weber ignored all of uranium industry's other problems. Western reports on the living conditions in the uranium provinces were dismissed as exaggerations: "The materials are obviously supplied by a few work-shy, lazy individuals anxious for adventure..."[233] Whatever Schumacher said could not, or simply should not, be right. The attitude was fixed by the Cold War and survived in East German propaganda for a long time. Criticism of the uranium industry was considered to almost be a mortal sin, as the Soviet Union was its ultimate target.

The propaganda battle over the uranium industry also involved a contest over the usage of certain terms. For instance, the term "Aue" was used as a synonym for the uranium industry. Whoever spoke of "Aue" had in mind the system of forced labor and of hard work in appalling conditions. The Aue labor exchange was the starting point of the miners' suffering: the "gate of tears." SED propaganda exerted itself to replace the term "Aue" with another, more positive, concept. Thuringian labor exchange proposed that the "...horror concept 'Aue' could be replaced in mass agitation by another one. One should avoid the word 'Aue' and speak only of the mining industry in Saxony. The term 'Aue,' coined by the class enemy, gives the public the creeps and should be substituted by the term 'ore-mining Saxony,' with an entirely different content..."[234] While the SED press confined itself to denying Western reports in 1947, the campaign for voluntary workers in 1948 received active support from the media. Interviews with leading party functionaries stressed the opportunities to earn more money. The president of the diet of Saxony, Otto Buchwitz, hinted at earnings as high as 1,200 and 1,300 marks a month, while the president of the DZVAS, Gustav Brack, spoke of even higher sums.[235] Apart from higher than average wages, the SED press campaign stressed extra food rations, the system of premiums, transition from compulsory to voluntary labor, as well as the cultural opportunities on offer to Wismut employees. No reference was ever made in the East German media to the size of the industry and the uses of uranium, including the contribution to the reparations account made by Wismut, were passed over in silence.

On 18 January 1948, an article on the Annaberg mines appeared in the *Tribune*, the trade union paper. Despite the complimentary tone of the report, the chairman of IG Bergbau, Paul Lahne, criticized the newspaper and insisted that it should have asked for permission to publish the article.[236] Fritz Apelt, the editor-in-chief of the *Tribune*, complained to the SED politburo and received a disappointing reply from Paul Merker: "You should not dismiss the demand of the trade union executive committee...out of hand. The comrades are acting without doubt at the request of friends who carry the responsibility for mining in the

Erzgebirge. Such unconfirmed news...could have various undesirable effects."[237]

The journalists discovered what the limit on freedom to choose their own subjects was. In the future, articles on the uranium industry could be written only after preliminary censorship. However, this rule was not applied with equal rigor in all media. A year later, in the spring of 1949, the East Berlin broadcasting station transmitted a report on the problems of the mining districts in Saxony. Labor exchanges which still used compulsory orders were criticized, and the reporter also referred to delays in the de-nazification process at Wismut and to the extension of the out-of-bounds territory to the outskirts of the town of Chemnitz.[238] The chief reporter of the broadcasting station, Skala, and his colleagues, had done their homework.

A few days after the broadcast, the department of mass agitation of the Politburo sent Walter Ulbricht a note pointing to the ban by the occupying power on reporting on the uranium industry, and criticized the broadcast for "negative attitudes."[239] The reporters defended themselves. Herbert Gessner wrote to the Politburo at length. He described the practice of forced labor in detail and enclosed declarations by miners made under oath. He concluded the letter: "I will not spread lies about Aue...that no compulsory orders are used any longer...Otherwise our attempt to hide the fact that Aue produces uranium seems to me entirely naive. Every schoolchild in the zone knows it."[240] Wilhelm Pieck probably read Gessner's letter, as he demanded that the "objectionable recruitment practices" should be discussed by the Politbureau.[241]

The management and the SED leadership continued insisting on the highest level of secrecy as far as Wismut AG was concerned. Even trivial pieces of information were regarded as state secrets. The Politburo succeeded in imposing the ban on reporting on Wismut, and the subject lay dormant for a long time, until *Neue Berliner Illustrierte* published a report on the daily lives of the miners early in 1958. The article contained information on the number of employees and on the location of the individual Wismut objects. The management condemned the article and the Politburo had to deal with the matter once again. Its new member, Erich Honecker, who was responsible for SED security policy, made his mark by advocating tough measures for the maintenance of secrecy at Wismut.[242]

Propaganda, the war of words, became closely intertwined with espionage, the most secret aspect of the Cold War. It was not all that surprising, as they both dealt with the same commodity: access to information and its uses. Many Cold War institutions were marked by the twofold nature of the contest and, on the Western side, the SPD Ostbüro was on the front line.

The SPD Ostbüro was founded in Hannover in February 1946 to look after social democrat refugees and to help social democrats in the Soviet zone of occupation. Apart from the outright opponents of forced unification between the Social Democrat and Communist parties. The social democrats who joined the SED had also hoped that the Ostbüro could serve as a shelter for social democracy until the end of the division of Germany.[243] The Ostbüro started to run a courier service between the two parts of Germany and supplied the comrades in the Soviet zone of occupation with social democrat newspapers, periodicals and other propaganda materials.

When the campaign to Stalinize the SED started in earnest in the autumn 1947, many former social democrats came under pressure. At worst, they were threatened with arrest by the Soviet authorities. In the conflicts inside the SED, "social-democratism" was commonly used as the label for the enemy within. The Ostbüro tried to warn the social democrats who came under threat and help them with flight from the Zone. Gradually there came into existence a network of social democrat underground groups, who regarded themselves as fighters against the unjust regime. They collected information on the political, economic and legal situation in the Soviet zone of occupation.

The borders between resistance against the existing regime and intelligence activities were fluid. The SPD executive knew that intelligence was being passed on to the Allied secret services, and that agreements on the exchange of information and on occasional assistance were concluded with the office for the protection of the constitution (Verfassungsschutz) and the federal information service.[244] In the year 1948, the number of Ostbüro agents in the Soviet zone stood at 2,000 people at least.[245] The cost of resistance was high, and thousands of persons were arrested, including social democrat couriers and agents.

From the beginning of 1947, a great deal of information on the rise and growth of Wismut reached Hannover.[246] The uranium industry offered a suitable conduit for confrontation with the SED regime and with the Soviet policy of occupation. In no other industrial sector was the Soviet gulag system, at least from the point of view of the outside world, so closely approximated as in the uranium industry. The Ostbüro was well informed, on the basis of detailed information from Wismut employees, about the problems facing the industry: the structure of Wismut AG, the layout of the mines, the system of forced labor, accidents, social conflicts, strikes, food supply problems as well as problems of rewards and of accommodation, the mistakes and arbitrary behavior by the pit managers or the functionaries of the SED. The Ostbüro also received reports on the meetings of the mining and district councils. An SPD executive published a memorandum at the end of 1949 drafted on the basis of the available information, entitled "Uranium Mining in the Soviet Zone."

The Ostbüro information was also made use of in the first detailed study of the uranium industry in the eastern zone by the British intelligence service,[247] which was passed on to the Americans.[248] The report contained comprehensive information about the number of employees, the location of Wismut objects and pits, as well as an estimate of production. Before the spring of 1948, fifteen objects had come into being and the report dealt in detail with the characteristics of their managers. Production figures and information on the uranium content of the ore were the least reliable parts of the report.

It was the quality and quantity of ore available to the Soviets from Wismut sources that were the most interesting items for the US and British intelligence services. For an assessment of the quality of the ore, it was necessary for the intelligence services to come into possession of its samples. The miners undertook the risky enterprise and smuggled the ore out of the workplace and handed it over to middlemen. They had political motives or financial gain as incentives. The miners invented new tricks for smuggling the ore; one of the miners brought it out on a trolley filled with sludge.[249] The price the ore fetched in West Berlin, about 1,000 DM for one kilogram, was also attractive.[250]

Despite strict controls, several samples reached the US military. The first documented case of this dates back to the beginning of 1949. The sample came from Object 1, pit 22, and weighed 2.4 kilograms with an average uranium oxide content of 22%.[251] Another sample, this time from pit 31, was of the same weight and it was examined in an US military laboratory. Its uranium oxide content was only 3%. A smaller piece of uranium ore from pit 22, which arrived about a month later, contained 11% uranium oxide.[252] Russian sources for the year 1949 indicate that 43 kilograms of pitchblende were stolen from Objects 1, 2 and 10.[253] After incoming reports and the analysis of samples of pitchblende, the Americans could no longer doubt the potential of Wismut AG to help the Soviets to close the uranium gap.

In the latter half of the 1950s, the flow of information from Wismut to the West became less frequent, virtually ceasing altogether after the construction of the Berlin wall in August 1961. The archives of the Ostbüro show this, though the silent struggle of the secret services continued. The Soviet secret service succeeded in preserving the "secrets of Wismut." Doing so required not only a lot of manpower, but also very tough methods.

In June and July 1955, several show trials took place at the House of Culture in Chemnitz (Karl Marx Stadt at the time), involving the alleged spies of the American secret services and of General Gehlen's intelligence organization. The defendants were accused, among other charges, of having betrayed Wismut production secrets. The SED party organization at Wismut organized meetings of employees on the occasion of the trials. The response proved to be disappointing, and there was little public interest in the trials.

The Miners Between Riot and Adjustment

In the meanwhile, the "wild years" at Wismut—a period lasting until the beginning of the 1950s—became legendary. The promise of high earnings attracted workers from all parts of the occupation zone to Wismut territory. After years of deprivation people tried to achieve a modest degree of comfort and amuse themselves as best they could. Contemporary eyewitnesses remember a gold rush mood: the Erzgebirge began to resemble the Wild West and the Klondike.

The lives of the miners, including hard work underground, communal accommodation, male society, special premiums and cheap liquor known as "miner's ruin" or "Kumpeltod," created a special culture. Many miners were regarded as hard drinkers and brawlers. "We are Wismut and none is better" (Wir sind die Wismut, uns kann keiner) was their motto. Because of their large numbers and the sharp definition of their community in contrast to the rest of the population, they regarded themselves as the "masters of the pitch." The actual masters however were the officers and the men of the occupying power. Face to face with them, the miners were powerless. When there was trouble, as was often the case in the early 1950s, the miners' anger turned against the "Russian stooges," members of the People's Police in the barracks. (Kassernierte Volkspolizei, or KVP)

On 5 September 1949, the mines and the railway police clashed at the Zwickau railway station. The policemen tried to pacify some rowdy miners and chase them away from the steps and the roofs of the carriages. The miners threw stones, pit lamps and hammers at the policemen, kicking them when they could.[254] The policemen, who fired a few warning shots, were forced to flee. An injured policeman was dragged to Hotel Wagner and some 2,000 people collected outside the hotel, dispersing after the security police arrived. The miners, who held the injured policeman captive, insisted that they would hand him over to the Soviet commander, because he had started shooting first.

The incidents at Zwickau were later examined by the police. The miners behaved the way they did because they were angry with their treatment by the local authorities, their long journeys to work, and were disappointed with the trade union and the party functionaries. They had no one to represent their interests, they were unhappy with the failures of the food supply system and they drank too much. Similar incidents took place at other railway stations—for instance, on 11 August 1950 at Johanngeorgenstadt.

In 1950–1951, when Wismut opted for tough personnel policies, it dismissed tens of thousands of miners at short notice. At the end of 1950, the workforces at Barenstein, Schwarzenberg, Schlettau, Oberschlema and Marienberg were greatly reduced. The papers stated that the miners were released at their own request. The dismissed miners were furious with Wismut, the party and the trade union functionaries. Their anger was reflected in routine reports on the mood of the population, prepared at the district police headquarters. The policemen reported that the dismissed miners included party members, activists and other loyal persons, whereas the former Nazis kept their jobs. The mood of the miners "was expressed in various ways, mostly at the expense of our party and trade union leadership at Wismut." Scathing references to "Russian methods" were frequently heard.[255]

Many miners who left the industry lost their long-service bonus, and Wismut tightened the purse strings for those who stayed on. Production norms were raised and Sundays were included in the working week. The mood among the miners was ugly and clashes between them and the police took place. On 21 March 1951, miners and members of the Peoples' Police clashed at an inn at Schneeberg. The miners tore up a picture of Stalin, chased away four policemen and freed a miner from custody at a nearby police station. A motorized police unit succeeded in getting the situation under control. The miners' actions were assessed as a political offence, rather than as rowdy behavior: the judge described it as enemy action. The leader of the riot was sentenced to eleven years' imprisonment.[256]

A few months later, miners' discontent at Saalfeld amounted to more than a disturbance.[257] The riot took place in the vicinity of the newly established Object Dietrichshutte, on the periphery of the uranium region, about 180 kilometers from Wismut headquarters at Chemnitz. A few hundred miners were transferred from Schlema to the new object and some of the dismissed miners were reemployed. The local people were not happy about the influx of Wismut employees and tended to be unfriendly, as they had good reasons to fear confiscations of their accommodation.

The tension grew when Wismut officials announced in summer 1951 that 2,000 further miners would come to Saalfeld and that they would need accommodation.[258] The steelworkers, particularly those at the Maxhutte nearby, felt themselves threatened and the tensions often ended in fights between the locals, with the steelworkers on one side and the uranium miners on the other. The police force was weak and could not stop people singing old Nazi songs.

At the Saalfeld market place in the evening on 16 August 1951, payday at Wismut, two miners started to fight and onlookers gathered to enjoy the spectacle. When the miners threatened a passerby, policemen who had been called to help arrested a miner and took him to the police station. The miners and locals who witnessed the arrest surrounded the police station to demand the release of their comrades. In the meantime, the crowd grew to some 3,000 people. The policemen were forbidden from using firearms and were ordered to release the miners in custody. The crowd believed that the police held other miners, and protests did not die down even after a delegation of miners was allowed to inspect the police station. Wismut lorries then brought stones to the market place and the crowd began to throw them at the station before finally invading the building. The prison behind the town hall was stormed, and two miners, convicts serving sentences, were freed. They had apparently been mistreated by the policemen. The crowd wrecked the prison, and people began shouting "Hang them, kill them!" The policemen escaped over the top of the roof. Five were injured, two of them seriously.

Soon after the assault on the prison, the first secretary of the SED in Thuringia, Erich Mückenberger, and the Minister of the Interior, Willy Gebhardt, arrived at Saalfeld. Trying in vain to calm down the crowd, they feared that they would be lynched themselves. After midnight, the triumphant miners left Saalfeld.

The tension continued the following day. Rumors of new arrests were in circulation. The miners returned to the market place and again demanded the release of the prisoners. Two women were arrested in the morning of 17 August, and strong police units arrived at Saalfeld, though they refrained from using force. SED functionaries succeeded in convincing the miners to take part in a meeting at a local hotel. The miners started returning to work and the situation was under control by the morning of 18 August. In the following weeks, several groups of miners were arrested by the security police, and twelve defendants were sentenced to terms of imprisonment of between 8 and 12 years in May 1952.[259]

The background of the miners' riots was provided by a poor food supply, despite all promises, and the poor housing conditions which most of the miners suffered from. Their economic interests were at stake, though the protests contained some moral elements. Mistreatment of the miners under arrest heightened the mood against the badly regarded Peoples' Police. The miners' action was focused on the prisoners' release, otherwise the miners made no political demands. The riot had no political aim, much less a directing agency abroad. It was nevertheless an implicit warning, which found expression less than two years later, in the revolt in the June 1953.

Wismut management reacted to the Saalfeld riots immediately, in August 1951. It established an office for complaints, one of the many measures introduced by the occupation authority and the SED leadership. After the riots, SED functionaries and trade union officials wanted to find the guilty men. Local officials were quick in naming them: "A large part of the employees consist of former Nazis, militarists or elements who oppose the laws of our order..."[260] Other reports offered similar analysis: SED leaders in Thuringia discerned "former militarists and fascists" at work and criticized the penetration of "alien class elements" (klassenfremden Elementen) into Wismut.[261] SED leaders regarded the miners' protests as a planned provocation, directed from abroad. They ignored the fact that the miners did not have anyone else other than the state authorities to protest against. Since the party had taken all aspects of life of the society under its control, it tended to politicize every kind of protest.

The SED Politbureau, as well as the occupying authorities, had forthright reports before them. "At several workers' settlements of Wismut AG, situated far from the pits, there exist no organs of the state, no people's police, no registration office of the inhabitants, too few shops, cinemas, libraries..."[262] The security organs in the mining districts were understaffed and their weakness was a matter for misgivings by the German and the Soviet authorities alike.

On 5 November 1951, the Russian Control Commission invited the SED leaders to report on the crisis in the uranium industry.[263] A long list of measures was agreed upon, including the partition of the Aue district, so as to provide better police control and smaller administrative units. The police and security apparatus in the mining districts was to be strengthened, propaganda activity improved, stricter control of all social organizations was promised, as well as personnel changes in the party and trade union organizations at Wismut. A new building program was to be launched and the miners were to be drawn into local administration and lead a richer cultural life.

The unrest at Saalfeld, unwelcome as it was, helped the Soviet military command and the top officials of the German party to give their full attention to the conditions in Wismut AG. The plans in 1952 for the construction of the miners' houses and of communal and cultural buildings were conceived on a generous scale. In sixteen towns 5,000 apartments, 1,500 individual houses, 12 schools, 12 post offices, 12 sports grounds, 5 swimming baths, 12 club houses, 14 canteens, 17 shops, 11 bath houses, 6 pioneer camps and 9 service stations of various kinds were to be built.[264] The list of requests was sent from Wismut to Walter Ulbricht on 28 November 1951. Instead of the 375 million marks required by Wismut, the State Planning Commission allowed 150 million, while the Ministry of Education was to finance the building of schools and club houses.

On 14 December 1951, Walter Ulbricht informed the chairman of the State Planning Commission about the understanding with Wismut.[265] Of the 150 million marks allowed for housing and communal buildings, 120 million were to come from Wismut AG and its social insurance account and 30 million from the state budget. At the planning commission, Rau was angered with the measure about which he had not been consulted.

On 15 and 16 February 1952, Walter Ulbricht took part in an extended meeting of the party secretariat at Wismut with the representatives of several institutions, including the Ministry of the Interior, the police and the FDGB.[266] Personnel changes at the Wismut SED Secretariat were agreed upon: the police force was to be strengthened, the inhabitants of the mining districts were to be re-registered and persons who came to the mining area from the West were to be removed. The replacement of town mayors was mentioned, in particular of the CDU mayor in Johanngeogenstadt, who was unpopular in SED circles. The effect of the SED resolution was limited. The miners were not impressed with the cultural program and remained loyal to their pubs.

Soon after the Saalfeld riots, the state planning commission appointed a "special commissioner for resettlement." He was to supervise the transfer of 3,000 miners from Johanngeorgenstadt and Oberschlema and carry it out as smoothly as possible. The Soviet occupation authorities expected of the commissioner "strictly confidential work."[267] Repeated protests against forced transfers had taken place, such as those in November 1950 when local people at Annaberg-Buchholtz, requested to vacate their houses in favor of Soviet employees of Wismut, complained to president Pieck.[268] In autumn 1951, the special commissioner was to prepare the evacuations demanded by Wismut and provide suitable compensation for the families concerned. The owners of building plots were to be compensated at 135% of their 1944 value. Between 1952 and 1955, 3,480 persons and 105 workshops were resettled and 330 building plots were bought. Despite improved compensations, the commissioner was unable to rely on the good will of the people who were affected by the transfers. No public protests took place, which was regarded by the state authorities as a success.

Politically motivated resettlement was a direct consequence of the Saalfeld riots. The occupation authorities demanded at the beginning of 1952 that the SED Politibureau should "remove criminal elements from the mining districts."[269] A short time before, the security regime in the Czechoslovak uranium districts had been tightened: the Russians apparently decided to impose a new and harsher regime on the whole Erzgebirge region.

For some time, the police had been compiling lists of "suspect persons." The register was based on information provided by agents and reports from policemen on patrol duty. On the basis of the reports, the Ministry of the Interior drafted a plan, of 15 December 1951, on the "cleansing of the Wismut territory of degraded elements."[270] People with criminal records, people without permanent residence or employment and prostitutes were to leave the uranium province.

In the early 1950s, the standard of living of Wismut workers was better than the rest of the population in the GDR. Even the accommodation situation began to improve after the Saalfeld riots. The Wismut workforce continued to be troubled, in the spring of 1953, by continuing dismissals and campaigns for voluntary increases in production norms. It may be that the threat of unemployment and the prospect of the loss of well paying jobs helped to discipline Wismut workers.

Discontent caused by low pay affected especially men employed in the garages and workshops of Wismut AG in Thuringia. The drivers and mechanics were paid considerably less than unskilled workers in production. Their pride was hurt and they were embittered. As early as

April 1953, the drivers threatened to strike. Party representatives at Wismut promised that wage demands would be dealt with, and the strike was averted.[271] The situation of the drivers and the mechanics remained virtually unchanged.

In spring 1953, the miners also came under the pressure of a new campaign for the increase of work norms. It did not help that the campaign was presented as a struggle for peace, under the slogan "every voluntary increase of the norm—a blow against the war-mongers!" Though the discontent of the miners with higher norms grew, they at least did not suffer wage reductions as did the industrial workers.

The setting aside of parts of the old town of Johanngeorgenstadt for Soviet personnel antagonized the local population, and on 15 June 1953 about a thousand people gathered in the market place to protest against the projected expulsions. The evacuation of the old town was slowed down, though the authorities did not give up the plan.

In other towns of the GDR, protests were also taking place, mostly against the increases in work norms. The SED tried to calm the situation by making new concessions, but the "new course" was regarded by the people as a declaration of bankruptcy. When the trade union newspaper *Tribüne* tried to justify the new work norms on 15 June 1953, the cup overflowed.

Demonstrations of the building workers in Berlin started in the morning on 16 June. They demanded withdrawal of the norms. The following day the protest movement spread over the whole territory of the GDR. The protests were accompanied by demands for the resignation of the government and for free elections. The SED regime was saved when the occupying authority declared martial law on 17 June and showed it was ready to use armed force against the protesters.[272]

In the Wismut districts, the protests were less intense than in the rest of GDR.[273] In West Germany, the press described violent protest actions by Wismut miners, destruction of equipment and dozens of dead people in the mining districts, but it was all pure invention.

Strikes too place in several Wismut objects on 17 and 18 June. The Wismut SED organization reacted more decisively than did most other district organizations. Without waiting for orders from Berlin, the SED leadership took extra security measures, with Soviet help. If their reports to the Central Committee are to be believed, then "model discipline" was the rule in Wismut objects during the critical days.[274] The party functionaries even endeavored to make the miners pledge themselves to meet higher production targets. According to SED officials, more than 50,000 employees had made the promises by 23 June.

While Wismut districts in Saxony remained, on the whole, calm, the situation in Gera, Weida and Jena in Thuringia became, on 17 and 18 June, more threatening. Every fifth employee in the Wismut objects in Thuringia laid down their tools. The starting point of the strike was the garage at Katzendorf near Ronneburg. The drivers and the mechanics drove to Weida, where they took part in the protests and asked other enterprises to follow suit. They went on to Gera, where a police unit failed to stop them. The policemen were disarmed by the workers before Soviet troops could come to their assistance.

Something rather disturbing occurred at Gera. The strikers pinned their hopes on the Wismut miners joining them, as the Wismut miners had a reputation for fearlessness. Some 6,000 people in Gera demanded the immediate abolition of State Security, increases in wages and the release of all political prisoners. The protest meetings were peaceful until the arrival of heavy Wismut vehicles. The protesters tried to storm the prison, and had to be stopped by Soviet troops from freeing the prisoners. Another clash took place between Wismut's employees and the police in the market place. The miners captured some weapons and destroyed them.

Only after the declaration of martial law and the arrival of Soviet troops did the miners return to Weida. They demolished several inns and attacked the police station once more. There was some shooting, and the Soviet troops cleared the public places and sealed the town off. By then the miners had left for Berga, where they attacked the police station. In Jena there were also many miners among the protesters. For a short time, the drivers' strike impaired production of uranium in the Ronneburg district.

Unrest kept breaking out. A few days after the uprising the district party leadership believed that it had the situation under control. The presence and the intervention by the Soviet troops were decisive in preventing the strike from spreading throughout the uranium district.

The whole Wismut territory was a part of a closed military zone and the Soviet armaments industry. Peoples' movement was closely monitored. The uranium districts could be entered only with a special pass. On their way to work, the miners themselves were controlled at checkpoints in the streets and on the trains. The mines themselves were patrolled by Red Army troops. The isolation of the mining districts prevented protests being carried from there to the outside.

Security measures in the uranium province were further tightened up on 17 June. Starting a strike on the company premises was pointless, even suicidal. The size of the mining districts made communications between the miners difficult and they knew that attempts to strike underground would have been regarded as a form of direct resistance to the occupying power. Furthermore, Wismut employees enjoyed higher wages than other industrial workers and they had something to lose. So did their wives, who had access to goods in short supply elsewhere.

After the defeat of the uprising, there came the reckoning with the "rabble rousers." In the following two weeks, 51 Wismut employees were arrested in connection with the strikes, of which 38 came from the Gera district.[275]

Despite its failure, the uprising was not entirely in vain. It remained, until the end of the existence of the GDR, an important harbinger of democracy and human rights. The occupying power and the SED were more inclined to make concessions after the uprising. The Soviet government announced in August 1953 that war reparations by the GDR would end on 1 January 1954. The increases in work norms were withdrawn, as were other unpopular measures.

At Wismut, the management issued order no. 245 in August, retroactively increasing the wages of the lower paid workers.[276] The unpopular Sunday shifts were rewarded with a premium of 50%. The projects to improve the miners' living standards, were speed up. The situation in Wismut was gradually stabilized and, by the end of 1953, the SED had regained its self-confidence to such an extent that it returned to its old practices.

Radiation Damage and Accidents

Miners' work has always been regarded as being dangerous, and working underground for Wismut AG was no exception to that rule. Because of material shortages and post-war chaos, it was hard to provide the mines with the necessary electricity, pumps, compressors, drills and mine props. As long as there was no compressed air available, nor pneumatic hammers, the workers had to improvise. Until autumn 1947, 41 working mines were under the supervision of one inspector and had only one ambulance at their disposal.[277] First aid necessities were unavailable, a rescue service did not exist, nor did regular safety controls.

Falling stones were the most frequent cause of accidents, especially after blasting. From time to time, soil subsidence occurred and fires broke out. Explosives, which were of poor quality, were the second most common cause of accidents. The inexperience of most of the miners and sometimes their carelessness, as well as the pressure on them for high production from the Soviet management and their own comrades, were also responsible for the growing accident rates. The journey to work itself could become an adventure. In 1947, the Reich railways took over the larger part of commuter traffic without having sufficient rolling stock at their disposal. Trains were overcrowded to such an extent that people traveled on the steps and the roofs of carriages and on the buffers. Jumping off moving trains was their favorite sport. At the end of 1948, Saxon government officials received reports of amputated limbs on a daily basis.[278] A former railway man at Schwarzenberg station remembered an accident in August 1948: "A shift train was just arriving from Johanngeorgenstadt on platform no. 2. Because the trains were very long, the engine driver had to go on, past the station. Many miners, standing on the steps of the carriages, did not wait for the train to stop and jumped off. One miner jumped sideways and hit a steel signal post with his face."[279]

Commuting started improving in the 1950s. In the spring 1950, a second track started being laid down on the line between Aue, Johanngeorgenstadt and Schwarzenberg. Other lines for Wismut commuters' traffic were constructed, rolling stock was improved and double-decker trains were introduced.

When they reached their places of work, the miners faced other dangers. They had to go down several hundred meters in unsafe conditions and falls down the shafts could be fatal. One of the miners, Norbert Schunemann, wrote in November 1947: "I myself work...at the level of 200 meters. Access to the shaft takes place on a perpendicular ladder fixed to the wall. The ascent from the deepest level (400 meters) to daylight takes almost two hours. It has been proposed only recently that the ladders should be provided with a barrier at the back, so that people would not fall down the shaft backwards. They usually fall into pools of ground water, which can be found on the 400 meter level. I know nothing about the whereabouts of the people who have fallen down. The highest number of fatal accidents we had in our shaft during a working week was fifteen people. Insufficient pit props and careless blasting are given as the reasons."[280]

One of the greatest dangers of the "wild years" of Wismut was the poor quality of explosives. The miners often entered the thick smoke caused by the explosion, so as to save time. "That was most dangerous! The yellow-brown smoke. We banged about in that. Every cutter (Hauer) in the squad would say 'we must have a look straightaway how it's come out'. And then we crawled around on all fours. And if the material reached as far as the ridge, that did not worry us. When the smoke brought dust with it, we were not worried either."[281]

Wismut's management dealt with frequent accidents by organizing courses in blasting.[282] Skilled miners from the black coal industry were brought in to Wismut, as well as members of the pioneer corps of the former Wehrmacht. As the specialized cutting and blasting gangs were built up, the situation began to improve in the early 1950s.

The Soviet board of directors tended to ascribe the majority of accidents to "sabotage." The fiction in these cases was hard to separate from reality. "The Russians' attitude was: there are no accidents! Every accident is somebody's fault. And the person responsible must be sought."[283] In the early years of Wismut, some politically motivated attempts to sabotage production did indeed take place. The miners, often acting on their own, tried to destroy the electrical system or mining machinery.[284] Wismut management and the SED party leadership had an explanation ready for the acts of sabotage: they were carried out by "Western agents." The management ascribed their mistakes to the work of the saboteurs. It also concealed accidents because it did not wish publicity to affect voluntary recruitment, nor did it want to upset employees or provide material for anti-Soviet propaganda. While Soviet managers blamed sabotage and subversion for the accidents, they never took the possibility of their own responsibility into account.

Nor was the SED leadership allowed insight into the situation with regard to accidents at Wismut. Party leaders relied on information from Max Weber. At a meeting in November 1947, Weber put the number of fatal accidents from the beginning of 1946 until October 1947 at 18, adding that "the mining of ore is no dolls' house."[285] It is impossible to tell whether Weber did not know the true situation, or whether he wanted to make the information acceptable for party leaders and assure them that the number of accidents in the uranium industry remained within reasonable limits.

Only a few weeks after the meeting, Weber's credibility suffered a setback. Several members of the SED Secretariat received Schunemann's letter, forwarded by the district committee at Zwickau, which referred to fifteen fatal accidents in one pit in a week. Paul Merker asked Weber to examine the information.

Weber expressed doubts about the reliability of Schunemann's report. "We have found out that in all the three districts, Aue, Annaberg and Marienberg, 19 fatal accidents took place in November. The number of fatalities and injuries were —in view of the number of workers employed in west Saxon mining, about 28,000 at the time—fortunately not very high."[286] Weber's note of January 1948 highlighted several problems. One was that the SED leadership gathered information about accident rates from letters of complaint. Weber played down the number of accidents and presented fatal accidents as an "acceptable risk." The murderous war, it is true, was a recent memory. Weber nevertheless helped to create an attitude of mind that ignored the risks faced by the miners. Accidents at Wismut AG were either not reported at all or they received only a fleeting mention.

Newspaper editors in the western zones used the information and rumors that reached the Ostbüro. They aimed to vilify the SED and the Soviet Union, and to deter people from working for Wismut. With regard to Wismut's accidents, Western reports tended to lead the readers seriously astray, and the number of victims could never be high enough for journalists. The accident in Johanngeorgenstadt in November 1949 made an especially deep impression on the press. In a late edition on 28 November, the *Telegraph* in West Berlin reported 2,500 dead. The Ostbüro in Bonn believed 600 dead more probable.[287]

What did actually happen in Johanngeorgenstadt? Rudolf Oeslner, engineer in charge of blasting, was asked on 24 November by Kirilenko, the Soviet political officer, to take care of the rescue work. Wismut's own rescue service was not yet fully operational. Looking back on the rescue work, Oelsner remembered: "I issued the following orders: 1. All galleries leading to the rock face (Grubenfeld) were to be closed down. 2. Rescue team from Zwickau black coal pit was to be brought over. 3. The arrival by rail and buses of the second shift was to be stopped. 4. (Make) further fire fighting and ventilation provisions. After the arrival of the Zwickau rescue team, fire-fighting orders were issued straightaway. About 13:30, director general Maltsev came and took over the command of the operation. Towards 14:00 hours I was asked to vacate the explosives store on level 78…as a precaution. Forty Soviet soldiers were put at my disposal…As the gallery was affected by fumes from the fire, several soldiers who were then working underground suffered gas poisoning. They were taken to the surface straightaway…The fire was put out during the weekend. About forty miners (including the Soviet troops) were poisoned during the fire fighting and were taken to the hospital for treatment. On the following day (25.11.49) the pump operator, Richard Reifenrath, was found dead. There were no other victims."[288] The "explosives disaster with many hundreds dead" was added to Cold

War legend, created by mass media which were unable to check their sources.

During 1948, the government of Saxony carried out inquiries into accidents in other branches of industry. Accident rates in the black coal industry were higher than at Wismut AG, Wismut held the lead in fatal accidents. According to the figures for 1 October 1948, there had been three fatalities in the black coal mines and 22 at Wismut: that is 0.07 and 0.38 deaths for per 1,000 employees.[289] Wismut AG tried to improve safety precautions at work by codifying regulations and instructing the workers, and the management issued detailed "Safety Precautions for the Mining Industry" as early as 1947.[290]

The works inspectors reacted to the rate of accidents sharply. In cases of security breaches, criminal proceedings were instituted against the offenders. The penalty was usually financial, in more serious cases Wismut employees were sentenced to imprisonment. In the first quarter of 1948, proceedings concerning breaches of safety regulations were instituted against two chief pit supervisors (Obersteiger), one blasting specialist, two miners and one electrician.

At the beginning of the 1950s, accident rates at Wismut started falling off. Wismut employed fewer workers and the training of a skilled workforce was completed. The technical equipment had also been improved; teams at the rock face were provided with supports for the drills and electric locomotives were available. Protective clothing was in greater supply, and the safety inspection service was upgraded. After the working conditions of the miners had improved and the workforce had stabilized, the pit disaster at Niederschlema struck.

On 15 July 1955 shortly before midnight, a fire broke out at pit 250. The fire was located in a blind shaft at the depth of 480 meters.[291] While turning off a ventilator, welder Günther Paersch noticed that one of the wires was still glowing.[292] While he searched for an electrician, the fire started. It spread very quickly to the adjacent parts of the mine, including the machine and the pump rooms.[293]

The night shifts at the Oberschlema and Niederschlema/Alberoda objects, comprising some 10,000 miners in all, were warned in time and ordered to go to the surface. Wolfgang Abendroth's team of five miners remained underground. They had not been warned and discovered the fire in the small hours of the night. By then, their escape route had been cut off by thick smoke and their only chance was to build a wall across the access route: "We built our own prison, without knowing whether we would ever be able to leave it."[294]

In the meanwhile, hectic rescue activities were undertaken. The pit supervisors were not certain how many miners were left in the place of the accident. On the morning of 16 July, the head of the rescue service, Max Markstein, left for the disaster area. He was an experienced man who had saved the lives of many miners. He was accompanied by supervisor (Steiger) Fritsche and the trade union security official Bruhl. They went down without any rescue apparatus. As they entered the shaft, they breathed in lethal monoxide gas.[295]

Rescue operations, initially conducted by the Wismut team, were hindered for hours, and ultimately days, by thick smoke. There were not enough oxygen masks to go round and the main Oberschlema rescue service was used in a chaotic manner. Its Soviet commanders issued conflicting orders and lost sight of the location of other rescue teams. Overzealous rescuers, fearing disciplinary consequences, were exposed to unnecessarily high risks. Rescue teams from other Wismut objects were sent underground immediately after their arrival, without adequate instructions. Their breathing apparatus held out at most for two hours. The teams worked separately and few men knew the layout of the mine. "It happened so that teams were often sent to blind shafts and nothing was done to ensure their way back. This caused the death of chief supervisors (Obersteiger) Röder and Jansch and four other colleagues in the rescue service would have also come to a bad end had...they not heard emergency signals of their comrades."[296]

On 18 July, the police rescue unit reported that 23 people were still missing. The management of the pit had failed to put out the fire with its own resources, yet it was reluctant to accept outside help.[297] For a time, there existed two command posts side by side: the pit management and the police rescue service. General Director Valentin Bogatov's 18 July order finally facilitated the use of outside help.

The five trapped miners were freed after some fifty hours. The fire was finally put out on 24 July. Only then were all the Wismut objects at Nieder- and Oberschlema, as well as at Alberoda (with the exception of blind shaft 208b, where the fire broke out) opened up again. The disaster resulted in 33 deaths: 14 miners, 7 assistants of the technical service, 1 policeman, 9 members of the rescue service and 2 management officials. More than 150 miners and rescue workers suffered from various degrees of poisoning. Forty years after the disaster, Werner Scheider, who was then deputy chief supervisor, described the causes and the extent of the catastrophe: "There were until then only a few fire protection measures in place. An underground fire service did not exist and people did not know how to use the rescue equipment. Every shaft had its own rescue team and there was no communication between them in the first hours. Every one of them worked blind..."[298]

The reactions of the management, the party and trade union leaders, the district headquarters of the Peoples' Police and of the Security Police, were all true to form. After the disaster was announced, communications between the leaders of the SED took place on the special telephone network. They were uncertain for a few days what kind of public position the party should take; or if it would take one at all. While the rescue operations were going on, the police and the state security started a probe of public opinion inside and outside Wismut.[299] As public pressure grew, it was agreed that the media should come into play. Conjectures that the fire was an "act of sabotage," as one of the first police reports assumed, appeared. On that assumption, twenty state security and criminal police officers started investigations on 19 July, collecting evidence from 59 miners.[300]

The task of informing the families of the victims fell to a member of the State Security and the SED district leadership. The victims' families received a provisional compensation of 1,000 marks, which was followed by comprehensive help, including the allocation of apartments, clothing and holidays, as well as further financial compensation from the Ostsee-Schwarzmeerversicherung.[301]

The Niederschlema catastrophe set off a review, by SDAG Wismut, of all the existing security systems. The rescue system was restructured, and its members better equipped and trained. Beginning in October 1956, the regulations for health and safety protection in the GDR became extended to Wismut as well.

No reliable statistics exist on the accident rate at Wismut. The results of research carried out in 1964 were destroyed for security reasons.[302] Its authors gave the figure of 376 fatalities for the years 1949 to the mid-1964, with the proviso that accidents that later resulted in death were not taken into consideration.[303] More reliable data is available only after 1965, and may be found in the documents of the deputy general director of safety at Wismut and of the mining authorities.

According to the available official records, it appears that between 1946 and 1990, when Wismut stopped uranium production, almost 42,000 accidents had taken place, of which many were serious and 772 fatal.[304] There are reasons to doubt the official figures and the true numbers of uranium victims were likely to have been considerably higher.[305]

Uranium mining, begun in chaotic circumstances by a largely inexperienced workforce, was carried out at a high human cost. The work was done under time constraints and pressure from the occupying authorities, and there was little opportunity for sufficient consideration of safety measures. Negligence and carelessness resulted in high accident rates. It took years to achieve a high safety standard in the Wismut objects. Yet the uranium industry at that time was no exception as far as accident rates were concerned. The threat of injury underground was similar in the Ruhrgebiet or in the black coal districts in Zwickau-Oelsnitz. The post-war years were exceptional for the German mining industry, both East and West, as they were marked by extensive plunder.[306] Accident risks in the uranium industry declined in the second half of the 1950s. After that, accident rates at Wismut fell under the average for the industry.

The long-term effects of radiation presented an additional health risk in uranium mining. Before 1945, No other region in the world had more experience with this invisible threat than did the Erzgebirge, and nowhere else was more done to prevent it. The "miners' sickness" of Schneeberg and Jáchymov has a place in the history of medicine. The first rule ever on tolerable radiation limits of was made by the Karlsbad mines office on 21 November 1940. Though it fell into virtual abeyance during wartime, it was later used by the head mines office at Freiberg as a model for the regulations applied in the Schneeberg district. On 15 June 1944, the Zwickau mines office issued "Safety Regulations for Trade Union Members in the Schneeberg Mining Industry."[307] The regulations consisted of thirteen paragraphs concerning regular medical examinations and detailed documentation, working hours and holidays, as well as safety rules above and below ground. Working time, including travel to work, was fixed at eight hours, and water drills were required. The upper limit set for radiation exposure was 3 Mache units. When this limit was exceeded, the miner was to be employed in a place with lower radiation for at least four weeks. The Karlsbad ruling of 1940, together with the Zwickau regulations four years later, were exemplary set of rules for the time.

Many studies concerning lung cancer at Schneeberg and Jáchymov were published in specialist magazines written in English before the Second World War. In 1944, Egon Lorenz reviewed 57 publications on the subject in the *Journal of the National Cancer Institute* in the USA.[308] After the end of the war, when extensive uranium mining activities were starting in the United States, American physicians turned to the discoveries made many years before in Saxony and Bohemia. The

US military and public authorities tried at first to minimize the dangers of uranium mining, or dismiss them as communist propaganda.

When it became apparent that the Soviets were interested in the Erzgebirge deposits, the medical profession warned of the hazards involved in the mining of uranium. Concern with the threat of lung cancer was first expressed by Professor Georg Wildführ, director of the Hygiene Research Institute in Dresden. In May 1947, he wrote several letters to the Saxon government, in which he summed up the state of research and advised caution. "Occupational hazards which can come into play in the mining for uranium are caused by the following three factors: dust, radiation and lead. The possibility of occupational damage is given by the combined effect of breathing dust and of radiation or emanations from the rock or the pit waters. One has thus to reckon in the case of the Schneeberg lung disease with the known effect of silicosis and of cancerous formations in the lung...it has to be taken into consideration that long-term impact of radioactive rays cause irreversible general damage, as is known from the sicknesses of people who work with radioactive materials."[309] Wildführ added that, based on his experience with Schneeberg disease, the first wave of silicosis and lung cancer cases could be expected only about ten years after the start of mining activities.

Dr. Seifert of the Department of Health used Wildführ's warning to explain the dangers of the new industry to the Minister for Labor and Social Welfare, Walter Gäbler.[310] Dr. Seifert recommended "an order concerning the classification of specific illnesses arising from uranium mining as occupational diseases."

In the meanwhile, the medical profession in the mining districts had more urgent problems to deal with, including the consequences of famine. Lack of suitable work clothing for the miners, as well as poorly heated accommodations, made matters worse. In 1947, up to 300 miners a month were unable to work because of sickness.[311] The doctors were also concerned about the rising numbers of cases of venereal disease and, because of the daily problems they faced, they had little time to consider the possibilities of the hidden, long-term dangers. They were nevertheless acquainted with such dangers in a circular concerning the "uranium mining sickness."

The Saxon government, as well as the SED Party leadership, took note of Professor Wildführ's warnings. The party leadership discussed the uranium industry's problems twice, in October and November 1947. Max Weber was invited to attend the meetings, probably on the recommendation of Helmut Lehmann. He downplayed the dangers of the uranium industry, and argued that not one of the 12,000 miners suffered from lung cancer.[312] In the minutes of the meeting, cries of "hear, hear" are recorded at this point, as Weber helped party leaders overcome their doubts. They put trust in their comrade and ignored the warnings from the Saxon physicians.

What would have happened had the leaders of the SED taken note of the Saxon doctors' warnings? It is hard to tell. The occupying power was unlikely to slow down the building the uranium industry. Timely improvements in safety standards, especially better underground ventilation, would have prevented many cases of lung cancer. The fact that the SED leaders did not invite medical experts to any of their meetings had fatal results.

It is not known whether the Soviet management of Wismut were aware of the threat of radiation. In annual reports between 1947 and 1953, protection against radiation is not mentioned once. The State Labor Inspectorate (Landesarbeitsinspektion) as well as the Saxon Department of Health demanded that Wismut undertake regular measurements of radon levels in the mines. Yet Wismut management never let the Saxon authorities know whether any measurements were carried out.[313]

Neither the Soviet management, nor the pit supervisors, warned the miners of the radiation risk; nor did the SED or the trade unions. The only people who cautioned the miners were the former spa doctors, or the local people who were aware of the history of the "Schneeberg sickness." Werner Schiffner, a miner, wrote that very few of them "knew the writings of Agricola, Paracelsus and of their successors. Yet all they needed to do was to look at the dates on gravestones in the cemeteries. Many a Wismut miner himself lost his father or uncle because of the 'Schneeberg lung cancer.' These experiences did not put most of them off exposing themselves to the danger of a shorter life."[314] Schiffner thus raised the fundamental problem of the early years of the uranium industry: the suppression of invisible dangers by the miners themselves. Many workers did not intend to stay at Wismut permanently—they simply wanted to earn some money quickly and leave. If they were aware of the threat, they ignored it.

Wismut management was, on the other hand, more inclined to deal with the danger of silicosis. The disability of "dusty lungs" (Staublunge) was long known in the coal industry. In 1948–1949, the Saxon works inspectorate tried to acquaint Wismut miners with the risk.[315] The inspectors assumed that about a half of the miners—out of the estimated total of some 70,000 workers in October 1948—were threatened with acute silicosis. The inspectorate suggested better ventilation as a preventive measure, as well as blasting at the end of the shift and the use of dust-masks. The inspectorate recommended that the regulations of 21 November 1940, or of 15 June 1944, should again come into force.[316]

The regulations from the Nazi period were left aside. Wismut management issued its own rules instead. The first order on safety issued in 1947 recommended the use of water-cooled drills, dust masks and improved ventilation.[317] The key point of the old regulations was missing from the rules issued after the war. It was the setting of limits on the maximum exposure to radiation. The Soviet management did not want to enter discussions on this sensitive point and avoided any hint of the existence of the hidden threat.

The new regulations were hard to follow. So as to start water-cooled drilling, several technical and material conditions had to be fulfilled. Water pipes with taps had to be laid down and the drills supplied. In addition the use of water-cooled drills would have made the work of the miners more difficult and would have slowed down production. According to eyewitness accounts, the transition to wet-drilling began as early as 1947 in a few of the Wismut objects, and was completed as late as 1950–1954.

The highest levels of radiation were suffered by miners who entered employment at Wismut between 1946 and 1955. As no radon exposure measurements had been carried out before 1954, their levels can be estimated only retrospectively.[318] Expert opinion indicates that until 1955, average exposure was extremely high, but later declined sharply thanks to better ventilation.[319] As late as 1957, samples of pit air were taken to the surface in football bladders and examined in laboratories for radon content.[320]

More than 90% of the miners who suffered silicosis or lung cancer belonged to the "first Wismut generation."[321] According to rough estimates, some 500,000 miners worked underground between 1946 and 1990. Until 1989, about 30,000 of them suffered from occupational disease: 15,000 from silicosis, 5,300 from lung cancer, 5,000 from stress-related damage and 4,700 from deafness caused by noise.

Is it possible to make an international comparison of these frightening figures? Can the Wismut operations be compared—as a daily newspaper attempted—with the explosions over Hiroshima and Nagasaki? Journalists referred in the article to research that compared the incidence of lung cancer in several large regions where uranium was mined with the cases of people who had survived attacks by atomic bombs.

Table 6. International Comparisons of the Incidence of Lung Cancer[322]

	Size of the Group	Number of Cases
Colorado	3,400	256
CSR	4,300	115
Eldorado	6,900	65
Ontario	13,000	87
Hiroshima/Nagasaki	76,000	638
Wismut	400,000	5,000

If methodological problems are set aside, a bewildering fact emerges: among the survivors of the bomb blasts there were fewer cases of cancer than among the uranium miners. If we accept the figures at their face value, then American uranium miners, per 1,000 workers, suffered the highest exposure to radiation. In fact, US doctors at the beginning of the 1950s estimated exposure to radon in the Colorado mines as having been ten times higher than the exposure found in the Erzgebirge before 1945.[323] Considering that some 400,000 people were employed by Wismut until 1950, and usually for a short time only, the tragedy in the Erzgebirge takes on a different dimension.

In Czechoslovakia, the incidence of lung cancer appeared to be surprisingly low. The first regular radon measurements were carried out in the Czech uranium industry as late as 1960,[324] and exposure levels at the time were much lower than the values established in US mines. Miners of the "first generation" workers in Jáchymov, particularly POWs and then political prisoners, had suffered a similar fate as the miners employed by Wismut AG. A study of site-specific lung cancer mortality in the 4,320 uranium miners at Jáchymov and Horní Slavkov, who had been followed up for an average of 25 years before the early 1990s, showed a four-fold excess of lung cancer over the national average. The study also hinted at a higher incidence of other cancers, especially among miners who had begun working in the uranium industry before reaching the age of 25.[325]

Eduard Goldstücker, a Communist Party member who was arrested during the purges of the 1950s, spent the last part of his sentence in Jáchymov. He was convinced that, during the few months he spent in the uranium mines, he was exposed to "unparalleled high radioactive radiation, without any protection. The accident rates were abhorrent, especially underground. Death by radiation was never registered. And this not only in the camps, but also later, when death started to claim its victims among the released prisoners, especially among those who had spent a longer time in Jáchymov."[326]

A Bavarian study from the late 1950s concerning 562 former POWs employed in Jáchymov showed that more than 8% of them suffered from cancer. (Other factors, including inferior nourishment or smoking habits, were not taken into account.) In comparison with the Bavarian average, cancer rates among the former miners were higher.[327]

In the GDR, one to three people per 10,000 inhabitants had cancer. In the district of Chemnitz, 44 cancer cases were officially registered. In Gera, 10 cases were officially registered. This means that in the two districts where SDAG Wismut operated and where most company employees lived, lung cancer rates were much higher than the national average.[328]

There was virtually no chance of recovery once the diagnosis of radiation-related cancer was made. Many miners died in great distress. "My father started working for Wismut when he was twenty. That was in 1950. They used dry drilling then. He worked at Schlema. He was tempted by the high wages and better food rations. He knew nothing about the danger of radiation...in 1963, my father was diagnosed as suffering from 30% silicosis. He was hospitalized more and more often. He met many miners there who had worked from the beginning. They were dying like flies. None of them was older than 55...He only shook his head now. He wanted at least to determine his own end. My father fell down from a window on the tenth floor. He knew nothing of radiation. Towards the end a doctor told my mother that it was lung cancer."[329]

A Final Remark on the Special Position of Wismut AG

The possibilities that German authorities or the German political parties could influence events at Wismut were, in the first years of the uranium industry, strictly limited. The Soviet military administration itself in Berlin-Karlshorst, with the SED Politburo in tow, only received fragmentary information about the developments in Erzgebirge. Wismut AG led its own life within the Soviet nuclear project. The Red Army and the SED were to be helpful to Wismut, especially in the matter of workforce recruitment, and were only informed about the uranium industry on a need-to-know basis. An SED executive came face to face with the uranium industry for the first time in October or November 1947. The impulse did not come from the Soviet side, but from the flood of complaints about the conditions in the mining districts. Kurt Schumacher's speech about the "uranium slaves" stung SED leaders. The reports by Helmut Lehmann and Max Weber may have also upset SED leaders, who were powerless to influence events in the uranium province.

Wismut AG caused severe difficulties for the party leadership. Forced employment combined with mass flights to the West injured the reputations both of the occupying power and the SED. In addition, the industry required enormous resources: timber, rails, drilling equipment and food supplies. Wismut AG was a company that rested largely on the shoulders of the German tax payers. The only consolation for the SED leaders was that the uranium deliveries were included as part of a reparation deal.

The greater the Soviet interest in Erzgebirge uranium, the more important it was for them to pursue a steady policy towards Germany and preserve the status quo. It was therefore less probable that Moscow would consider alternatives to the division of Germany, before and after the establishment of the two German states in May and in October 1949. Before the first successful test of Stalin's bomb on 29 August 1949, it was unlikely that the Soviets could afford to do without more than a half of their resources of uranium. The development of the uranium industries in Czechoslovakia and in the Soviet zone of occupation gave the Erzgebirge, in 1947 or 1948 at the latest, a vastly increased significance in the geopolitical deliberations in Moscow.

The SED easily came to terms with the separate existence of the Soviet zone of occupation from the rest of Germany. The interests of the occupying power and of the SED were not identical, but by the time the GDR was established, the two parties shared similar positions on the question of uranium production, at the latest. The uranium industry was no longer a "necessary evil" for the SED, but a pledge for its very existence.

After the foundation of the GDR, the first written agreement on the uranium industry was signed, and the GDR thus achieved a certain limited sovereignty. The SED's ability to influence developments at Wismut still remained very restricted. Even after the conclusion of the treaty on the establishment of SDAG Wismut in August 1953, nothing changed much. After the conclusion of the "Protocol on the Stationing of Soviet Troops on the Territory of GDR" in September 1955, the costs of occupation were carried jointly by the GDR and the Soviet Union, as were the costs of the uranium industry. In both cases, sharing the costs represented an undertaking by the GDR to go on carrying on payments resembling reparations.[330] The protocol of September 1955 left a number of important questions open, including the final amount of the cost of the stationing of Soviet troops, the question of which jurisdiction did the Soviet troops come under, as well as a "public declaration on the question of deliveries by Wismut and the payments for them."[331] Negotiations at the government level took place in Moscow in July 1956. On 12 March 1957, an agreement on the stationing of Soviet troops was concluded, which remained valid as long as the GDR persisted.

Beginning in 1955, regular negotiations on occupation costs were connected with calculations of the costs of the uranium industry. An unfavorable exchange rate (finally it was 1 ruble for 5.50 GDR marks) was fixed for German contributions. The SED leaders regarded the cost of the uranium industry—in view of its equal ranking with the costs of occupation—as "supplementary defense subsidies." Between 1954 and 1958, the GDR paid about 6.45 billion marks for the costs of occupation and and about 4.1 billion for SDAG Wismut.[332] Ulbricht became increasingly confident about using those special payments in negotiations with the Soviet partners, and insisted on the high value of the GDR for the communist bloc.

Around the mid 1950s, a new perspective opened before the German uranium industry. The peaceful uses of nuclear power promised to overcome the chronic shortages of electric power in the GDR. The SED leaders decided in favor of nuclear power stations. Political and material conditions had changed during the decade after the war, and a debate about the future direction of science and technology took place. In the center of the discussion, there stood mechanization, automatization and atomic energy. Chemistry was added later. The green light for the peaceful use of atomic energy was given in 1955. A number of prominent scientists returned from the Soviet Union that year, where they had taken part, either voluntarily or under compulsion, in the nuclear program. At the same time, there took place a conference on the peaceful uses of atomic energy in Geneva.

The decision to build atomic power stations in the GDR was influenced by the existence of the local uranium industry. The agreement on the formation of a Soviet–German joint stock company on 22 August 1953 ended the absolute dependence of uranium production on Soviet dictate. It made it possible for politicians and scientists to contemplate setting aside a part of the production for their country's own uses. The requirement by the GDR to produce uranium for its own purposes was first written into the agreement on 7 December 1962.

Notes

[1] Dimitrii Filipovich, Vladimir Sakharov, "Deutsches Uran für die sowjetische Atombombe," *Der Anschitt*, 2/3 1998, 83f.

[2] Decision no. 740-29; cf. W. I. Wetrow, "Die Bildung von Betrieben für die Gewinnung und Aufbereitung von Uranerz," in Kruglov, *Die sowjetische Atombombe*, Moscow 1995, draft translation into German, 19.

[3] Dimitrii Filipovich, Vladimir Sakharov, 85.

[4] Rob Roeling, De grote Trek naar het Diitse Ertsgebergte, thesis for Rotterdam university, 1991, 19.

[5] Chronik der Wismut, CD-Rom, Chemnitz 1999, Teil 2, 1202.

[6] cf Werner Schiffner, Agricola und die Wismut, 53 and "Martin Vogel schlug als erster Uran," *Dialog*, 1996, 13, 28.

[7] Chronik der Wismut, 1210.

[8] SÄHStA Dresden, LRS, MfW Nr. 1524, Schreiben des Treuhanders für das Vermögen der ehem. Sachsenerz Bergwerks AG an die Landesregierung, vom 13. März 1948.

[9] SÄHStA Dresden, LRS, MfW, Nr. 1510, Verzeichniss der Unternehmen, die zu Sowjetischen Aktiongesellschaften gehören (April 1947).

[10] Martin Vogel schlug als erster Uran, in: Dialog 1996/13, 28.

[11] Der Uranbergbau in der Sowjetzone, Bonn 1950, 3.

[12] Chronik der Wismut, Teil 2, 1244.

[13] Volksstimme, 30 June 1956.

[14] Mario Kaden, Wismut—die "wilde" Zeit, Beitrage zur Geschichte des Landkreises Annaberg, Annaberg 1994, 76 et seq.

[15] Der Uranbergbau in der Sowjetzone, 5.

[16] SÄHStA Dresden, LRS, MfW, Nr. 1 510.

[17] W. I. Wetrow, op cit., 19.

[18] Rainer Karlsch and Johannes Bähr, "Die Sowjetischen Aktiengesellschaften (SAG) in der SBZ/DDR," in Karl Lauschke and Thomas Welskopp (eds.), *Mikropolitik im Unternehmen*, Essen 1994.

[19] Jan Foitzik, Inventar der Befehle der Obersten Chefs der Sowjetischen Militäradministration in Deutschland (SMAD), 1945–1949, Munich 1995, 133; Sächsisches Hauptstaatsarchiv, Landesregierung Sachsen, Ministerpräsident, Nr. 147, Bl 4.

[20] Unternehmensarchiv der Wismut GmbH, microfilm Nr. 438, Materialy po oformelenye A/O Wismut, Urkunde der Stadt Aue, 10 February 1948.

[21] Kaden, op. cit., 12.

[22] Sächsisches Hauptstaatsarchiv, Landesregierung Sachsen, Ministerium des Innern, Nr. 303.

[23] cf. Wismut archive, microfilm Nr. 438, Protokoll Nr. 1.

[24] Gerhard Rohner, "Finanzierung der Wismut, MSS," *BAK*, B 137, Nr. 1355.

[25] Rolf Badstübner and Wilfried Loth (eds), Aufzeichnungen zur Deutschlandpolitik 1945–1953, Berlin 1994, 318.

26 Jochen Laufer, "Politik und Praxis der sowjetischen Demontagen 1941–1950," in Rainer Karlsch and Jochen Laufer (eds.), *Sowjetische Demontagen in Deutschland*, Berlin 2001.
27 Assembled from BArch Berlin-Lichterfelde, DN 1, VS04/81, Nr. 5, Analyse über die Entwicklung der Finanzwirtschaft der DDR 1951–55, vom 9.9.1954.
28 cf. Ray C. Ringholz, *Uranium Frenzy, Boom and Bust on the Colorado Plateau*, New York 1989, 269.
29 Rainer Karlsch, *Allein bezahlt?*, 197 et seq.
30 Rudolf Lange, "Gründung und Entwicklung der SDAG Wismut," *in Chronik der Wismut GmbH*, Vol. 1, Chemnitz 1997.
31 Rainer Karlsch, "Ungleiche Partner—Vertragliche und finanzielle Probleme der Uranlieferungen der DDR," in Rainer Karlsch and Harm Schröter, *Strahlende Vergangenheit*, 279.
32 Wismut GmbH Archiv, Abschlussdokumentation, Anlage zu Abschnitt 2
33 Anton Antonov-Ovseienko, "Der Weg nach oben, Skizzen zu einem Berija-Portrat," in Vladimir F. Nekrasow, *Berija, Henker in Stalins Diensten*, Augsburg 1997, 142.
34 Filipovich and Sakharov, *Nemetskii uran*, 34.
35 Nikolai Grishin, *The Saxony Uranium Mining Operation ("Vismut")*, 130 et seq.
36 GARF Moscow, fond 9401, op 1, g 4152, Bl 140-2; letter from Maltsev and Malygin to Serov of 24.2.47.
37 Grishin, op. cit., 131.
38 *Die SED. Geschichte, Organisation, Politik, Ein Handbuch*, Berlin 1997, 21 et seq.
39 Klaus Schröder, *Der SED Staat*, Munich 1998, 60.
40 SÄHStA Aussenstelle Chemnitz, GPL Wismut, IV 2/3/2 and 2/3/11; Sekretariatsitzung des Kreisvorstandes Siegmar 4.1.49; Protokoll der Sitzung der Gebietsparteileitung der SED, 8.3.54.
41 SÄHStA Dresden, BPA, Nr. A/1193.
42 SÄHStA Dresden, BPA, A/1193, Zentrale Betriebsgruppenleitung der SED in Bergbau.
43 Zur Geschichte der Gebietsparteiorganisation Wismut der SED, 22.
44 BArch Berlin-Lichterfelde, SAPMO, IGW, Nr. 10, Sekretariatssitzung, 17 February 1949.
45 SÄHStA Dresden, BPA, Nr. A/1140; Die innerparteiliche Lage im Kreisvorstand Aue, Bericht von 17.5.50.
46 Ralf Engeln, *Uransklaven oder Sonnensucher*, 153.
47 SPD-PV, Ostbüro der SPD, Nr. 0072 B (Nr. 02367), Bericht vom 11.11.52
48 Ralf Engeln, "Die industriellen Beziehungen im Uranbergbau der SAG Wismut," in Karlsch and Schröter, op. cit., 193.
49 Ralf Engeln, op. cit., 194.
50 SÄHStA Dresden, BPA, A/796, Sitzung des Landesvorstandes, 18.1.52.
51 Wolfgang Kiessling, *Partner in "Narrenparadies," Der Freudenkreis um Noel Field und Paul Merker*, Berlin 1994.

[52] Rainer Karlsch, "Ungleiche Partner," in Rainer Karlsch and Harm Schröter, (eds.), Strahlende Vergangenheit, 263–300.
[53] Mike Reichert, *Kernenergiewirtschaft in der DDR*, 113 et seq.
[54] Karl Schirdewan, *Aufstand gegen Ulbricht*, 92.
[55] Hermann Weber, *Geschichte der DDR*, 26.
[56] Archiv der Friedrich Ebert Stiftung, Ostbüro der SPD, Nr. 0072B (Nr. 02363).
[57] BArch Berlin-Lichterfelde, SAPMO, J IV 2/2-5/4, Sitzung des Politbüros am 11. Marz 1958.
[58] Günther Mai, *Der Allierte Kontrollrat in Deutschland 1945–1948, Allierte Einheit—deutsche Teilung?*, Munich 1995, 372.
[59] Idem., 370.
[60] Wolfgang Zank, *Wirtschaft und Arbeit*, 104.
[61] BArch Berlin-Lichterfelde, DQ-2, Nr. 6, Bl. 31ff; Im ersten Geschäftsjahr erlassene Verordnungen, Verfügungen und Richtlinien betreffend Arbeitslenkung.
[62] Ralf R. Leinweber, *Das Recht auf Arbeit im Sozialismus*, 80 et seq.
[63] Günter Mai, *Der Allierte Kontrollrat*, 381 et seq.
[64] Werner Abelshauser, *Der Ruhrkohlenbergbau seit 1945*, 30 et seq.
[65] Ralf Engeln, *Uransklaven oder Sonnensucher*, 70.
[66] Engeln, *Uransklaven oder Sonnensucher*, 153.
[67] Rob Roeling, *Arbeiter im Uranbergbau*, 103.
[68] Roeling, *Arbeiter im Uranbergbau*, 111.
[69] BArch Berlin-Lichterfelde, SAPMO, DY 30 IV, 2/2 027/25 (Büro Lehmann), Schreiben des Präsidenten DVAS, Brack, an den Zentralvorstand der SED, 20.9.47.
[70] SÄHStA Dresden, LRS, MfAS, Nr. 391.
[71] BArch Berlin-Lichterfelde, SAPMO, FBS 336/13 238, Sekretariat Lehmann, Schreiben vom 4.9.47.
[72] The Secret Mines of Russia's Germany, Life 29/1950, 74.
[73] BArch Berlin-Lichterfelde, SAPMO, DY 30 IV, 2/2 027/25 (Büro Lehmann), Brief an den Ministerpräsidenten des Landes Thüringen 6. Dezember 1947.
[74] *Idem.*, Abschrift von der LR Sachsen, Besprechung über die Unterbringung und Versorgung der Arbeitskräfte im Erzbergbau Aue vom 7.8.47.
[75] BArch Berlin-Lichterfelde, SAPMO, DY 30 IV, 2/2 027/25 (Büro Lehmann), DWK, HVAS, Bericht über eine Reise nach Dresden und Aue in der Zeit vom 14.9.–16.9.48.
[76] SÄHStA Dresden, BPA, SED-LL, A/778, Bl 263 Beschluss-Protokoll vom 13.11.46.
[77] BArch Berlin-Lichterfelde, SAPMO, DQ-2, Nr. 2 138.
[78] *Idem.*, draft of an article by Jenny Matern, 14.8.47.
[79] SÄHStA Dresden, LRS, MAS, Nr. 397, Brief des Präsidenten Brack an die Landesregierungen, vom 18.10.48.
[80] BArch Berlin-Lichterfelde, SAPMO, DQ-2, Nr. 1 995, Schreiben von DWK an die SMAD, Morenow, 16.11.48.

81 Ralf Engeln, *Uransklaven oder Sonnensucher*, 116.
82 BÄrch Berlin-Lichterfelde, SAPMO, DQ-2, Nr. 1 995, Protokoll vom 10.6.48
83 Ralf Engeln, *Uransklaven oder Sonnensucher*, 160.
84 Archiv der Wismut GmbH, Jahresbericht für 1949, in Russian on microfiche Nr. 410.
85 Norman Naimark, *The Russians in Germany*, 244.
86 Dmitri N. Filippovich and Vladimir V. Zacharow, Deutsches Uran für die Sowjetische Atombombe. Zur frühen Geschichte der Sächsischen Bergbauverwaltung und der Staatlichen Sowjetischen Aktiengesellschaft, "Wismut," in *Der Anschnitt. Zeitschrift für Kunst und Kultur im Bergbau*, 50 Jg., 2–3/1998, 82–95.
87 Jan Lipinsky, "Mobilität zwischen den Lagern," in Alexander von Plato (ed.), *Sowjetische Speziallager in Deutschland 1945–1950*, vol. 1, 228.
88 Lipinsky, op. cit., 230.
89 Zur Geschichte der Gebietsparteiorgansation der Wismut, Chemnitz 1988, 12.
90 BArch Berlin-Lichterfelde, SAPMO, DY 30 IV, 2/2 027/25 (Büro Lehmann), Plakat "Deutschland erwartet euch!"
91 *Idem.*, Bericht aus dem Heimkehrerlager Fürstenwalde.
92 SÄHStA Dresden, Kreistag/Kreisrat Annaberg, Nr. 568/3.
93 BArch Berlin-Lichterfelde, SAPMO, DY 30 IV, 2/2 027/25 (Büro Lehmann); DZVAS: Bericht über die am 15.8.47 durchgeführte Dienstreise nach Frankfürt/O.
94 Mannfred Jahn, "Vertriebene Deutsche als attraktive Zielgruppe der Arbeitskraftelenkung in den Uranbergbau," *Der Anschnitt*, 2/3, 1998.
95 SÄHStA Dresden, BPA, Nr. 685, Schreiben von Weber an den LV der SED Sachsen von 20.3.47.
96 SÄHStA Dresden, LRS, MdI, Nr. 2645, Schreiben an die Aussenstelle Erzbergbau, 16.6.48.
97 *Idem.* Schreiben des MAS, HA, Umsiedler, Sonderabteilung Erzberbau, 15.6.48.
98 Reimar Paul, *Das Wismut Erbe*, 147.
99 Archiv Friedrich Ebert Stiftung, Ostbüro, 0072 B (Nr. 02362) CCG, press release 24.8.50.
100 BArch Berlin-Lichterfelde, SAPMO, DY 30 IV, 2/2 027/25 (Büro Lehmann), Bericht über eine Reise nach Dresden und Aue in der Zeit von 14.9.–16.9.48.
101 *Idem.*
102 BArch Berlin-Lichterfelde, SAPMO, DY 30 IV, 2/2 027/25 (Büro Lehmann), Katastrophale Verhältnisse im Erzbergbaugebiet Aue, 30.8.48.
103 Chronik der Wismut, CD Rom, part 1, 745.
104 GARF Moscow, Fond 9401, Op 2, D 207, p 3; report of the Minister of the Interior of the USSR, S. N. Kruglov to Beria, 20 March 1948.
105 SÄHStA Dresden, MAS, Nr. 62, Aktennotiz Frau Dr Heinze, 14.1.47.
106 *Idem.*, Schreiben des Landesarbeiteramtes an Minister Selbmann, 7.11.47.
107 SÄHStA Dresden, MAS, Nr. 303, Bl 104, Beschwerdebrief von Karl C an das Landesarbeitsamt, 16.12.46.

[108] SÄHStA, BPA A/1193, Bergwerks Objektleiter Sitzung, 24.11.47.
[109] SÄHStA, Aussenstelle Chemnitz, GPL Wismut, IV 2/3/11, Protokoll der Sitzung des KV Siegmar, 4.1.49.
[110] SÄHStA Dresden, BPA A/1193, Schreiben der Ortsgruppe der SED Silberstrasse an Oberstleutnant Einbinder, 24.12.48.
[111] *Idem.*
[112] SÄHStA Dresden, BPA A/783, Bl 274, Sitzung, 6.9.48.
[113] BArch Berlin-Lichterfelde, SAPMO, Do 11, Nr. 20, Bl 24; HVDVP: Kontrollbericht über VPKA Aue, 5.12.51.
[114] *Idem.*
[115] BArch Berlin-Lichterfelde, SAPMO, Do 11, Nr. 20, HVDVP, Bl 40; Kontrollbericht über das VPKA Marienberg, 4.12.51.
[116] Archiv der Fridrich-Ebert Stiftung (FES), Ostbüro, 0072 B (Nr. 02358), Bericht vom Mitte 1949.
[117] Dmitrii N. Filipovich, Vladimir V. Sakharov, *Deutsches Uran für die sowjetische Atombombe*, 90.
[118] FES, Ostbüro der SPD, 0072 B (Nr. 02358), Bericht von Schudy, undated.
[119] Mario Kaden, "Kriminalität, Polizei, Justiz- und Sicherheitsapparat in der 'Uranprovinz' 1946 bis 1958," in Karlsch and Schröter, op. cit., 151.
[120] FES, Ostbüro, 0072 B (Nr. 02367); SPD-Pressemitteilungen, 1.7.54.
[121] BArch Berlin-Lichterfelde, SAPMO, DO 1, Nr. 11/546; Schreiben des Stellv. Chef der VP, Seifert an A. F. Kabonov, 27.9.50.
[122] Der Uranbergbau in der Sowjetzone, 24.
[123] BArch Berlin-Lichterfelde, SAPMO, DO 1, HDVP 11/509, Jahresbericht der BDVP (BS) Wismut für 1957.
[124] *Idem.*, Nr. 11/50.
[125] Jan Foitzik, Inventar der Befehle der SMAD, Munich 1995.
[126] BArch Berlin-Lichterfelde, SAPMO, DQ-2, Nr. 2105; Schreiben des MfAG an alle Landesregierungen, 18.11.49.
[127] *Idem.*, Schreiben des Landesregierung Sachsen an die DWK, 30.8.49.
[128] *Idem.*, DO 11, Nr. 932, Bl 2105, Schreiben, Seifert an Maltsev, 17.1.50.
[129] *Idem.*
[130] *Idem.*, Bericht über die Vorsprache von Vertretern MdI et al bei Oberleutenant Einbinder, 11.2.50.
[131] *Idem.*, Nr. 933, Bl 29 et seq, Verfahrensweise bei Umtausch von Personalausweisen für Bergarbeiter der Wismut AG.
[132] Gotthard Bretschneider, *Kein Schutz von Willkür? Zur Arbeit der Gewerkschaft in der SAG Wismut 1946–1953*, Oberinitz 1998, Vol. 1, 8 et seq.
[133] A copy of the letter to the Russian trade unions was kindly made available to the authors by Gotthard Bretschneider from his private archive.
[134] GARF Moscow, 9401-2-173 Bl 19, written report by Kusnetsov to Molotov of 13 May 1947.
[135] Rainer Karlsch and Vladimir V. Sakharov, "Ein Gulag im Erzgebirge? Besatzer un Besiegte beim Aufbau der Wismut AG" in Deutschland-Archiv, 1/1999, 16–34.

136 BArch Berlin-Lichterfelde, SAPMO, FBS/123/16 502 (Nachlass Grotewohl), Bericht von 24.5.47; one document refers to 23.4 as the date of the meeting, another to 26.4.
137 Report on the meeting between Maltsev and German trade unionists on 26 April 1947, a copy of which was kindly provided for the authors by Herr Bretschneider.
138 Gotthard Bretschneider, Kein Schutz von Willkür?, 12.
139 BArch Berlin-Lichterfelde, SAPMO, DY 30 IV, 2/2 027/25 (Büro Lehmann).
140 *Idem.*, FBS 93/1172 (Nachlass Pieck), Schreiben der Landesleitung Sachsen, "Ein zusammengedrangter Bericht von dem Erzbergwerk" vom 24.3.47.
141 BArch Berlin-Lichterfelde, SAPMO, FBS/123/16 502 (Nachlass Grotewohl), Vorlage von Helmut Lehmann für das Zentralsekretariat der SED vom 12.8.47.
142 BArch Berlin-Lichterfelde, SAPMO, DY 30, IV2/2.1/177, Arbeit in Uranbergbau (Paul Merker), 15.8.47.
143 Erich W. Gniffke, *Jahre mit Ulbricht*, 292.
144 *Idem.*, Tagung des Parteivorstandes, 15/16.10.47.
145 Peter Hübner, *Konsens, Konflikt und Kompromiss. Soziale Arbeiterinteressen und Sozialpolitik in der SBZ/DDR 1945–1970*, Berlin 1995.
146 Privatarchiv Gotthard Bretschneider, Amt für Arbeit Bleicherode, Niederschrift vom 30.12.49.
147 FES Bonn, SPD-PV Ostbüro, 0072 B(Nr. 02358); Bericht vom 6.8.47.
148 FES, Ostburo 0072 B (Nr. 02358), Arbeitskraftevermittlung für den Erzbergbau, 17.4.50.
149 *Idem.*, Bericht, 27.11,49.
150 BArch Berlin, Nachlass Pieck, NL36/736.
151 BArch Berlin-Lichterfelde, SAPMO DY30 2/2/83, Sitzung des Politbüros, 11.4.50.
152 Ulrich Gill, *Der Freie Deutsche Gewerkschaftsbund (FDGB). Theorie–Geschichte–Organisation–Funktionen–Kritik*, Opladen, 1989.
153 Privatarchiv Gotthard Brettschneider, Bundesvorstand des FDGB: Empfehlungen zur Überwindung der Fluktuation bei der Wismut AG, 30.12.49.
154 *Idem.*, Schreiben des Bundesvorstand des FDGB an Minister Heinrich Rau, 13.1.50.
155 Private archive Bretschneider, Bundesvorstand des FDGB, Referat Erzbergbau, Aktennotiz vom 24.2.50.
156 BArch Berlin-Lichterfelde, Q-2, Nr. 2105.
157 Jahresbericht der Generaldirektion für 1949 (in Russian), UA der Wismut GmbH, microfiche.
158 SÄHStA Dresden, BPA, A/685, Bl 103; SED Landesleitung, Abt Wirtschaft, Aufstellung der gestellten und entlassenen Arbeitskräfte im 1. Hj 1950.
159 Ralf Engeln, *Uransklaven oder Sonnensucher*, 111.
160 SÄHStA Dresden, MfW, Nr. 819/1.
161 BArch Berlin-Lichterfelde, SAPMO, FBS 123/16502 (NL90/359); Schreiben des Landesregierung von Mecklenburg and die DVAS, 4.9.47.

[162] BArch Berlin-Lichterfelde, IV 2/2027/25, Sitzung der Aue Kommission, 11.5.50; *Idem.*, SAPMO, NL 36 (Pieck), Nr. 736, Besprechung mit Marschall Tschuikow, 6.4.50.
[163] Rob Roeling, in Karlsch and Schröter, op. cit., 112.
[164] SÄHStA Dresden, LRS, MfW, Nr. 1175/1, GVS, Aktennotiz von Minister Goschütz für Minister Hoffmann, 14.7.51.
[165] UA der Wismut GmbH; the authors are grateful to Herr Rainer Kohlisch for drawing their attention to the commentary on the annual report.
[166] Ralf Engeln, *Uransklaven oder Sonnensucher*, 46.
[167] Mario Kaden, Uranprovinz, Zeitungen der Wismut erinnern sich, Marienberg 2000, 36.
[168] Rainer Karlsch, "Kohle, Chaos und Kartoffeln," in Jurgen Engert (ed), *Die wirren Jahre,* Berlin 1996, 105ff.
[169] Rob Roeling, *Arbeiter im Uranbergbau*, 113.
[170] Quoted in Paul Reimar, *Wismut Erbe*, 165.
[171] Werner Abelshauser, *Der Ruhrkohlenbergbau seit 1945*, Munich 1984, 42.
[172] Peter Hübner, *Konsens, Konflikt und Kompromiss. Soziale Arbeiterinteressen und Sozialpolitik in der SBZ/DDR*, Berlin 1995, 19.
[173] Ralf Engeln, op. cit., 202.
[174] BArch Berlin-Lichterfelde, DG-2, Nr. 6, Verordnung des Generaldirektors der Wismut, 20.7.49.
[175] Ralf Engeln, op. cit., 234.
[176] BArch Berlin-Lichterfelde, SAPMO, DO 11, Nr. 20; Kontrollbericht über VPKA Aue, 5.12.51.
[177] Rainer Potratz, "Zur Entfernung deklassierter Elemente..." in Karlsch, Schröter, op. cit., 211.
[178] BArch Berlin-Lichterfelde, SAPMO, DO 11, Nr. 20, Bl 147 et seq; Vorlage für das Sekretariat des Politbüros der SED "Reorganisation des Kreises Aue," 29 November 1951.
[179] SÄHStA Dresden, LRS, MdI, Nr. 2645; Schreiben des Kreisrates von Annaberg und die Landesregierung, 21.2.48.
[180] *Idem.*, Niederschrift über die Besprechung am 16. Januar 1948.
[181] *Idem.*, Schreiben des MAS an die Sonderabteilung Erzbergbau, 28.10.47.
[182] *Idem.*, Schreiben der Sonderabteilung Erzbergbau and die Hauptabteilung Umsiedler beim MAS, 20.12.48.
[183] *Idem.*, Aktennotiz, 6.1.49.
[184] *Idem.*, Notiz anlässlich einer Besprechung bei der SMA Sachsen am 29.3.49.
[185] BArch Berlin-Lichterfelde, SAPMO, FBS 123/16355 (NL 90/424) Bl 16ff, Staatliche Vervaltung an Grotewohl, 13.2.50.
[186] Chronik der Wismut, CD-Rom, Chemnitz 1999, part 2.2.1, 4.
[187] BArch Berlin-Lichterfelde, SAPMO, FBS 123/16497 (NL 90/306); protokoll einer Besprechung während der 5. Gebietsparteikonferenz der PO Wismut, 15.12.51.
[188] Wismut Archiv, MF, Nr. 439, Bl 19-26; Vertrag zwischen der Wismut AG und dem Ministerium für Aufbau der DDR (russisch), 1.10.52.

[189] GARF, Bericht über die Tätigkeit der Rechtsabteilung SMA/Sachsen für das IV Quartal 1948.
[190] *Idem.*, Bericht für das I. Halbjahr 1949 vom 14.7.49.
[191] Der Uranbergbau in die Sowjetzone, 28 et seq.
[192] *Idem.*, 29.
[193] BArch Berlin-Lichterfelde, SAMPMO, DQ 2, Nr. 1995, Schreiben des Arbeitsamtes Salzwedel an die Landesregierung Sachsen-Anhalt, 24.5.48.
[194] *Idem.*, Nr. 2105, DWK Hausmitteilung, 20.7.49.
[195] Karl Wilhelm Fricke, "Kampf dem Klassenfeind," in Alexander Fischer (ed.), *Studien zur Geschichte der SBZ/DDR*, Berlin 1993, 166.
[196] *Idem.*, 179.
[197] Peter Erler, "Zur Tätigkeit der Sowjetischen Militartribunale (SMT) in der SBZ/DDR," in Sergei Mironenko, Lutz Niethammer, Alexander von Plato (eds.), *Sowjetische Speziallager in Deutschland*, Berlin 1998, vol. 1, 173.
[198] Privatarchiv Mario Kaden, Rehabilitierungbescheid of 25 May 1994.
[199] *Idem.*
[200] Interview with Ludwig Hecker, Neudorf.
[201] GARF, Moscow, OSL Rybakov, military commander of the Zwickau district, protocol of 12 August 1949.
[202] Amt für Arbeit Bleicherode. Niederschrift vom 30.12.49, Privatarchiv Gotthard Bretschneider.
[203] *Life*, 1950, 73–83.
[204] Uranbergbau in der Sowjetzone, 67 et seq.
[205] SÄHStA Dresden, LRS, MdJ Nr. 96; cf Bonn, Ostbüro der SPD, 0072B (Nr. 02358), Bericht vom 13.12.48.
[206] SÄHStA Dresden, LRS Nr. 65, MdJ; Notiz von September 1949, Einweisung von Strafgefängenen in den Arbeitsprozess.
[207] SÄHStA Dresden, LRS, MAS Nr. 191, Richtlinie der DJV und der DZVAS vom 1.9.47.
[208] SÄHStA Dresden, LRS, MAS Nr. 192, Schreiben des Ministeriums für Arbeit und Gesundheitswesen an die Landesregierung vom 13.5.50.
[209] SÄHStA Dresden, LRS, MdJ, Nr. 866 und Nr. 866/1; only a few cases were reported.
[210] Dmitrii N. Filipovich, Vladimir V. Sakharov "Deutsches Uran für die sowjetische Atombombe," 88 et seq.
[211] FES Bonn, SPD-PV, Ostbüro der SPD, 0072B (Nr. 02366), Berich vom 29.5.52.
[212] Gotthard Bretschneider, "Warum enstand die SAG/DSAG Wismut?," *Schlema* 1999, 18.
[213] FES Bonn, SPD-PV, Ostbüro der SPD, 0072B (Nr. 02367), Bericht vom 18.9.54.
[214] Ostbüro, 0072 B (Nr. 02367), Bericht (ohne Datum) über die Arbeit der Bergpolizei und des NKVD.
[215] *Idem.*
[216] Neues Deutschland, 12 July 1955, "Riasverbrecher im Erzbergbau unschädlich gemacht."

[217] SÄHStA Dresden, LRS, MdJ, Nr. 873, Bl 16RS; Schreiben des Amtgerichtes Chemnitz am MdJ vom 4.10.49.

[218] SÄHStA Dresden, LRS, MdJ, Nr. 866, Bl 122, Schreiben Staatsanwaltschaft beim Landgericht an das MdJ vom 26.2.48.

[219] SÄHStA Dresden, LRS, MdJ, Nr. 873, Bl 168; Revision der Gerichtsgefangnisse im Bergarbeit vom 22.10.49.

[220] GARF, Maltsev to Dubrowski, 1.2.49.

[221] Mario Kaden, *Kriminalität*, 154.

[222] SÄHStA Dresden, LRS, MfW, 1530/1, Schreiben des MAS an die Arbeitsgerichte vom 9.4.49.

[223] Kaden, op. cit., 153.

[224] SÄHStA Dresden, LRS, MfW, 1530/1, Schreiben der DKW an die Landesregierung Sachsen vom 17.8.49.

[225] *Idem.*, MAS, Vermerk vom 5.12.49.

[226] SÄHStA Dresden, LRS, MfW, 1530/1, Protokoll über eine Besprechung... vom 26.5.50.

[227] SÄHStA Dresden, LRS, MfW, Nr. 1530.

[228] Bretschneider private archive, a letter from FDGB in Saxony to FDGB central office, Hans Jendretsky, 1 August 1947.

[229] cf. Peter Merseburger, Der schwierige Deutsche, Kurt Schumacher; Stuttgart 1995, 345.

[230] BArch Berlin-Lichterfelde, SAPMO, FBS 194/20776; Tagung des Parteivorstandes am 12/13.11.47.

[231] DVAS Pressedienst, 5 November 1947, Neue Falschmeldungen über Aue.

[232] BArch Berlin-Lichterfelde, FBS 123/16502 (NL90/359), Die Wahrheit über den westsächsischen Erzbergbau.

[233] *Berliner Zeitung*, Nr. 57/1948, Im sächsischen Erzbergbau.

[234] Der Uranbergbau in die Sowjetzone, 83.

[235] *Berliner Zeitung*, Nr. 57/1948, Im sachsischen Erzbergbau.

[236] BArch Berlin-Lichterfelde, SAPMO, IV 2/2027/25 (Büro Lehmann), Schreiben von Paul Merker an Fritz Apelt, 25.2.48.

[237] *Idem.*, Merker an Apelt, 1.3.48.

[238] *Idem.*, Skala Bericht (Berliner Rundfunk), 10 August 1949.

[239] *Idem.*, Information der Abt. Massenagitation an Ulbricht, 27.5.49.

[240] *Idem.*, Schreiben von Herbert Gessner (Berliner Rundfunk), 10.8.49.

[241] *Idem.*, Schreiben von Pieck an Ulbricht, 22.8.49.

[242] *Idem.*, Sitzung des Politbüros der SED am 11.3.58.

[243] Werner Buschford, *Das Ostbüro der SPD*. The measure of cooperation of the Ostbüro with the British and the American secret services was not fully appreciated by Buschford. Ehrhart Neubert, *Geschichte der Opposition in der DDR 1949–1989*, 93

[244] Peter Merseburger, *Kurt Schumacher*, 341; and Karl Wilhelm Fricke, "Organisation und Tätigkeit der DDR Nachrichtendienste," in Wolfgang Krieger, Jürgen Weber (eds.), *Spionage für den Frieden? Nachrichtendienste in Deutschland wahrend des Kalten Krieges*, 221.

[245] Das Ostbüro der SPD, Sendung des Wesdeutschen Rundfunks 1997.

246 Bericht vom 13.12.48, FES Bonn, Ostbüro der SPD, 0072 B (Nr. 02358).
247 Friedrich Ebert Stiftung Archiv, Bonn, Ostbüro der SPD, Nr. 0072 B (02365), Bericht vom 13.12.48.
248 Institut für Zeitgeschichte, OMGUS, AGTS 44/b, Report RP-226-48.
249 BArch Berlin-Lichterfelde, DO 11, Nr. 11/546.
250 BArch Berlin-Lichterfelde, DO 11, Nr. 11/546, Schreiben des BS Wismut an die HDVP, 2.4.52.
251 Institut für Zeitgeschichte, OMGUS, AGTS 38/1a, Uranium Ore Sample Obtained from Johanngeorgenstadt, January 1949.
252 Institut für Zeitgeschichte, OMGUS, AGTS 38/1a, Uranium Ore Sample Obtained from Johanngeorgenstadt, 18 April 1949.
253 Russian State Military Archive (RGVA), Fond 32925, op 1, d 178, l. 25–27.
254 SÄHStA, Dresden, BPA, Nr. A/799. Bl 132; Bericht der Landesbehörde der VP, 9.9.49.
255 SÄHStA, Dresden, BPA, Nr. A/1235; Bericht des VPKA Annaberg, 28.9.50.
256 Untersuchungs-Vorgang wegen Aufruhr in Schneeberg am 21.3.51, Archiv des Bundesbeauftragten für die Unterlagen des ehemaligen Staatssicherheitsdienstes der DDR (BStU), HA/St, Nr. 29/52.
257 Heidi Roth, Torsten Diedrich, "Wir sind Kumpel—uns kann keiner. Der 17 Juni 1953 in der SAG Wismut," in Karlsch and Schröter (eds.), *Strahlende Vergangenheit*, 233 et seq.
258 Andrew Port, "Mates in Agents' Clothing, The Wismut Upheaval of August 1951," in *German History*, 1998/1.
259 BStU, HA/St, Nr. 29/52.
260 BArch Berlin-Lichterfelde, SAPMO, FBS 363/15 418 (NL 182/986), Bericht über die Ereignisse in Saalfeld, 20.8.51.
261 BArch Berlin-Lichterfelde, SAPMO, FBS 123/16 497 (NL 90/306), Bericht der Landesleitung der SED Thüringen, 22.8.51, über die Ergebnisse der Überprüfung der Lage der Parteiorganisation der Wismut AG im Kreisgebiet Saalfeld.
262 BArch Berlin-Lichterfelde, SAPMO, FBS 363/15 418 (NL 182/986), Bericht vom 10.11.51.
263 Rolf Badstübner, Wilfrid Loth (eds.), *Wilhelm Pieck—Aufzeichnungen zur Deutschlandpolitik 1945–1953*, 378.
264 BArch Berlin-Lichterfelde, SAPMO, FBS 363/15 418 (NL 182/986), Aufstellung der notwendigen Wohnungs-Kommunal-und Kulturbauvorhaben für 1952, 21.11.51.
265 BArch Berlin-Lichterfelde, SAPMO, FBS 363/15 418 (NL 182/986), Schreiben von Ulbricht an Rau, 14.12.51.
266 BArch Berlin-Lichterfelde, SAPMO, NL 182 Nr. 987 (FBS 15 222) Bl. 30ff.
267 BArch Berlin-Lichterfelde, C-20, Nr. 105, Bericht des Sonderkommissars für Siedlungsfragen, 30.7.56.
268 SÄHStA Dresden, LRS, Ministerpräsident, Nr.334, Resolution der Hausgemeinschaft Haldenstrasse, 4.11.50.
269 Badstubner and Loth, op. cit., 378.
270 Rainer Potrass, *Zur Entfernung "deklassierter Elemente,"* in Karlsch and Schröter, op cit., 221.

[271] Heidi Roth, der 17. Juni 1953 in Sachsen, Köln 1999, 372.
[272] Torsten Diedrich, *Der 17. Juni 1953 in DDR*, Berlin 1991; Armin Mitter, Stefan Wolle (eds.), *Der Tag X-17. Juni 1953*, Berlin 1955.
[273] Roth and Diedrich in Karlsch and Schröter (eds.), op. cit., 228 et seq.
[274] Heidi Roth, op. cit., 318.
[275] Heidi Roth, Torsten Diedrich, *Wir sind Kumpel—uns kann keiner*, 253.
[276] Ralf Engeln, *Uransklaven und Sonnensucher*, 224.
[277] SÄHStA Dresden, Bezirksparteiarchiv (BPA) A/685, Bericht von Max Weber an den LV der SED Sachsen, 6.10.47.
[278] Mario Kaden, *Wismut—die "wilde" Zeit*, 34 et seq.
[279] Reimar Paul, *Das Wismuterbe*, 24.
[280] BArch Berlin-Lichterfelde, SAPMO, DY 30/IV 2/2.027/25 (Büro Lehmann), Schreiben des Bergarbeiters Norbert Schunemann und das Amt für Arbeit in Horsmar, 11.11.47.
[281] Werner Schiffner, in RADIZ-Information 10/96, Gesprächsrunde ehemaliger Wismut-Angehoriger, Schlema 1993, 39 et seq.
[282] Günther Klose, *Sprengwesen, in Chronik der Wismut*, Chemnitz 1997, Teil 1.
[283] Kaden Sammlung, Zeitungsinterview Werner L.
[284] Mikhail Semiriaga, *Kak my upravliali Germanii*, Moscow 1995, 137.
[285] BArch Berlin-Lichterfelde, SAPMO, FBS 194/20776, Tagung des Parteivorstandes der SED, 12-13.11.47.
[286] *Idem.*, Schreiben von Max Weber und das Amt für Arbeit und Sozialfürsorge Muhlhousen, Nebenstelle Horsmar, 28.1.48.
[287] FES Ostbüro, 0072B (Nr. 02362), Bericht vom 9.1.50.
[288] Zeitzeugenprotokoll von Rudolf Oelsner (Aue), March 1998.
[289] SÄHStA Dresden, LRS, MAS Nr. 216, Unfallstatistik nach Wirtschaftsgruppen.
[290] Wismut AG Archiv, Staatliche AG Wismut Sicherheitsvorschriften für die Bergwerkindustrie 1947, 1949; Chemnitz 1949.
[291] Much of the information was supplied by Herr Schiffner of Aue. BArch Berlin-Lichterfelde, IG Wismut Nr. 1501, Grubenungluck in Niederschlema, Protokoll über Vorkomnisse, 16 July 1955.
[292] *Neues Deutschland*, 14.7.55.
[293] *Volksstimme*, 20.7.55.
[294] *Freie Presse*, 18.7.55.
[295] *Volksstimme*, 22.7.55.
[296] SÄHStA Dresden, MdI, Gebietskommando DVP (BS) Wismut, Nr. 27/17, Bericht über der Einsatz des Grubenrettungskommando...in Niederschlema.
[297] *Idem.*, Bericht über die Massnahmen zur Bekampfung des Untertagsbrandes in der SDAG Wismut, 23.7.55.
[298] Neues Deutschland, 14.7.95.
[299] SÄHStA Dresden, MdI, Gebietskommando DVP (BS) Wismut, Nr. 27/17, Stimmungsbericht, 17.7.55.
[300] *Idem.*, Abschlussbericht zum Grubenbrand in Niederschlema, 1.8.55.
[301] BArch Berlin-Lichterfelde, IG Wismut Nr. 1501, Bericht über die Massnahmen des ZV der IG Wismut, 2.8.55.

302 Letter from Gotthart Bretscheider to the author, 8.10.97.
303 Chronik der Wismut GmbH, CD-Rom part 1, 598.
304 *Idem.*, 599.
305 Rainer Karlsch, "Die Erzbergbau ist keine Puppenstube," in *Unfälle bei der Wismut AG*, Berlin 1998.
306 Norbert Ranft, *Vom Objekt zum Subjekt, Montanmitbestimmung, Sozialklima und Strukturwandel im Bergbau seit 1945*, Bochum 1988, 336; Mark Roseman, *Recasting the Ruhr, Manpower, Economic Recovery and Labour Relations*, New York, Oxford 1992, 43.
307 SÄHStA Dresden, LRS, MAS, Nr. 2174/1, Schreiben des Bergamtes Zwickau an die Gewerkschaft Schneeberger Bergbau, vom 15.6.44, and "Polizeiverordnung für Radiumbergwerke," no date.
308 Egon Lorenz, "Radioactivity and Lung Cancer: a critical review of lung cancer in the mines of Schneeberg and Joachimsthal," *Journal of the National Cancer Institute*, 1944, no. 5.
309 SÄHStA Dresden, MAS Nr. 2174/1, Schreiben von Professor Wildführ an Ministerialrat Dr Seifert, 27.5.47.
310 *Idem.*, Dr. Seifert an Minister Gabler, 25.6.47.
311 *Idem.*, Dr. Schratz betr. Gesundheitsverhaltnisse im Erzbergbau, 16.11.47.
312 BArch Berlin-Lichterfelde, SAPMO, FBS 194/20776, Tagung des Parteivorstandes, 15-16.10.47.
313 SÄHStA Dresden, MAS Nr. 2174/1, Bl 28.
314 Werner Schiffner, Agricola und die Wismut, 110.
315 SÄHStA Dresden, MAS Nr. 2174/1, Bl 12, Schreiben des Landesarbeitinspektion an Minister Ziller, 16.10.48.
316 *Idem.*, Bl 15, Schreiben des Landesinspektion, 2.12.48.
317 *Wismut Chronik*, Teil 1, 477.
318 I. Bruske-Hohlfeld et al., *Abschlussbericht zum Forschungsvorhaben "Stichprobenerhebung und Auswertung von Personaldaten der Wismut,"* Neuherberg 1995, 15.
319 G. Eigenwillig, E. Ettenhuber (eds.), *Strahlenexposition und Strahleninduzierte Berufskrankheiten im Urabergbau am Beispiel Wismut*.
320 *Chronik*, Teil 2, 1432.
321 Rolf Selig, op. cit., 374.
322 K. Martignoni, Epidemiologische Studien im Rahmen des Projektes "Altlastenkataster" in *Bundesamt für Strahlenschutz*, 2. Biophysikalische Arbeitstagung in Schlema, Berlin 1992, 83.
323 Raye C. Ringholz, *Uranium Frenzy*, 93.
324 J. Švec, "Lung Cancer in Uranium Miners and Long-term Exposure," *Health Physics* 1976, 433
325 Ladislav Tomášek, Sarah C. Darby, Anthony J. Swerdlow, Václav Plaček, Emil Kunz, "Radon Exposure and Cancers Other than Lung Cancer among Uranium Miners in West Bohemia," *The Lancet*, vol. 341, 10.4.93.
326 Eduard Goldstücker, Prozesse, 251.
327 Hans Ulrich Walter, *Epidemiologische Untersuchung über die Lebenserwartung und Mortalität von in der Nachkriegszeit in Joachimsthaler Uranbergwerken beschäftigten Personen*, Inaugural-Dissertation, Munich 1982.

[328] Rolf Seelig, op. cit., 380.
[329] Michael Beleites, *Altlast Wismut*, 121.
[330] Rainer Karlsch, "Belastung durch bewaffnete Organe," in *Materialen für der Enquete-Komission "Überwindung der Folgen der SED-Diktatur im Prozess der deutschen Einheit*, Vol. III/2, Baden-Baden 1999, 1500–1584.
[331] BArch Berlin-Lichterfelde, SAPMO, NL 90/471, Notiz von Minister Grotewohl vom September 1955.
[332] Rainer Karlsch, *Belastungen durch bewaffnete Organe*, 1571.

Concluding Remarks

While we were working on the study of uranium in the Erzgebirge with Rainer Karlsch, it seemed to us that its role in the historiography of the Cold War had been underestimated. By the middle of 1948 at the latest, Stalin and the Soviet leaders knew that the Erzgebirge was the richest, most accessible source of the precious ore. The strategic importance of the region greatly increased.

The Soviet nuclear project, including of course the search for uranium, was under the control of Beria and of the Soviet security agencies. This circumstance alone left a lasting mark on the development of both the Czechoslovak and the German uranium industries. During the great uranium rush that followed the Second World War, the Soviets put aside even elementary political and diplomatic caution. They started to use, under the guidance of Beria's agencies, methods of economic mobilization that they had employed in the distant regions of Siberia and central Asia. On both sides of the border, a line that ran across the peaks and the valleys of the Erzgebirge, Soviet methods, including the use of forced labor and manpower to make up for the technological deficit, were employed.

In Czechoslovakia as well as in the Soviet zone of occupation in Germany, the uranium industry became the flagship of Soviet influence. They taught the people and the local communist leaders that they could stand up to Soviet imperial interests at their own peril. They instructed them in Soviet ways of doing and making things, including economic planning regardless of the needs of the people. While Western Europe tried to repair the damage of the war, Stalin mobilized the resources of his new empire in an attempt to challenge the American nuclear supremacy.

The secret engagement between the Soviet and the Western intelligence services was, in the Erzgebirge region, at its most vigorous, as was the open battle between the Eastern and Western machineries of propaganda. It was to be expected that Stalin would try and bind the two countries into the Soviet bloc as tightly as possible. As long as the communist regimes lasted and uranium was mined, that is from the end of the Second World War until 1989, the Czechs supplied the Soviet Union with nearly 100,000 tons of pure uranium, while the East Germans sent more than twice the amount. At least until the first Soviet atomic bomb was tested on 29 August 1949, the Erzgebirge supplied the Soviet nuclear project with most of the uranium it required.

During the greatest uranium shortage, the Soviets, assisted by the local communist regimes, applied strong pressure on the Czechs and the Germans alike. The pressure abated more slowly in Czechoslovakia than the GDR, perhaps because the Czech communists were keener students of Soviet methods than the Germans.

A personal reminiscence is perhaps fitting here, as it helps illuminate the contrasts between the development of the Czech and German uranium industries. On the occasion of the opening of a 2005 exhibition based on the German version of this book in Chomutov, the retired German miners, former employees of Wismut AG, were accompanied by their families and wore festive dress uniforms. The Czech employees of Jáchymov n p (state owned company), on the other hand, were represented by a group of former political prisoners. The Czechoslovak communist leaders who admired the Soviet Union and gradually came to fear it, found it easy or imperative to follow the Soviet example. The local uranium industry thus found it difficult to shed its Stalinist past, even after the Soviets loosened their control over Jáchymov n p.

The Germans, on the other hand, who had less to lose after the disastrous war and less reason to be grateful to the Russians, built an efficient industry which managed to at least avoid the worst excesses of Beria's command methods. Wismut AG had a turbulent history prior to the German control of the mine. It became a secret enclosure of Soviet authority, a state within a state; yet instead of stark compulsion of any part of its workforce, it managed to develop a system of ingenious incentives. The German managers later served Moscow as reliably as their Soviet predecessors, and the uranium industry covered the largest part of the German reparations payments.

The Czech and the German uranium industries, the fastest growing sectors of the economy of the two countries, reached their zenith in the early 1950s. In 1950, Wismut AG employed some 200,000 people. The Soviet-sponsored uranium rush was different from the other precious metal rushes in the region. No other parts of the GDR or of Czechoslovakia experienced such violent transformations as did the uranium provinces. They became strictly guarded security zones, where the demands of the industry reigned supreme. The countryside, the villages and whole parts of ancient towns fell victim to the uranium rush. There was no time to repair the damaged countryside, remove the vast slagheaps, or fill in the mudflats and brackish lakes created by the extensive mining activities.

While the Czechoslovak government developed its own system of uranium gulags after 1948, hundreds of Wismut miners suffered imprisonment or, in some cases, death sentences, imposed by the special military tribunals for petty offences. In addition, deliberately concealed health hazards remained. One the whole, voluntary miners were well-rewarded, while at the same time being grossly deceived. Despite local experience with the "Schneeberg sickness," Soviet management kept quiet about the radiation risks. Due to their ignorance and the fact that they underestimated the risks, thousands of miners subsequently suffered from silicosis and lung cancer.

After the hectic growth of the uranium industries in the Soviet zone of occupation and in Czechoslovakia, the development of the sector took a more peaceful course in the 1960's. Uranium ore was no longer in short supply and the potential for using it for peaceful purposes had emerged. The "wild years" of Wismut AG ended long before the Soviet–German joint-stock company was founded in 1954. In Czechoslovakia, the last political prisoners had disappeared from the lists of workers of Jáchymov n p only by the end of 1961.

In the late 1970s, uranium mining in Czechoslovakia and the GDR went into decline. The richest deposits had been exhausted and the cost of production increased. It led, in the case of Wismut, to conflicts between Soviet and German parties as to the division of the growing burden. While the Germans pleaded for controlled production, which would protect the remaining resources, their Soviet partners went on insisting on the highest possible production. The basic consensus on the continuation of production was however never questioned.

The Chernobyl disaster in 1986 preceded the break-up of the Soviet bloc by some three years. The end of the Cold War brought the end of uranium mining in GDR and its virtual winding down in Czechoslovakia. When environmental damage in Saxony and Thüringia was surveyed after the reunification of Germany in October 1990, it was calculated that the redevelopment of the countryside would take at least fifteen years and cost about 13 billion marks. In Czechoslovakia (and since 1993, the Czech Republic), the environmental damage was fortunately smaller, since the state did not dispose of sufficient resources to repair it.

In the Erzgebirge on the German and the Czech sides and in the Příbram district in south Bohemia as well, the uranium rush followed the much earlier silver trails. In this case it lasted, as most rushes for precious metals did, for some four decades. It changed the lives of hundreds of thousands of people, for many of them, unfortunately, for the worse.

Appendix 1

Translation of the Czechoslovak–Soviet Uranium Treaty of 23 November 1945*

Strictly confidential
AGREEMENT between the government of the Union of Soviet Socialist Republics and the government of the Czechoslovak republic concerning the expansion of the mining of ores and concentrates in Czechoslovakia, containing radium and other radioactive elements, as well as their supplies to the Union of Soviet Socialist Republics.

The government of the Union of Soviet Socialist Republics and the government of the Czechoslovak republic have agreed as follows:

Part 1.
The Czechoslovak government organizes a state enterprise for the exploration and exploitation of all deposits containing radium and radioactive elements, which are the property of the Czechoslovak state.

---

* A note on bibliography: since the opening of the archives of the communist period, Karel Kaplan and Vladimír Pacl published, in Prague in 1993, *Tajný prostor Jáchymov*, containing the full text of the treaty. The publication came too late for David Holloway, who wrote in *Stalin and the Bomb, The Soviet Union and Atomic Energy 1939–1956*, (Yale University Press 1994, 108–109) that "At the end of March 1945 the Czechoslovak government in exile, headed by President Edvard Beneš, travelled from London to Moscow, on its way home to Prague. While it was in Moscow it signed a secret agreement giving the Soviet Union the right to mine uranium ore in Czechoslovakia and transport it to the Soviet Union." Holloway quotes as his source Jiři Kašpárek's article in The Russian Review (1952, no. 2, 98–99) as well as a letter from Eugen Loebl, deputy minister of foreign trade, who had apparently tried to "get the Soviet Union to pay world prices for Czechoslovak uranium." In a series of articles on Czechoslovak foreign policy, in Mezinárodní politika 1991, Jaroslav Šedivý describes, in chapter 3, entitled "Jak se velký bratr vybavoval," how he had in the 1960's looked for the treaty in the archives of the foreign ministry. It was missing there, but after about two years it was found at the embassy in Cairo. It was apparently taken there by an ambassador who defected in 1948 and who left a copy of the treaty behind in a hurry. In both German and English, Zbyněk Zeman published an article entitled "Der tschechoslowakisch-sowjetische Uranvertrag von 23 November 1945" in a book entitled *Strahlende Vergangenheit* (edited by Rainer Karlsch and Harm Schröter and published in Germany by the Scripta Mercaturae Verlag in 1996) and an article on a similar subject in *Transition*, a monthly magazine of the Open Media Research Institute, in May 1996.

Part 2.

The Czechoslovak government will do its best to secure the maximum increase of the extraction of ore and concentrates, containing radium and other radioactive elements.

Part 3.

The government of the Soviet Union will provide comprehensive technical aid for the research and exploitation of the above-mentioned deposits. This aid will consist firstly of dispatching experts for the organisation of the search and industrial examination of the deposits and for the mining operations for the ore and concentrates, as well as in the supply of the necessary equipment and material.

Part 4.

Both governments will form a permanent Czechoslovak–Soviet commission based in Prague and consisting of four members (two from each government). The commission will have the following tasks:

a) to work out directives for the purposes of the extension of the work in geological exploration and the increase of extraction of the ore and concentrates.

b) to consider plans for the extraction of the ore and concentrates, while the basic plans have to be ready in time and for the period of at least five years, with a gradual increase of the plan in case the results of geological exploration will provide the basis for it.

c) to solve all questions arising in the framework of the fulfilment of the agreement in the case of technical assistance and supplies.

d) to fix the price for the ore and concentrates and for radium in accord with paragraph 5 of this agreement on the basis of production costs (*svestojnych nakladu*) and adding a normal percentage of profit.

The work of the commission will be regulated by a statute provided by itself. Decisions of the commission will be binding with the agreement of the two sides. In case of an disagreement between the Czechoslovak and Soviet members, the matter will be resolved by the two governments.

Part 5.

The Czechoslovak-Soviet commission will decide in the sense of paragraph 4 which part of the mined ore and concentrates will remain in Czechoslovakia for its necessary economic and scientific needs. The rest of the mined ore and concentrates, containing radium and other radioactive elements will be handed over to the Union of Soviet Socialist Republics, while 50% of the radium will be returned to Czechoslovakia, as long as it will be extracted from the ore and concentrates supplied by Czechoslovakia for processing in the USSR.

Reciprocal accountancy, resulting from the supply of the ore and concentrates for processing in the Soviet Union and the return of radium to Czechoslovakia will be carried out on the basis of the prices for the ore and concentrates and the prices for radium, fixed with the agreement of both governments, with payments of the differences so arising either in supplies of goods or in money (ve valute) according to the two parties' agreement.

Part 6.
The Soviet party agrees with providing one expert in the rank of the technical director among the experts sent to Czechoslovakia, one expert in the rank of chief engineer and one expert in the rank of the head of technical control of the Jáchymov enterprise.

Part 7.
The two parties undertake to exchange scientific data concerning the use of ores and concentrates containing radium and radioactive elements.

Part 8.
This agreement comes into force after signature and is valid for a period of twenty years.

Drafted in Prague on 23 November 1945 in two authentic copies, each in the Czech and in the Russian language, both texts having the same validity.

On the basis of full powers of the government of the Czechoslovak republic H. Ripka
On the basis of full powers of the government of the Union of Soviet Socialists Republics I. Bakulin

Strictly confidential

PROTOCOL to the agreement between the government of the Czechoslovak republic and the government of the Union of Soviet Socialist

Republics concerning the expansion of the mining of ores and concentrates in Czechoslovakia, containing radium and other radioactive elements, as well as their supplies to the Union of Soviet Socialist Republics.

In connection with the signing in Prague today of an agreement between the government of the Czechoslovak republic and the government of the Union of Soviet Socialist Republics concerning the expansion of the mining of ores and concentrates in Czechoslovakia, containing radium and other radioactive elements, as well as their supplies to the Union of Soviet Socialist Republics it was agreed that:

1. From the total amount of the mined ores and concentrates containing radium and other radioactive elements, for the duration of the first five years of the above-mentioned agreement, Czechoslovakia will retain for its economic and scientific needs up to 10% of these ores and concentrates.

2. The government of the Czechoslovak republic will provide according to the needs of the Soviet members of the Czechoslovak–Soviet commission suitable accommodation in Jáchymov and Prague.

3. For the purpose of ensuring the confidentiality of the extraction of the ores and concentrates containing radium and other radioactive elements as well as their supplies to the Union of Soviet Socialist Republics the Czechoslovak–Soviet commission will provide corresponding regulations in the Jáchymov and other possible enterprises in Czechoslovakia.

4. The agreement above between the government of the Czechoslovak republic and the government of the Union of Soviet Socialist Republics is strictly confidential.

Drafted in Prague on 23 November 1945 in two authentic copies, each in the Czech and in the Russian language, both texts having the same validity.

With the full powers of the government of the Czechoslovak republic H. Ripka
With the full powers of the government of the Union of Soviet Socialists Republics I. Bakulin

Appendix 2

The reply was drafted on 18 January 1946, on Masaryk's personal writing paper, from 58 Westminister Gardens, Marsham Street, London S W 1. The first draft was written in blue ink and was corrected in black. Corrections are given here in square brackets. It read:

"Please note, in connection with your fears and rebuke, that

1. I have always spoken, during the war, about the nationalization and control of the armament industries, I am convinced of it [addition: and I have never been contradicted].
2. Concerning uranium I said that we hope that our uranium will be used for peaceful purposes. The Soviet delegation openly applauded that part of my speech. I did not talk to Reuters about inspection in Jáchymov mines, I said that the quantity and quality of our radium has long been known and that we have no cause to be secretive about it [changed to: and we have never kept it secret].
2A. I have not done anything adversely to affect [addition: on the contrary] our treaty with the Soviets, which of course remains secret. We hope that there will be no war and that neither the Soviets nor we will have to use atomic weapons. The Soviets here have not gained an unfavorable impression and they understood my expression of the hope that no war will take place. [addition: There will be no unpleasant repercussions here, unless you start them at home.]
3. I made the whole speech in full and absolute agreement with the delegation, which supplied me with all the material, and the passage in particular concerning uranium was not drafted by me. All the members had twenty-four hours and approved the speech unanimously.
4. I regret if [addition: because of ignorance of the atmosphere here] a big affair, unpleasant for you, is built up from this. Please take note of the fact that I am ready immediately [addition: more than ready] to resign the leadership of the delegation and my post. All the members of the Soviet delegation congratulated me [addition: especially Gromyko and Kuznetsov]. Manuilsky, the Poles and the Yugoslavs the same.
5. I take full responsibility for the speech and I beg you not to extend the blame to other members of the delegation, who supplied me with ideas that made you so cross. I should humbly advise you to remain calm, if I may be so bold.

6. [The following passage was deleted] I repeat that if I have not your trust, I cannot go on and I would be glad to leave for other [the rest is illegible].
7. [Addition: you have the full text of my speech.]
    The Russians came to me after the speech and talked to me about the general secretaryship [of the United Nations] If nobody at home uses the incident against me, everything will turn out well.

Archives

Germany
Archiv der Friedrich-Ebert-Stiftung Bonn (FES), Ostbüro der SPD
Archiv des Bundesbeauftragten für die Unterlagen des ehemaligen
   Staatssicherheitsdienstes der DDR (BStU)
Bundesarchiv (BArch) Berlin-Lichterfelde und Koblenz
Geologisches Archiv (GA) der Wismut GmbH
Institut für Zeitgeschichte München (IfZ)
Landesarchiv Berlin
Radon-Dokumentations- und Informationszentrum Schlema e. V. (RADIZ)
Sächsisches Hauptstaatsarchiv (SÄHStA) Dresden und Außenstelle Chemnitz
Stadtarchiv Schneeberg
Stiftung Archiv der Parteien und Massenorganisationen der DDR (SAPMO) im
   Bundesarchiv Berlin-Lichterfelde
Sudetendeutsches Archiv München
Unternehmensarchiv (UA) der Wismut GmbH.

Privatarchiv Karl Aurand
Privatarchiv Gotthard Bretschneider
Privatarchiv Werner Schiffner

Czech Republic
Archiv Českého geologického ústavu (CGÚ), Prague
Archiv DIAMO, Příbram
Státní ústřední archív (SÚA), now Národní archiv (NA), Prague
Archiv ministerstva vnitra (AMV), Prague
Archiv ministerstva zahraničních věcí (AMZV), Prague
Archiv Národního musea (ANM), Prague
Archiv Ústavu T. G. Masaryka, ve správě MSU—A AV ČR, (AÚTGM), Prague
Okresní archiv, Karlovy Vary-Rybáře
Vojenský historický archiv (VHA), Prague
Private archive of Dr. Hettler

Russia
State archive of the Russian Federation (GARF), Moscow
Russian Economical State Archive Economy (RGAE), Moscow
Russian State Military Archive (RGVA) Podolsk

Great Britain
Public Record Office (PRO), now National Archiv (NA), London.

USA
Hoover Institution Archives (HIA), Stanford
National Archives (NA), Washington

Bibliography

30 let uranového průmyslu (30 years of the uranium industry), Prague, 1975.

Abele, Johannes. *„Wachhund des Atomzeitalters." Geigerzähler in der Geschichte des Strahlenschutzes* ("Watch dogs of the atomic age." The Geiger counter in the history of radiation protection), Dissertation, Munich, 1998.

Abelshauser, Werner. *Der Ruhrkohlenbergbau seit 1945, Wiederaufbau, Krise, Anpassung* (The Ruhr coal mining industry since 1945. Recovery, Crisis, Adaption), Munich, 1984.

Albrecht, Ulrich, Andreas Heinemann-Grüder, and Arnd Wellmann. *Die Spezialisten. Deutsche Naturwissenschaftler und Techniker in der Sowjetunion nach 1945* (The specialists. German natural scientists and technicians in the Soviet Union after 1945), Berlin, 1992.

Albright, Joseph, Marcia Kunstel. *Bombshell: The Secret Story of America's Unknown Atomic Spy Conspiracy.* New York, 1997.

Andrew, Christopher, Vassilii Mitrokhin. *The Mitrokhin Archive*, London 1999.

Antonov-Ovsejenko, Anton. „Der Weg nach oben. Skizzen zu einem Berija-Porträt" (The way to the top. Sketches for a Beria portrait) in Nekrassov, Vladimir F., ed. *Berija. Henker in Stalins Diensten* (Beria. Executioner in Stalin's duty), Augsburg, 1997.

Ardenne, Manfred von. *Ich bin ihnen begegnet* (I've met them), Düsseldorf, 1997.

Ardenne, Manfred von. *Sechzig Jahre für Forschung und Fortschritt*, (Sixty years of research and progress), Berlin, 1988.

Aurand, Karl, Werner Schüttmann. *Die Geschichte der Außenstelle Oberschlema des Kaiser-Wilhelm-Instituts für Biophysik Frankfurt/M.* (The history of the branch office of the Kaiser-Wilhelm-Institute for Biophysics Frankfurt in Oberschlema), Salzgitter, 1991.

Badstübner, Rolf, Wilfried Loth, eds. *Wilhelm Pieck – Aufzeichnungen zur Deutschlandpolitik* (Wilhelm Pieck – Notes on German policy), Berlin, 1994.

Běhounek, František. „Über die Verhältnisse der Radioaktivität im Uranpecherzbergbaurevier von St. Joachimsthal in Böhmen" (On the rates of radioactivity in the uranium mining district of St. Joachimsthal in Bohemia), *Physikalische Zeitschrift* 28/1927.

Beleites, Michael. *Altlast Wismut, Ausnahmezustand, Umweltkatastrophe und das Sanierungsproblem im deutschen Uranbergbau* (Disused dump Bismuth Corporation. State of emergency, environmental disaster and the problem of cleaning up in the German uranium mining industry), Frankfurt/M., 1992.

Beyer, Klaus. *Arbeitsschutz und technische Sicherheit im erzgebirgischen Uranerzbergbau – einige kritische Gedanken zu einem heißen Thema* (Safety and technical standards at work in the uranium mining industry in the Erzgebirge – some critical thoughts on a hot topic), Schlema, 2001.

Bogsch, Walter. *Der Marienberger Bergbau seit der zweiten Hälfte des 16. Jahrhunderts* (Mining in Marienberg since the second half of the 16th century), Cologne, 1966.

Bretschneider, Gotthard. *Kein Schutz vor Willkür? Zur Arbeit der Gewerkschaft in der SAG Wismut 1946–1953* (No protection from arbitrariness? The work of the trade union in the Soviet stock company Wismut 1946–1965), Schlema, 1997.

Bretschneider, Gotthard. Warum *entstand die SAG/SDAG Wismut?* (Why was the Wismut corporation founded?), Schlema, 1999.

Brooks, Geoffrey. *Hitler's Nuclear Weapons*, London, 1992.

Brüske-Hohlfeld, I., M. Möhner, and H. E. Wichmann. „Abschlußbericht zum Forschungsvorhaben 'Stichprobenerhebung und Auswertung von Personaldaten der Wismut' für den Hauptverband der Gewerblichen Berufsgenossenschaften e. V" (Final report about the research project „Samples and evaluation of personal files of the Wismut "for the trade association having liability for industrial safety and insurance"), *GSF-Bericht* 17/95, Neuherberg, 1995.

Buchler, Walter, ed. *Dreihundert Jahre Buchler. Die Unternehmen einer Familie 1651–1958* (300 years of Buchler. The companies of a family 1651–1958), Braunschweig, 1958.

Buschfort, Werner. *Das Ostbüro der SPD—von der Gründung bis zur Berlin-Krise* (The "Eastern bureau" of the Social Democratic Party. From its foundation up to the Berlin crisis), Munich, 1991.

Caufield, Catherine. *Das strahlende Zeitalter. Von der Entdeckung der Röntgenstrahlen bis Tschernobyl* (Multiple Exposures: Chronicles of the Radiation Age), Munich, 1994.

Khariton, Iulii B. "Memories," *The Bulletin of Atomic Scientists*, May 1993.

Chronik der Wismut GmbH (Chronicle of Wismut Limited Company), CD-Rom, Chemnitz,1998.

Chronik der Wismut GmbH (Chronicle of Wismut Limited Company, Part I), Chemnitz, 1997.

Chronik der Wismut GmbH (Chronicle of Wismut Limited Company, Part II), CD-Rom, Chemnitz, 1999.

Churchill, Winston S. *The Second World War: Triumph and Tragedy*, vol. 6., London, 1954.

„Der Uranbergbau in der Sowjetzone" (Uranium mining in the Soviet Zone), Vorstand der SPD, ed. *Denkschriften* no. 27, Hannover o. J. [1950].

Die SED. Geschichte, Organisation, Politik. Ein Handbuch (The Socialist Unity Party of Germany. History, Organisation, Politics. A Handbook), Berlin, 1997.

Diedrich, Torsten. *Der 17. Juni 1953 in der DDR. Bewaffnete Gewalt gegen das Volk* (The 17th June 1953 in the German Democratic Republic. Armed forces against the people), Berlin, 1991.

Drápala, Milan. *Na ztracené vartě Západu: Antologie české nesocialistické publicistiky z let 1945–1948* (The lost watchtower of the west. Anthology of Czech nonsocialist poitical journalism 1945–1948), Prague 2000.

Dvořák, Tomáš. "Těžba uranu versus „očista" pohraničí" (Uranium mining versus cleanup of the borderlands), *Soudobé dějiny* 3–4/2005.

Eichholtz, Dietrich. *Geschichte der deutschen Kriegswirtschaft* (History of the German war economy), vol. II, Berlin, 1985.

Eigenwillig, G., E. Ettenhuber, eds. *Strahlenexposition und Strahleninduzierte Berufskrankheiten im Uranbergbau am Beispiel Wismut* (Radioactive contamination and occupational diseases in the uranium mining of Wismut corporation), Cologne, 2000.

Elsner, Gine, Karl-Heinz Karbe. *Von Jachymov nach Haigerloch. Der Weg des Urans für die Bombe* (From Jachymov to Haigerloch. The pathway of uranium to the bomb), Hamburg, 1999.

Engeln, Ralf. "Die industriellen Beziehungen im Uranbergbau der SAG Wismut" (Industrial relations in the uranium mining of Soviet stock company Wismut) in Karlsch, Rainer, Harm Schröter, eds. „Strahlende Vergangenheit." Studien zur Geschichte des Uranbergbaus der Wismut ("Radiant past." Studies on the history of uranium mining by the Wismut corporation), St. Katharinen, 1996.

Engeln, Ralf. *Uransklaven oder Sonnensucher? Die Sowjetische AG Wismut in der SBZ/DDR 1946–1953* (Uranium slaves or Sun searchers? The Soviet Wismut corporation in the Soviet Zone of Occupation and GDR 1946–1953), Essen, 2001.

Erler, Peter. "Zur Tätigkeit der Sowjetischen Militärtribunale (SMT) in der SBZ/DDR" (Soviet military tribunals in the Soviet Zone of Occupation and GDR) in Mironenko, Sergei, Lutz Niethammer, and Alexander von Plato, eds. *Sowjetische Speziallager in Deutschland* (Soviet special camps in Germany), vol. 1, Berlin, 1998.

Faensen, Hubert. *Hightech für Hitler. Die Hakeburg—Vom Forschungszentrum zur Kaderschmiede*

(High tech for Hitler. The Hakeburg- Form a brain trust to a centre for political stuff), Berlin, 2001.

Fierlinger, Zdeněk. *Ve službách ČSR. I.* (In duty for the Czechoslovakian Republic), vol. I, Prague, 1951.

Filippovich, Dmitrii N., Vladimir V. Sakharov. "Deutsches Uran für die sowjetische Atombombe" (German uranium for the Soviet atomic bomb) in *Der Anschnitt*, 50. Jg., 2–3/1998.

Fischer, Helmut J. "Hitler und die Atombombe. Bericht eines Zeitzeugen" (Hitler and the atomic bomb. Report of an eyewitness), Asendorf, 1987.

Fischer, Helmut J. *Hitlers Apparat. Namen, Ämter, Kompetenzen* (Hitler's system. Names, positions, competences), Kiel, 1988.

Foitzik, Jan. *Inventar der Befehle des Obersten Chefs der Sowjetischen Militäradministration in Deutschland (SMAD) 1945–1949* (Inventory of the Orders of the chief of the Soviet military administration in Germany (SVAG) 1945–1949), Munich, 1995.

Frank, Mario. *Walter Ulbricht. Eine deutsche Biographie* (Walter Ulbricht. A German biography), Berlin, 2001.

Franklin, William. *Zonal Boundaries and Access Routes to Berlin, World Politics*, October 1963.

Fricke, Karl Wilhelm. "Organisation und Tätigkeit der DDR-Nachrichtendienste" (Organisation and activities of the GDR's intelligence services) in Krieger, Wolfgang, Jürgen Weber, eds., *Spionage für den Frieden? Nachrichtendienste in Deutschland während des Kalten Krieges* (Spying for peace? Intelligence services in Germany in the Cold War years), Munich, 1997.

Fricke, Karl Wilhelm. "Kampf dem Klassenfeind" (The fight against the class enemy) in Fischer, Alexander ed., *Studien zur Geschichte der SBZ/DDR* (Studies on the history of the Soviet Zone of Occupation and GDR), Berlin, 1993.

Friedman, Otto. *The Break Up of Czech Democracy*. London, 1950.

Friedrich, Ulrike. *Die Außenstelle des KWI für Biophysik (Frankfurt/M.) im Radonbad Oberschlema* (The branch office of the Kaiser-Wilhelm-Institute for Biophysics Frankfurt/M. at the radon spa Oberschlema), dissertation, Freie Universität Berlin, 1997.

Fuchsloch, Norman. "Forschungen zur Uranprospektion an der Bergakademie Freiberg im Auftrag der Sowjetunion" (Research on uranium prospecting performed at the Bergakademie Freiberg and commissioned by the Soviets) in *Der Anschnitt*, 50. Jg., 2–3/1998.

Gilbert, Martin. *The Road to Victory: Winston S. Churchill 1941–1945*. London, 1986.

Gilbert, Martin: *Never Despair: Winston S. Churchill* 1945–1965, London, 1988.

Gill, Ulrich. *Der Freie Deutsche Gewerkschaftsbund (FDGB). Theorie–Geschichte–Organisation–Funktionen–Kritik* (The Free German Federation of Unions. Theory. History. Organisation. Function. Critics.), Opladen, 1989.

Gniffke, Erich W. *Jahre mit Ulbricht* (Years with Ulbricht), Cologne, 1966.

Goebbels, Joseph. *Tagebücher aus den Jahren 1942–1943* (Diary of the years 1942–1943), Zurich 1948.

Goldstücker, Eduard. *Prozesse. Erfahrungen eines Mitteleuropäers* (Processes. Experiences of a Central European), Munich, 1989.

Goudsmit, Samuel. *Alsos*, New York, 1947.

Grieß, Rainer. *Die Rationen-Gesellschaft. Versorgungskampf und Vergleichsmentalität: Leipzig, München und Köln nach dem Kriege* (The ration-society. Struggle for supplies and the comparative mentality: Leipzig, Munich and Cologne after the war), Münster, 1991.

Grishin, Nikolai. "The Saxony Uranium Mining Operation ('Wismut')," in Slusser, Robert, ed., *Soviet Economic Policy in Postwar Germany. A Collection of Papers by Former Officials*, New York, 1953.

Groves, Leslie Richard. *Now it Can be Told: the Story of the Manhattan Project*, London, 1963.

Harriman, Averell, and Elie Abel. Special Envaj to Churchill and Stalin, 1941–1946. New York 1975.

Heinemann-Grüder, Andreas. *Die sowjetische Atombombe* (The Soviet atomic bomb), Münster, 1992.

Hejl, Vilém, and Karel Kaplan. *Zpráva o organizovaném násilí* (Report about organised violence), Toronto, 1986.

Helmreich, Jonathan E. *Gathering Rare Ores: The Diplomacy of Uranium Acquisition 1939–1954*. Princeton, 1986.

Henke, Klaus-Dietmar. *Die amerikanische Besetzung Deutschlands* (The American occupation of Germany), Munich, 1995.

Henshall, Philip. *The Nuclear Axis: Germany, Japan and the Atom Bomb Race 1939–1945*. London, 2000.

Herbert, Ulrich. *Fremdarbeiter. Politik und Praxis des „Ausländer-Einsatzes" in der Kriegswirtschaft des Dritten Reiches* (Foreign workers. Politics and practice of the "Ausländer-Einsatz" in the war economy of the Third Reich), Berlin, 1985.

Herken, Gregg. *The Winning Weapon: The Atomic Bomb in the Cold War 1945–1950*. New York, 1980.

Heym, Stefan, *Schwarzenberg*, Frankfurt am Main, 1984.

Hildebrandt, H. J. "Die Kernenergie im System der Elektrizitätsversorgung der DDR" (Nuclear energy as part of the electrical energy system in the GDR) in *Energietechnik*, April 1957.

Hinsley, F. H., et al. *British Intelligence in the Second World War*, London, 1979–1990.

Hoffmann, Dieter, ed., *Operation Epsilon: Die Farm-Hall-Protokolle oder die Angst der Alliierten vor der deutschen Atombombe* (The Farm-Hall protocols or the fear of the Allies about a German atomic bomb), Berlin, 1993.

Holloway, David. *Stalin and the Bomb: the Soviet Union and Atomic Energy 1939–1956*. New Haven–London, 1994.

Holloway, David. *The Soviet Union and the Arms Race*. New Haven, 1983.

Horák, Vladimír. *Paměti královského horního města Jáchymova a jeho stříbrných a uranových dolů* (The monuments of the royal mining city Jachymov and its silver and uranium mines), Prague, 1993.

Houdek, K. *Vývoj a současné problémy uranového průmyslu v ČSSR* (Development and present problems of uranium industry in the Czechoslovakian Socialist Republic), Prague, 1969.

Hübner, Peter. *Konsens, Konflikt und Kompromiss: Soziale Arbeiterinteressen und Sozialpolitik in der SBZ/DDR 1945–1970* (Consensus, conflict and compromise: Social interests of workers and social policy in Soviet Zone of Occupation/GDR 1945-1970), Berlin, 1995.

Irving, David. *Der Traum von der deutschen Atombombe* (First published in English under the title: The virus house), Hamburg, 1969.

Jahn, Manfred. "Vertriebene Deutsche als attraktive Zielgruppe der Arbeitskräftelenkung in den Uranbergbau" (Expelled Germans as potential work forces for the uranium mines) in *Der Anschnitt*, 50. Jg., 2–3/1998.

Jehles, Ursula. *Der Bergbau im Erzgebirge und die Arbeitsmedizin* (Uranium mining in the Ore mountains and occupational medicine and health care), Erlangen, 1961.

Jentsch, Frieder. "Begegnungen mit einem Giganten" (Meetings with a giant) in *Ans Licht gebracht. Begegnungen und Erinnerungen* (Up to the light. Meetings and memories), Chemnitz, 1998.

Johnson, David Alan. *Germany's Spies and Saboteurs*. New York, 1998.

Kaden, Mario. *Auswertung von Zeitzeugenbefragungen* (Evaluation of Eyewitness protocols), Manuscript, Annaberg, 1999.

Kaden, Mario. "Kriminalität, Polizei-, Justiz- und Sicherheitsapparat in der 'Uranprovinz' 1946 bis 1958" (Criminality and the Police, Justice and Secret Service in the "Uranium province" 1946 to 1958) in Karlsch, Rainer Harm, Schröter, eds., *"Strahlende Vergangenheit". Studien zur Geschichte des Uranbergbaus der Wismut* ("Radiant past". Studies on the history of uranium mining by the Wismut corporation), St. Katharinen, 1996.

Kaden, Mario. *Uranprovinz: Zeitzeugen der Wismut erinnern sich* (Uranium province: Eyewitness reports), Marienberg, 2000.

Kaden, Mario. *Wismut–die "wilde" Zeit. Beiträge zur Geschichte des Landkreises Annaberg* (The "wild years" of Wismut corporation. Studies on the history of the district of Annaberg), Annaberg, 1994.

Kaplan, Karel. *Der kurze Marsch. Kommunistische Machtübernahme in der Tschechoslowakei 1945–1948* (The short march. Communist takeover of Czechoslovakia 1945–1948), Veröffentlichungen des Collegium Carolinum, Bd. 33, Munich, 1981.

Kaplan, Karel. *Zpráva o zavraždění generálního tajemníka* (Report about the murdering of the General secretary), Prague, 1992.

Kaplan, Karel, Vladimír Pacl. *Tajný prostor Jáchymov* (The secret place Jáchymov), Prague, 1993.

Karlsch, Rainer. *"Der Erzbergbau ist keine Puppenstube". Unfälle bei der Wismut AG* ("Uranium mines are not a dollhouse." Accidents in the Wismut corporation), Berlin, 1998.

Karlsch, Rainer. *Allein bezahlt? Die Reparationsleistungen der SBZ/DDR 1945–1953* (Paid on its own? The reparations of the SBZ/GDR 1945–1953), Berlin, 1993.

Karlsch, Rainer. "Belastungen durch bewaffnete Organe" (The costs of military and paramilitary organizations) in *Materialien der Enquete-Kommission "Überwindung der Folgen der SED-Diktatur im Prozess der deutschen Einheit,"* vol. 2, Baden-Baden, 1999.

Karlsch, Rainer. "Der Uranwettlauf 1939 bis 1949" (The hunt for uranium ore 1939 to 1949) in *Der Anschnitt*, 50. Jg., 2–3/1998.

Karlsch, Rainer. "Kohle, Chaos und Kartoffeln" (Coal, chaos and potatoes) in Jürgen Engert ed., *Die wirren Jahre. Deutschland 1945–1948* (The confused years. Germany 1945–1948), Berlin, 1996.

Karlsch, Rainer. „Ungleiche Partner—Die vertraglichen Grundlagen des Uranbergbaus und der Verrechnung der Uranerzlieferungen" (Unequal partners – The treaties for the uranium mining and the price for the uranium ore shipments) in Karlsch, Rainer, Harm Schröter eds, *"Strahlende Vergangenheit". Studien zur Geschichte des Uranbergbaus der Wismut* ("Radiant past". Studies on the history of uranium mining by the Wismut corporation), St. Katharinen, 1996.

Karlsch, Rainer, Johannes Bähr. "Die Sowjetischen Aktiengesellschaften (SAG) in der SBZ/DDR" (The Soviet stock companies in the SBZ/GDR) in Lauschke, Karl, Thomas Welskopp eds., *Mikropolitik im Unternehmen: Arbeitsbeziehungen und Machtstrukturen in industriellen Grossbetrieben des 20. Jahrhunderts* (Micro policy in the corporation: industrial relations and power structures in large industrial enterprises in the 20th century), Essen, 1994.

Karlsch, Rainer, Vladimir Zakharov. "Ein Gulag im Erzgebirge? Besatzer und Besiegte beim Aufbau der Wismut AG" (A GULAG in the Ore mountains?

Occupiers and the defeat in the building up of the Wismut corporation) in *Deutschland-Archiv*, 32. Jg., 1/1999.

Kennan, George. *Memoirs 1925–1950*. London, 1968.

Kennan, George.Memoirs *1950–1963*. London, 1973.

Kießling, Wolfgang. *Partner im "Narrenparadies." Der Freundeskreis um Noel Field und Paul Merker* ("Partners in the paradise of fools." The circle of friends around Noel Field and Paul Merker), Berlin, 1994.

Klapper, Lothar. *Die Gründung: Festschrift 500 Jahre Annaberg* (The beginnings: 500 years Annaberg), Annaberg, 1996.

Knyschweskij, Pavel. *Moskaus Beute: Wie Vermögen, Kulturgüter und Intelligenz nach 1945 aus Deutschland geraubt wurden* (Moscow' s booty. The plunder of values, cultural goods and intelligence after WWII in Germany), Munich, 1995.

Kořán, J. *Sláva a pád českého hornictví* (Rise and Fall of the Czechoslovakian mining industry), Příbram, 1984.

Korbel, Madeleine Jana. *Zdeněk Fierlinger's Role in the Communization of Czechoslovakia: the Profile of a Fellow-Traveller*. MA Dissertation, Wellesley College, 1959.

Kosatík, Pavel, Michal Kolář. *Jan Masaryk*, Prague, 1998.

Kowalczuk, Ilko-Sascha, Armin Mitter, and Stefan Wolle, eds. *Der Tag X – 17. Juni 1953. Die "Innere Staatsgründung" der DDR als Ergebnis der Krise 1952/53* (The X Day – 17th June 1953. The real foundation of the GDR as a result of the crisis of 1952/53), Berlin, 1995.

Kramer, Mark. "Documenting the Early Soviet Nuclear Weapons Program." *Bulletin of the Cold War International History Project*, Nr. 6/7 (Washington 1995).

Kramish, Arthur. *The Griffin*, Boston, 1986.

Kratochvíl, Antonín. *Čaluji*, Prague, 1990.

Kratochvíl, Josef. *Topografická mineralogie Čech* (Topographical mineralogy of Bohemia), Prague, 1936.

Kroker, Eyelyn, Michael Farrenkopf. *Grubenunglücke im deutschsprachigen Raum* (Mining accidents in Germany and Austria), Bochum, 1999.

Kruglov, A. K., A. M. Petrossjanz. "Erste Forschungsinstitute, Konstruktionsbüros und Projektierungsorganisationen, die an der Entwicklung der Atomindustrie gearbeitet haben" (Research institutes, construction and project bureaux, which worked for the soviet atomic industry) in Ministerium der Russischen Föderation für Atomenergie eds., *Die Entwicklung der ersten sowjetischen Atombombe* (The development of the first Soviet atomic bomb), Moscow, 1995.

Kunetka, James W. *City of Fire: Los Alamos and the Birth of the Atomic Age 1943–1945*, New York, 1978.

Laufer, Jochen. "Die UdSSR und die Zonenteilung Deutschlands (1943/44)" (The USSR and the zonal division of Germany 1943/44), *Zeitschrift für Geschichtswissenschaft*, 43. Jg., 4/1995.

Laufer, Jochen. *Politik und Praxis der sowjetischen Demontagen 1941–1950* (Politics and praxis of Soviet dismantling 1941–1950), Manuskript, Potsdam, 2001.

Leinweber, Ralf R. *Das Recht auf Arbeit im Sozialismus. Die Herausbildung einer Politik des Rechts auf Arbeit in der SBZ/DDR 1945 bis 1961* (The right to a working place in Socialism. The beginnings of a policy on the right to a working place in the SBZ/GDR 1945 to 1961), Marburg, 1983.

Lipinsky, Jan. "Mobilität zwischen den Lagern" (Mobility between the special camps) in Plato, Alexander von ed., *Sowjetische Speziallager in Deutschland 1945–1950* (Soviet special camps in Germany 1945–1950), Berlin, 1998.

Lorenz, Egon. "Radioactivity and Lung Cancer: A Critical Review of Lung Cancer in the Mines of Schneeberg and Joachimsthal," *Journal of the National Cancer Institute*, 5/1944.

Löwy, Julius. "Über die Joachimsthaler Bergkrankheit (Vorläufige Mitteilung)" (About the Joachimsthal illness [Preliminary findings]), *Medizinische Klinik*, Vol. 25, 25. January 1929.

Mai, Gunther. *Der Alliierte Kontrollrat in Deutschland 1945–1948. Alliierte Einheit—deutsche Teilung?* (The Allied control council in Germany 1945–1948, Allied unity—German division?), Munich, 1995.

Martignoni, K. "Epidemiologische Studien im Rahmen des Projektes 'Altlastenkataster'" (Epidemiological Studies for the "disused dump register project") in Bundesamt für Strahlenschutz ed., *2. Biophysikalische Arbeitstagung in Schlema*, Berlin, 1992.

Melzer, Richard. *Breakdown: How the Secret of the Atomic Bomb Was Stolen During World War Two*, Santa Fe, 2000.

Merseburger, Peter. *Der schwierige Deutsche. Kurt Schumacher* (The complicated German. Kurt Schumacher), Stuttgart, 1995.

Müller, Wolfgang D. *Geschichte der Kernenergiewirtschaft in der Bundesrepublik Deutschland* (The history of nuclear industry in the Federal Republic of Germany), vol. I, Stuttgart, 1990, vol. II, Stuttgart, 1996.

Nagel, Günter. "Oranienburg als Vorstation auf dem Weg zur deutschen Atombombe" (Oranienburg as a pre-station to German atomic bomb), *Potsdamer Neuste Nachrichten*, 17. November 2000.

Naimark, Norman M. *The Russians in Germany: A History of the Soviet Zone of Occupation 1945–1949*, Cambridge, MA, 1995.

Neubert, Ehrhart. *Geschichte der Opposition in der DDR 1949–1989* (The history of opposition in the GDR 1949–1989), Bonn, 1997.

Pacific War Research Society, ed. *The Day Man Lost*, Tokyo–New York, 1972.

Paleček, Pavel. *Ministr Hubert Ripka a jeho osobní archiv* (Minister Huber Ripka and his personal archive), Brno, 2000.

Pash, Boris T. *The Alsos Mission*, New York, 1969.

Patton, George S. *War, as I Knew it*, Boston, MA, 1947.

Paul, Reimar. *Das Wismut-Erbe. Geschichte und Folgen des Uranbergbaus in Thüringen und Sachsen* (The Wismut inheritance. History and consequences of the uranium mining in Thuringia and Saxony), Göttingen, 1991.

Pecka, Karel. *Motáky nezvěstnému* (Kites to the missing one), Brno, 1990.

Peterson, Heinrich. *Zur Geschichte der Glasfarben-Erzeugung in Joachimsthal* (The history of the production of glass dyes in Joachimsthal), Vienna, 1894.

Petráčová, Ludmila. "Vězeňské tábory v jáchymovských uranových dolech 1949–1961" (Prison camps in the Joachimsthaler uranium mining district 1949–1961), *Sborník archivních prací*, 2, XLIV, Prague, 1994.

Petrov, Nikita V. "Pervoi Predsedatiel KGB Ivan Serov" (The first chief of the KGB Ivan Serov), *Otchestvenia Istoria*, 5/1997.

Picugin, Vladimir. "Aus der Geschichte des sowjetischen Atomprojektes" (On the history of the Soviet atomic project) in Karlsch, Rainer, Harm Schröter, eds., *"Strahlende Vergangenheit." Studien zur Geschichte des Uranbergbaus der Wismut* ("Radiant past." Studies on the history of uranium mining by the Wismut corporation), St. Katharinen, 1996.

Pluskal, Oscar. *Vývoj československého uranového průmyslu* (On the history of the Czechoslovak uranium industy), Prague, 1998.

Pluskal, Oscar. *Poválečná historie jáchymovského uranu* (Post-war history of the Jachymov uranium), Special Papers, Czechs Geological Institute, Prague, 1998.

Podewin, Norbert. *Walter Ulbricht: Eine neue Biographie* (Walter Ulbricht: A new biography), Berlin, 1995.

Port, Andrew. "When Workers Rumbled: The Wismut Upheaval of August 1951 in East Germany," *Social History*, 22/2 (1997), 145–173.

Potratz, Rainer. "Zur Entfernung „deklassierter Elemente." "Die Ausweisungen aus den Uranbergbaukreisen 1952–1954" (The removal of "antisocial persons". The Expulsions from the uranium mining districts 1952–1954) in Karlsch, Rainer, Harm Schröter, eds., *"Strahlende Vergangenheit." Studien zur Geschichte des Uranbergbaus der Wismut* ("Radiant past". Studies on the history of uranium mining by the Wismut corporation), St. Katharinen, 1996.

Powers, Thomas. *Heisenbergs Krieg: Die Geheimgeschichte der deutschen Atombombe* (Heisenberg's war. The Secret History of the German Bomb), Hamburg, 1993.

Rajewsky, Boris. "Bericht über die Schneeberger Untersuchungen" (Report about the Schneeberg investigations) in *Zeitschrift für Krebsforschung*, 49 (1939).

Ranft, Norbert. *Vom Objekt zum Subjekt: Montanmitbestimmung, Sozialklima und Strukturwandel im Bergbau seit 1945* (From object to subject: co-determination, social climate and structural change in the mining industry since 1945), Bochum, 1988.

Reichert, Mike. *Kernenergiewirtschaft in der DDR: Entwicklungsbedingungen, konzeptioneller Anspruch und Realisierungsgrad (1955–1990)* (Atomic industry in the GDR: Genesis, concepts and realisation 1955–1990), St. Katharinen, 1999.

Remdt, Gerhard, Günter Wermusch. *Rätsel Jonastal: Die Geschichte des letzten "Führerhauptquartiers."* (Secret Jonastal. The history of Hitler's last Headquarters), Berlin, 1992.

Rhodes, Richard. *The Making of the Atomic Bomb*, New York, 1986.

Riehl, Nikolaus. *Zehn Jahre im goldenen Käfig: Erlebnisse beim Aufbau der sowjetischen Uran-Industrie* (Ten years in a gilded cage: Experiences in the development of the Soviet uranium industry), Stuttgart, 1988.

Ringholz, Raye C. *Uranium Frenzy: Boom and Bust on the Colorado Plateau*, New York, 1989.

Riabev, L. D., ed. *Atomnii projekt SSR* (The Soviet atomic project), Moscow, 2002.

Roch, W. *Annaberg 1496–1946: Chronik.* (Annaberg 1496–1946. A chronicle), Annaberg, 1946.

Roeling, Rob. "Arbeiter im Uranbergbau" (Workers in the uranium mines) in Karlsch, Rainer, Harm Schröter, eds., *"Strahlende Vergangenheit." Studien zur Geschichte des Uranbergbaus der Wismut* ("Radiant past." Studies on the history of uranium mining by the Wismut corporation), St. Katharinen, 1996.

Rose, Paul Lawrence. *Heisenberg and the Nazi Atomic Project: A Study in German Culture*, Berkeley, CA–London, 1998.

Roseman, Mark. *Recasting the Ruhr 1945–1958: Manpower, Economic Recovery and Labour Relations*, New York–Oxford, 1992.

Roth, Heidi. *Der 17. Juni 1953 in Sachsen* (The 17th June 1953 in Saxony), Cologne, 1999.

Roth, Heidi, Torsten Diedrich. "'Wir sind Kumpel – uns kann keiner.' Der 17. Juni 1953 in der SAG Wismut" ("We are miners—we are strong." The 17th June 1953 in the Wismut corporation) in Karlsch, Rainer, Harm Schröter, eds., *„Strahlende Vergangenheit." Studien zur Geschichte des Uranbergbaus der Wismut* ("Radiant past." Studies on the history of uranium mining by the Wismut corporation), St. Katharinen, 1996.

Sborník přednášek a dokumentů, Stálá mezinárodni konference o zločinech komunismu (Edition with documents and lessons. International permanent commission for the communist crimes), Prague, 1991.

Scalia, Joseph M. *Germany's Last Mission to Japan: The Failed Voyage of U-234*, Annapolis, 2000.

Schenck, E. G., W. von Nathusius. "Extreme Lebensverhältnisse und ihre Folgen" (Extreme living conditions and their consequences), *Schriftenreihe des ärztlich-wissenschaftlichen Beirates des Verbandes der Heimkehrer Deutschlands e. V.*, Cologne, 1958.

Schiffner, Werner. *Agricola und die Wismut: Ein Zeitbogen vom Bergkgeschrey auf Sylber zum Berggetöse ums Uran im Erzgebirge* (Agricola and the Vismit corporation: A report from the silver mining to the uranium mining in the ore mountains), Leipzig, 1994.

Schirdewan, Karl. *Aufstand gegen Ulbricht: Im Kampf um politische Kurskorrektur, gegen stalinistische, dogmatische Politik* (Uprising against Ulbricht: The struggle for another politics against Stalinism), Berlin, 1994.

Schmidt, Reinhard. "Vorgeschichte, Beginn und Frühzeit der Urangewinnung im Erzgebirge" (Prehistory, origins and early years of the uranium mining in the Erzgebirge) in Karlsch, Rainer, Harm Schröter, ed., *"Strahlende Vergangenheit." Studien zur Geschichte des Uranbergbaus der Wismut* ("Radiant past." Studies on the history of uranium mining by the Wismut corporation), St. Katharinen, 1996.

Schreiber, R. "St. Joachimsthal, die Stadt der Radioaktivität" (St. Joachimsthal, town of radioactivity) in *Bericht der Oberhessischen Gesellschaft für Natur- und Heilkunde. Neue Folge: Naturwissenschaftliche Abteilung*, vol. 32, Gießen, 1962.

Schroeder, Klaus. *Der SED-Staat: Partei, Staat und Gesellschaft 1949–1990* (The state of the Socialist Unity Party: Party, State and Society 1949–1990), Munich, 1998.

Schulte, Jan Erik. *Zwangsarbeit und Vernichtung: Das Wirtschaftsimperium der SS* (Slave labor and Genocide: The economic empire of the SS), Paderborn, 2001.

Schüttmann, Werner. "Die Geschichte des 'Schneeberger Lungenkrebses'" (The history of the Schneeberger lung cancer) in *Der Anschnitt*, 2/3 1998.

Schüttmann, Werner, Karl Aurand. "Die Geschichte der Außenstelle Oberschlema des Kaiser-Wilhelm-Instituts für Biophysik Frankfurt am Main" (The history of the branch office of the Kaiser-Wilhelm-Institute for biophysics Frankfurt/M. in Oberschlema), *Schriftenreihe des Bundesamtes für Strahlenschutz*, Salzgitter, 1991.

Schüttmann, Werner, Helmut Schnatz. "Ein erster Schritt zum Kalten Krieg?

Der amerikanische Luftangriff auf Oranienburg am 15. März 1945" (A first step towards the cold war? The American air raid on Oranienburg at 15th March 1945) in *Der Anschnitt*, 50. Jg., 2–3/1998.

SDAG Wismut, ed. *Seilfahrt. Auf den Spuren des sächsischen Uranerzbergbaus* (Elevator into the mine. On the track of the uranium mining in Saxony), Haltern, 1990.

Seelig, Rolf. "Gesundheitliche Folgen des Uranbergbaus" (The uranium mining and its consequences for health) in Karlsch, Rainer, Harm Schröter, eds., *"Strahlende Vergangenheit." Studien zur Geschichte des Uranbergbaus der Wismut* ("Radiant past." Studies on the history of uranium mining by the Wismut corporation), St. Katharinen, 1996.

Semiriaga, Mikhail. *Kak my upraviali Germaniie* (How we have governed Germany), Moscow, 1995.

Semmig, Hellmuth. *Die wirtschaftliche Entwicklung der Exulantensiedlung "Johanngeorgenstadt" von der Gründung 1654 bis zum Stadtbrand 1867* (The economic development of Johanngeordenstadt from its foundation 1654 up to the great fire 1867), Dresden, 1931.

Šesták, Miroslav, Emil Voráček, eds. *Evropa mezi Německem a Ruskem, Historický ústav* (Europe between Germany and Russia), Prague, 2000.

Švec, J. "Lung cancer in uranium miners and long-term exposure," *Health Physics*, 1976.

Sieber, Siegfried. *Um Aue, Schwarzenberg und Johanngeorgenstadt* (Around Aue, Schwarzenberg and Johanngeorgenstadt), Berlin, 1972.

Slánská, Josefa. *Zpráva o mém muži* (Report about my husband), Prague, 1990.

Smelser, Ronald, ed. *Die SS: Elite unter dem Totenkopf. 30 Lebensläufe* (The SS: Elite under the death's head. 30 curriculum vitas), Paderborn, 2000.

Šorf, František. "30 let naší geologické činnosti na Příbramsku" (30 years of geological work in the Pribram uranium mining district), *Atom*, 17, 1977/18.

Šorf, František. *Uran*, Příbram, 1982.

Soukup, Lumír. *Chvíle s Janem Masarykem* (Years with Jan Masarky), Prague, 1994.

Speer, Albert. *Erinnerungen* (Memoirs), Berlin, 1970.

Stange, Thomas. *Die Genese des Instituts für Hochenergiephysik der Deutschen Akademie der Wissenschaften zu Berlin (1940–1970)* (The genesis of the Institute for High energy physics of the German Academy of science in Berlin 1940–1970). Hamburg, 1998.

Stettner, Ralf. *"Archipel Gulag." Stalins Lager*, Munich, 1997.

Sudoplatov, Pavel A. *Special Tasks*, Boston, 1994.

Táborský, Eduard. *Prezident Beneš mezi Západem a Východem* (President Beneš between West and East), Prague, 1993.

Tomeš, Josef, ed. *Československý biografický slovník (CBS) 20. století* (Czechoslovakian biographical lexicon of the 20th century), Prague, 1999.

Tomek, Prokop. *Československý uran 1945–1989* (Czechoslovakian uranium 1945–1989), Prague, 1999.

Tomíček, Rudolf. *Těžba uranu v Horním Slavkově.* (Uranium mining at Horni Slavko), Sokolov, 2000.

Valenta, Vladimír. "Po stopách uranového hornictví na Příbramsku" (Tracing the uranium mining in Příbramsko) in *Podbrdsko, Sborník státního okresního archivu v Příbrami* (Miscellany of the State Regional Archive in Příbram), vol. IV, 1997.

Vaněk, Vladimír. *Jan Masaryk.* Prague, 1994.

Veselý, T. "Stavba a význam jednotlivých žilných uzlů uranového ložiska Jáchymov" (Structure and importance of different uranium shafts for the uranium deposit Jachymov), *Sborník geologických věd, řada LG c 28,* Prague, 1986.

Wächtler, Eberhard, Ottfried Wagenbreth. *Bergbau im Erzgebirge: Technische Denkmale und Geschichte* (Mining in the Ore Mountains: Technical monuments and history), Leipzig, 1990.

Walker, Mark. *Nazi Science: Myth, Truth and the German Atomic Bomb*, New York–London, 1995.

Walter, Hans Ulrich. *Epidemiologische Untersuchung über die Lebenserwartung und Mortalität von in der Nachkriegszeit in Joachimsthaler Uranbergwerken beschäftigten Personen* (Epidemiological research on the life expectancy and mortality of miners from Joachimsthal), Dissertation, Munich, 1982.

Werth, Nicola. "Ein Staat gegen sein Volk" (A state against its people) in Courtois, Stephane, ed., *Das Schwarzbuch des Kommunismus. Unterdrückung, Verbrechen und Terror* (The black book of communism. Repression, crime and terror), Munich, 1998.

Wetrow, W. I. "Bildung von Betrieben für die Gewinnung und Aufbereitung von Uranerz" (Foundation of enterprises for the mining and refining of uranium ore) in Ministerium der Russischen Föderation für Atomenergie, ed., *Die Entwicklung der ersten sowjetischen Atombombe* (The development of the first Soviet atomic bomb), Moscow, 1995.

Wheeler-Bennett, John, Anthony Nicholls. *The Semblance of Peace*, London, 1972.

Wilcox, Robert K. *Japan's Secret War*, New York, 1995.

Williams, Robert Chadwell. *Klaus Fuchs: Atomic Spy*, Cambridge, 1987.

Zank, Wolfgang. *Wirtschaft und Arbeit in Ostdeutschland 1945–1949: Probleme des Wiederaufbaus in der Sowjetischen Besatzungszone Deutschlands* (Economy and work in East Germany 1945–1949: Problems of the economic recovery of the Soviet Zone of Occupation), Munich, 1987.

Zeman, Zbyněk. "Der tschechoslowakisch-sowjetische Uranvertrag vom 23. November 1945" (The Czechoslovakian-Soviet uranium treaty of 23th November 1945) in Karlsch, Rainer, Harm Schröter, eds., *"Strahlende Vergangenheit." Studien zur Geschichte des Uranbergbaus der Wismut* ("Radiant past." Studies on the history of uranium mining by the Wismut corporation), St. Katharinen, 1996.

Zeman, Zbyněk, Antonín Klimek. *The Life of Edvard Beneš 1884–1984: Czechoslovakia in Peace and War*, Oxford, 1997.

Zubok, Vladislav, Constantine Pleshakov. *Inside the Kremlin's Cold War: From Stalin to Khrushchev*, Cambridge, MA, 1996.

Zur Geschichte der Gebietsparteiorganisation Wismut der SED (History of the Socialist Unity Party organisation at Wismut corporation), Chemnitz, 1988.

Name Index

Where no first name is given, it is because the first name was not found by the authors.

Abakumov, Viktor, 174
Aeckerlin, Gustav, 28, 29
Agricola (Georg Bauer), 41, 49, 252
Alexander, tsar, 3
Alexandrov, Semion, 27, 28, 69, 71, 72, 75, 99
Altschuler, Lev, 26
Anderle, Josef, 151, 152
Anderson, John, 4
Apelt, Fritz, 231
Arakatsu, Bunsaku, 15
Ardenne, Manfred von, 9n, 12, 25
Attlee, Clement, 21

Bacílek, Karol, 120n
Bakulin, Ivan, 71–75, 277, 278
Bamborough, O., 108
Baraniuk, 173
Baudyš, V., 116
Becquerel, Henri, 43, 44
Běhounek, František, 51
Bělehrádek, Jan, 85, 89–91
Belousov, 124
Beneš, Edvard, 46, 63–69, 71, 73–75, 80, 82–84, 86, 90, 91, 109, 110, 275n,
Beria, Lavrentii, 16–18, 25–27, 76, 80, 92, 96, 111, 162, 173, 174, 186, 196, 200, 203, 271, 272
Bevin, Ernest, 107
Boček, Karel, 70
Bogatov, Valentin, 171, 248
Bohr, Niels, 10, 15
Bonaparte, Napoleon, 3
Bothe, Walter, 8, 9
Böhme, Kurt, 176–178, 180
Brack, Gustav, 187, 231
Bradley, Omar, 22, 23
Brandt, Artur, 51, 53–55
Brandt, Willy, 58

Bräutigam, Alois, 179, 182
Brecht, Bertold, 211
Brouāek, 119
Bruhl, 248
Bub, 150
Bulander, 147
Byrnes, James, 84

Cairncross, John, 16, 18
Calvert, Horace C., 13
Čepicka, Alexei, 114
Čermák, František, 119
Chesnokov, N., 75
Chmela, Stanislav, 144, 148
Chuikov, Vasily, 204
Churchill, Winston, 3, 4, 6, 7, 18, 22–24, 106, 108, 161
Clementis, Vlado, 87
Čmelák, 110, 142, 144, 148, 149
Cockroft, John, 7
Curie, Joliot, 5, 14
Curie, Pierre, 44
Curie-Sklodowska, Marie, 44

Dahlem, Franz, 181
Dashkievich, 93, 97
Dejean, M., 108
Diebner, Kurt, 8, 10–12
Dimitrov, Georgi, 27
Doolittle, James, 20
Doubek, 150
Dubrovski (general), 226
DuPont, Henry, 6
Ďuriš, Július, 74
Dušek, 142

Eden, Anthony, 4, 21, 23, 68
Einbinder, 173, 176, 194, 199
Einstein, Albert, 5, 11, 15
Eisenhower, Dwight, 22

Elliot, 147
Elstner, 99
Esau, Abraham, 8

Fermi, Enrico, 7
Fieker, Gerhard, 221
Fierlinger, Zdenŭk, 63–75, 80, 81, 83, 84, 87, 89, 91, 95n, 97, 98, 108
Flerov, Georgi, 16
Franco, Francisco, 65
Franz, Alfred, 178
Frederick, the III, duke of Saxe-Gotha, 42
Frederick, the Wise, 37
Frejka, Ludvik, 150
Friedman, Otto, 80, 81
Friedrich, Richard, 43
Frisch, Otto, 5
Fritsche, 248
Forst, Karel, 152
Fuchs, Klaus, 18, 19

Gäbler, Walter, 251
Gebhardt, Willy, 237
Gehlen, Reinhard, 235
Gerlach, Walther, 12
Gessner, Herbert, 232
Getseva, 98
Gide, André, 64
Giese, Friedrich, 45
Glaser, Kurt, 79, 80
Goebbels, Joseph, 10, 52
Gold, Harry, 18
Goldstücker, Eduard, 254
Gorsky, Anatolii, 16
Gottwald, Klement, 68, 81, 87, 94, 97, 102, 110, 111, 114, 121, 139, 141, 144, 145, 149, 150
Goudsmit, Samuel, 13, 14
Göring, Hermann, 47
Greenglass, David, 19
Grischek, 58
Gromyko, Andrei 279
Grotewohl, Otto, 168, 175, 202, 203, 217
Groves, Leslie, 6, 13, 14, 19, 21, 24, 25, 109

Hacha, Emil 66
Hahn, Otto, 5, 8, 12, 14, 16, 47
Hahner, Karl, 166
Hall, Theodore, 18, 19
Hambro, Charles, 108, 109
Hankey, Maurice, 16
Hanzlíček, 150
Harteck, Paul, 8
Havelková, Bohumila, 146
Hecker, Ludwig, 222
Hegner, Bohuslav, 110, 111, 144, 148, 149
Heiner, Max, 50–52
Heinze, 184
Heisenberg, Werner, 8–12, 14, 15, 17
Henckel, Johann, 49
Henlein, Konrad, 46, 47, 51–53
Hennecke, Adolf, 214
Hertz, Gustav, 25
Hettler, Josef, 151, 152
Heym, Stefan, 24
Himmler, Heinrich, 13, 47
Hitler, Adolf, 3, 5, 9n, 10–14, 22, 46, 47, 53, 63, 64, 66, 67, 177, 180
Hoffmann, Gerhard, 9
Honecker, Erich, 171, 232
Horák, Jiří, 73, 90, 91, 109, 110
Horáková, Milada, 139
Houterman, Fritz, 9
Huebner, Clarence, 23
Hund, Magnus, 49

Ihwe, Egon, 19, 21

Jansch, 248
Johann Georg, the I, 37, 38

Kammler, Hans, 13
Kapitsa, 17
Karlsch, Rainer, 13, 271
Kašpárek, Jirí, 98, 275n
Keitel, Wilhelm 22
Kessler, Robert, 176
Khariton, Iulii, 17
Khaustov, N., 29, 162
Khozianov, Nikolai,139

Khrushchev, Nikita, 128
Kirilenko, 246
Kirsche, Hans, 221
Klaproth, Heinrich, 43
Kloss, Milan, 119
Koblic, Odolen, 95
Kobulov, Bogdan, 168, 173, 196
Kohlerová, 145
Kolař, Jan, 79
Konoplov, 123
Kopřiva, Ladislav, 120n, 140, 141
Koryma (luitenent), 93
Košulic, 129, 137
Koutek, J., 99, 100
Kovář, Václav, 93, 95–98, 102, 148
Krasnikov, 98
Kratochvíl, J., 99
Krebs, Adolf, 29
Kreiter, Vladimir, 27, 28
Krivonosov, 94, 99
Kroll, Fritz, 174
Kruglov, Sergei, 189
Kruml, 122
Kuhn, Josef, 42
Kurchatov, Igor, 17–19
Kuznetsov, Vladimir, 200, 279
Kutuzov, Mikhail, 3

Lahne, Paul, 231
Laue, Max von, 9
Laušman, Bohumil, 70, 71, 75, 95, 96, 98
Leeb, Wilhelm Ritter von, 10
Lehmann, Helmut, 201, 202, 251, 256
Lejsek, Bohuslav, 144
Lenin, Vladimir, 63, 64, 67, 175, 180, 212
Leppi, Richard, 179, 180
Likhachev, Dmitry, 139
Liška, Vladimír, 105
Lorenz, Egon, 250
Löwy, Julius, 51
Luc, Robert, 108
Luck, Wilhelm, 214

Mache, Heinrich, 42
Maclean, Donald, 18
Maisky, Ivan, 67
Makarov, Fyodor, 139
Malenkov, Georgy, 139
Maltsev, Mikhail, 162, 164, 171–174, 184, 199–201, 203, 205, 207, 224, 226, 227, 28, 230, 246
Malygin, 195
Manuilsky, Dmitry, 279
Margrave Henry, the Enlightened, 37
Markstein, Max, 248
Marshall, George, 19
Martínek, Karel, 119, 144
Masaryk, Jan, 68, 75, 79–91, 109, 279
Masaryk, Thomas, 82
Mathesius, Johannes, 49
Mathias, the II, 42
Merker, Paul, 181, 202, 231, 245
Merkulov, Vsevolod, 168, 200
Mesiakova, 123
Meyer, Stefan, 42
Mikhailov, Vladimir, 69, 99
Milch, Erhard, 10, 11
Molotov, Vyacheslav, 17, 18, 66, 73, 91, 200, 203
Morenov, Pavel, 206
Morgan, J., 21
Morozov, A., 93
Mückenberger, Erich, 237
Mutschmann, Martin, 53
Müller, H., 50

Nathusius, Wolfgang von, 58
Neusser, 42
Nicols, Philipp, 83
Nishina, Yoshio, 15
Nosek, Václav, 140
Novotný, Antonín, 137

Oelsner, Oskar, 166, 246
Oelssner, Fred, 182
Ohnesorge, Wilhelm, 12, 13
Oppenheimer, Robert, 6, 17

Orlov, Alexander, 27
Osubka-Morawski, Edward, 68
Otterbein, Georg, 12

Paersch, Günther, 247
Paracelsus (Theophrast Bombast von Hohenheim), 49, 252
Pash, Boris, 13, 14
Patton, George, 22, 23
Patzschke, Kurt, 48, 53, 54, 99
Pavel, Josef, 140
Pavlenko, 94
Pavlov, Alexander, 94
Pecka, Karel, 120
Peierls, Rudolf, 5
Pervukhin, Mikhail, 17, 18
Peter the I, 26
Philby, John, 18
Pieck, Wilhelm, 168, 169, 202, 204, 230, 232, 240
Pika, Heliodor, 147
Pixa, Kamil, 150
Plaček, Štěpán, 139
Pohl, Otakar, 93, 98, 119, 144, 148
Pomp, Herbert, 176
Pose, Heinz, 25
Prchala, Antonín, 151
Pytlík, Josef, 118

Rada, Svatopluk, 81, 93, 96–98, 105, 110, 111, 140–142, 145–151
Radecker, Horst, 214
Raikin, 195
Rais, Stefan, 116
Rajewsky, Boris, 29, 54
Rajk, László, 139
Razdan, 129
Razhev, 144
Reifenrath, Richard, 246
Reiter, Hans, 56
Riehl, Nikolaus, 8, 21, 25
Ripka, Hubert, 79–81, 88, 89, 277, 278
Roentgen, Wilhelm, 43
Rohner, Gerhard, 167
Rommel, Erwin, 196

Roosevelt, Franklin D., 4–7, 18, 108
Rosbaud, Paul, 13
Ross, 147
Röder, Günther, 178, 179, 248
Rutherford, Ernest, 15, 44, 50

Sauckel, Fritz, 47
Scheffler, Carl, 49
Schiffner, Carl, 42, 43, 50
Schiffner, Werner, 252
Schindler, Antonín, 92, 93
Schirdewan, Karl, 182
Schlick, Caspar Count, 39, 40
Schlick, Stephan Count, 39, 40
Schmidt, Johannes, 27, 174, 175
Scheider, Werner, 248
Schreiber, Fritz, 201
Schröder, Klaus, 174, 182
Schudy, Gotthard, 196, 198, 225
Schumacher, Friedrich, 40
Schumacher, Kurt, 229–231, 256
Selbmann, Fritz, 171, 176, 180–182, 202
Semichastny, V., 149
Sengier, Edgar, 8, 77
Seydewitz, Max, 216
Sigismund, emperor, 39
Simin, 144, 149
Skala, 232
Skobeltsyn, 17
Slánský, Rudolf, 110, 111, 139–141, 144–146, 149, 150
Sling, Otto, 150
Slutski, 17
Sokolovski, 200
Solzhenitsyn, Alexander, 26
Soukup, Lumír, 85–88, 99
Spaatz, Carl, 21
Speer, Albert, 10, 11
Šrámek, Jan, 66, 74, 87
Stalin, Joseph, 3, 4, 7, 13, 16, 17, 19, 22–26, 63–65, 69, 76, 78, 83, 84, 107, 108, 114, 118, 121, 123, 128, 139, 140, 145, 146, 151, 161, 165, 168, 172, 175, 181, 236, 256, 271
Stepanov, A., 104

Stirsky, 123
Stránský, Jaroslav, 74, 87, 88
Strassman, Fritz, 5, 12
Strauss, Franz Josef, 181
Strittmatter, Erwin, 214
Sukhanov, Dmitry, 93, 195–197
Sum, Antonín, 90
Suslov, 188
Šváb, Karel, 140, 141
Švenek, Jaroslav, 151, 152
Svoboda, 110, 140
Szilard, Leo, 5

Tannenbaum, Antonín, 141
Taussigova, Jarmila, 141
Thiessen, Peter, 25
Tikhonov, Vyacheslav, 139
Trotsky, Leo, 64
Truman, Harry, 3, 4, 23, 24, 82, 138, 161
Tulpanov, 188

Ulbricht, Walter, 168, 171, 178, 179, 181, 182, 202, 203, 232, 239, 257

Vančata, Jaroslav, 138
Vaněk, Vladimír, 142
Vesely, Konrad, 178
Vogel, Martin, 163, 164
Vojíř, 129, 137, 138

Volmer, Max, 25
Volodin, 184
Voloshchuk, S., 93

Wabra, Ernst, 176
Wajda, Andrzej, 213
Webb, Beatrice, 64
Webb, Sidney, 64
Weber, Max, 184, 202, 203, 216, 229, 230, 245, 246, 251, 252, 256
Weidenberg, Axel, 221
Weidig, Max, 43
Weizsäcker, Karl Friedrich von, 9, 11
Wells, H. G., 64
Wenig, Josef, 174, 214
Wildführ, Georg, 251
Wollweber, Ernst, 182

Yezhov, Nikolai, 172

Zaimov, Vladimir, 147
Zalud, 110
Zápotocký, Antonín, 102, 112
Zaveniagin, Avramii, 25, 162, 200
Zhukov, Georgy, 3
Ziller, Gerhard, 182
Zorin, Valerian, 71, 87, 90
Zoubek, V., 99, 100
Zubov, A., 103, 104

For Product Safety Concerns and Information please contact our EU representative GPSR@taylorandfrancis.com Taylor & Francis Verlag GmbH, Kaufingerstraße 24, 80331 München, Germany